Science, Africa and Europe

Historically, scientists and experts have played a prominent role in shaping the relationship between Europe and Africa. Starting with travel writers and missionary intellectuals in the 17th century, European savants have engaged in the study of nature and society in Africa. Knowledge about realms of the world like Africa provided a foil against which Europeans came to view themselves as members of enlightened and modern civilisations. Science and technology also offered crucial tools with which to administer, represent and legitimate power relations in a new global world but the knowledge drawn from contacts with people in far-off places provided Europeans with information and ideas that contributed in everyday ways to the scientific revolution and that provided explorers with the intellectual and social capital needed to develop science into modern disciplines at home in the metropole. This book poses questions about the changing role of European science and expert knowledge from early colonial times to post-colonial times. How did science shape understanding of Africa in Europe and how was scientific knowledge shaped, adapted and redefined in African contexts?

Martin Lengwiler is Professor in the Department of History at the University of Basel, Switzerland.

Nigel Penn is Professor in the Department of Historical Studies at the University of Cape Town, South Africa.

Patrick Harries was Professor of African History at the University of Basel, Switzerland and Emeritus Professor in the Department of Historical Studies at the University of Cape Town, South Africa. He died in 2016.

Routledge Studies in Science, Technology and Society

The Ethics of Ordinary Technology
Michel Puech

Imagined Futures in Science, Technology and Society
Edited by Gert Verschraegen, Frédéric Vandermoere, Luc Braeckmans and Barbara Segaert

Adolescents and Their Social Media Narratives
A Digital Coming of Age
Jill Walsh

Scientific Imperialism
Another Facet of Interdisciplinarity
Edited by Uskali Mäki, Adrian Walsh and Manuela Fernández Pinto

Future Courses of Human Societies
Critical Reflections from the Natural and Social Sciences
Edited by Kléber Ghimire

Science, Africa and Europe
Processing Information and Creating Knowledge
Edited by Martin Lengwiler, Nigel Penn and Patrick Harries

The Sociology of "Structural Disaster"
Beyond Fukushima
Miwao Matsumoto

The Cultural Authority of Science
Comparing across Europe, Asia, Africa and the Americas
Edited by Bauer, MW, Pansegrau, P and Shukla, R

For the full list of books in the series: www.routledge.com/Routledge-Studies-in-Science-Technology-and-Society/book-series/SE0054

Science, Africa and Europe

Processing Information and Creating Knowledge

Edited by
Martin Lengwiler, Nigel Penn and Patrick Harries

Routledge
Taylor & Francis Group
LONDON AND NEW YORK

First published 2019
by Routledge
2 Park Square, Milton Park, Abingdon, Oxon OX14 4RN

and by Routledge
52 Vanderbilt Avenue, New York, NY 10017

First issued in paperback 2020

Routledge is an imprint of the Taylor & Francis Group, an informa business

British Library Cataloguing-in-Publication Data
A catalogue record for this book is available from the British Library

Library of Congress Cataloging-in-Publication Data
A catalog record for this book has been requested

ISBN 13: 978-0-367-58471-9 (pbk)
ISBN 13: 978-0-8153-7831-0 (hbk)

Typeset in Times New Roman
by Apex CoVantage, LLC

Contents

Figures

Contributors

Sonia Abun-Nasr, Dr. phil., is the Director of the Cantonal Library Vadiana St. Gallen, Switzerland. Her research focuses on the colonial history of Ghana and mission history and, in recent years, on topics in the field of library and information science.

Adrien Delmas, PhD, is a Researcher at the Insitut des mondes africains, Paris, France (CNRS UMR 8171) and Director of the Centre Jacques Berque, Rabat, Morocco (USR 3136). He has published on travel writing in the early modern world and also engages in research on African medieval history (11th to 17th centuries).

Marcel Dreier, Dr. phil. des., is Managing Director of the Fund for Development and Partnership in Africa (fepa) in Basel, Switzerland. His research focuses on the history of transnational development cooperation and the history of health systems in Eastern and Southern Africa.

Patrick Grogan is a PhD student and member of the Basel Graduate School of History at the University of Basel, Switzerland. His research focuses on early 19th-century German naturalists in southern Africa.

Tanja Hammel, Dr. phil. des., is a Scientific Collaborator at the Department of History of the University of Zurich, Switzerland. Her research focuses on the social history of science and knowledge in colonial contexts. She is particularly interested in visual history and the history of women.

Patrick Harries (1950–2016) was Professor for African History in the Department of History at the University of Basel, Switzerland, until his retirement in 2015. His research focused on the history of southern Africa, the history of missions and the history of science in Africa.

Heinrich Hartmann, Dr. phil., is a Senior Lecturer (Privatdozent) at the Faculty for Humanities and Social Sciences at the University of Basel, Switzerland. In his research, he focuses on the history of the social sciences and anthropology, as well as the history of development thought in a transnational perspective.

Dag Henrichsen, Dr. phil., is a Namibian Historian at the Basler Afrika Bibliographien and Lecturer at the Department of History, University of Basel,

Switzerland. His research and publications focus on central Namibian history, colonial histories of science as well as audiovisual and archives studies.

Martin Lengwiler is Professor in the Department of History at the University of Basel, Switzerland. His research focuses on European History in a transnational and global perspective. He specialises in the history of insurance and the welfare state.

Lukas Meier, Dr. phil., is Deputy Managing Director of the R. Geigy Foundation (Swiss TPH) at Basel, Switzerland. His research interest includes the history of health and science, development aid and ecology.

Sandra Näf-Gloor, MA, is working at the University of Zurich, Switzerland, focusing on school-based violence prevention. She graduated from the University of Basel, with an MA thesis on Hinrich Lichtenstein.

Nigel Penn is Professor in the Department of Historical Studies at the University of Cape Town, South Africa. His research focuses on topics in the indigenous and colonial frontiers of southern Africa as well as on microhistories of crime and punishment in the early colonial Cape.

Pascal Schmid, Dr. phil., is an academic associate at the Centre for African Studies of the University of Basel, Switzerland. His research interests include the history of Swiss relations to Africa as well as the development of health care and higher education.

Daniel Speich Chassé is Professor at the Department of History of the University of Lucerne, Switzerland. His research focuses on knowledge in global modernisation, in particular the history of global statistics.

Preface

Tribute to Patrick Harries (1950–2016)

This book is dedicated to, and inspired by, Patrick Harries, who first conceived of the project a few years before his sudden death in June 2016. Patrick had been Professor of African History at the University of Basel between 2001 and his retirement in 2015. Before that he had been an Associate Professor in the Department of History at the University of Cape Town. His work, and life, thus spanned the two continents of Africa and Europe and, as an historian, he always sought fresh ways of understanding the links between them.

Patrick's early work emerged from his PhD research. He obtained his doctorate from the School of Oriental and African Studies (SOAS), University of London in 1983 with a thesis on migrant workers from Mozambique in South Africa. This found full expression in his book, *Work, Culture and Identity: Migrant Labourers in Mozambique and South Africa, 1860–1910* (1994). Born and educated in Cape Town, Patrick grew up under apartheid and like many of his generation sought both to understand and remove the racist regime and its ideology. SOAS, at the time, was the intellectual home of a radical Marxist critique of South African history and Patrick's work was in some ways typical of a generation of historical materialists who identified the Mineral Revolution in South Africa as the most crucial event in southern Africa's political economy. Like others, he too concentrated on the rise of the South African working class, stressing the role played in this by the mines and industries of Kimberley and the Witwatersrand. What distinguished his work from his peers, however, was that he brought the sensitivities of a cultural anthropologist to his account of the transformation of Mozambican peasants into industrial workers and that he explained how these men were not necessarily the passive victims of capitalism but created a new and vibrant culture for themselves in the streets and compounds of Kimberley and Johannesburg. Though his model might have been E.P. Thompson's *Making of the English Working Class*, Patrick brought a uniquely well-informed knowledge of cultural anthropology to his work and his stress on "culture, identity and interpretation" marked a breakaway from the more mechanistic Marxism of some of his colleagues. His focus on a transnational, rather than a national theme was also fairly novel for his time and place and presaged his later interest in cross cultural and cross border currents.

Thus far, Patrick's interest in science might be thought to have been confined to history, historical materialism and anthropology. But it was whilst mining the sources of the history of Mozambique migrants that Patrick encountered the archives of the Swiss Mission in Lausanne and Neuchatel. Here he came across the polymath figure of the missionary Henri-Alexandre Junod who had worked amongst the Thonga of southern Mozambique in the late 19th and early 20th centuries. Junod was primarily famous as an anthropologist, author of the influential *The Life of a South African Tribe* (1912–3), but he was also an entomologist, a botanist and a linguist, as well as being curious and informed about many other branches of scientific knowledge. Patrick became fascinated by Junod, particularly by the ways in which he sought to understand or represent Africa and Africans. What systems of classification informed Junod as he attempted to define the boundaries of a people, a language, a culture or an ethnic identity? How closely were these systems of classification influenced by the natural sciences and the scientific organisation of plants and insects? As Patrick pondered these questions he produced a stream of articles that reflected a growing awareness of the global "literary turn" in historical studies, which stressed the idea that traditions and identity might be inventions, and that Western systems of knowledge might be discourses masking a colonising project of power.

Once he moved to Basel in 2001 he set about the self-imposed task of making the city the centre of a vibrant network of African studies. On his newly established chair for African History, financed by the *Carl Schlettwein Foundation*, he was among the founders of the *Center for African Studies* and a key figure for positioning his department as a leading place in the field of African history. He won major funding grants and worked tirelessly to raise money in order to host conferences and to bring scholars of African history to Basel. He taught three to four new courses a semester and personally supervised over 17 PhD students and 30 Master theses. He also began to shape what would become his second major book, a book with Junod at its heart but so much more than a biography. He sought to write a study of the evolution of modern European or Western scientific knowledge about Africa, about how Africa and Africans were constructed as objects of scientific knowledge, but also how this knowledge then altered or transformed European scientific consciousness itself. The book appeared in 2007 as *Butterflies and Barbarians: Swiss Missionaries & Systems of Knowledge in South-East Africa*. The title reflects that the book is as much a study of the intellectual and cultural world of Swiss missionaries as it is about Africa and one of the book's ironic twists is to turn the gaze of the cultural anthropologist onto the Protestant evangelists of the Jura. Junod's world is explored before Africa. Once the mental and material universe of the Swiss is dealt with we are better equipped to deal with their perceptions of Africa which appeared to them to be almost the exact opposite of Switzerland. This is not to say that Junod is portrayed as an insensitive and inflexible colonialist seeking to impose the discourse of Western science upon his objects of study. The Junod who emerges from Patrick's account is open and intelligent, fascinated by the diversity of life and undogmatic in his beliefs. It would be no exaggeration to say that Junod's encounter with Africa broadened both his understanding and humanity, with science not so much providing the answers as presenting new things at which to marvel.

As soon as his work on Junod was completed Patrick turned his attention to more general work on the cultural history of knowledge production and the history of science. Africa remained central to his vision and so too did the science of anthropology although he began to look more deeply into the roots of other "field science", that is, sciences where data collection and observation happened in the outdoors rather than in a laboratory. The present collection of essays reflects this enthusiasm, and is a result of his engagement with the individual scholars – some of them ex-students – featured here. At the time of his death Patrick had great plans to continue working in the field of the history of science and had, as usual, applied for funding so as to generate a hub of students and scholars around him and his project, the history of science in southern Africa. Sadly, of course, this project has not been fulfilled and the present slim volume by no means does justice to either Patrick's vision or his achievements. It should, instead, be regarded as a modest tribute to him, as a historian, a colleague and a teacher who had a rare ability to inspire and encourage others by his intellectual enthusiasm. It is also important to mention that, like Junod, Patrick was always interested in many things at once, spanning from African history, missionary history, the history of science, to a post-colonial history of Switzerland. One of these things, which he intended to work on in his retirement at Cape Town, was the history of Mozambican slaves and their descendants, "Mozbiekers" in Afrikaans. This was a subject that would have returned him to his earliest historical studies, those which dealt with the creation of identity amongst African migrant or involuntary labourers in a world shaped by colonial capitalism. Though it may seem that, intellectually, Patrick had come a full circle, in truth he had probably only completed the first revolution in an intellectual voyage that had only just begun. Like a field scientist, he was still busy collecting his material, a servant of that science that, in Junod's words

> goes out across the continents, gathering its rich harvest of facts, studying geographic and climatic phenomena, collecting new animal forms, observing the customs and languages of primitive races, all in order to one day reconstruct the admirable set of facts, to understand if not the reason behind, at least the way in which humans and things are arranged on our marvelous planet.

Nigel Penn, Martin Lengwiler
Basel and Cape Town, March 2018

1 Science between Africa and Europe

Creating knowledge and connecting worlds (introduction)

Martin Lengwiler and Nigel Penn

This book investigates the role of science in the relationship between Europe and Africa and in creating knowledge about Africa. Since at least the 17th century, scientists and experts have played a prominent role in shaping the view Europeans had on Africa. Starting with missionary intellectuals and travel writers, generations of European savants have engaged in the study of nature and society in Africa. Knowledge about Africa and other non-European realms of the world provided a foil against which Europeans came to view themselves an enlightened and modern civilisation. Science and technology also offered crucial instruments with which to administer, represent and legitimate power relations in a world of empires and, after decolonisation, in a world of a globalised capitalism. In the colonial empires of the 19th and 20th centuries, science and technology was an intellectual resource for reform policies, in order to develop and modernise the colonies and their infrastructure. Also in the context of post-independent African nations, science and technology kept this role for modernisation policies, in different areas such as education, health or infrastructure. Knowledge about Africa finally also reflected back on European institutions. The information and specimen drawn from contacts with people in far-off places provided Europeans with ideas and objects that contributed in multiple ways to the scientific revolution and that provided explorers with the intellectual and social capital needed to develop science into modern disciplines at home in the metropoles. Traces of this African heritage can be found in publications, but also in museums, university collections and the private estates of scientists and travellers.[1]

The contributions to this volume investigate the history of European knowledge on Africa from the early colonial to the post-colonial era. The timespan witnessed the establishment of academic institutions and knowledge-based bureaucratic organisations both in Europe and later in Africa. The chapters analyse how science and expert knowledge shaped the understanding of Africa since the early colonial period. How was scientific knowledge shaped, adapted and redefined in African contexts in the 18th and 19th centuries? How did it permeate the administration of colonial empires in the late 19th and early 20th centuries? How did the colonial context feedback on the development of science and technology in Europe? And how did the entanglement between science and colonialism change with decolonisation?

We use the term of "science" in a broad sense: it includes fundamental sciences as well as applied sciences, encompassing a vast range of activities like collecting material, cataloguing, arranging and exhibiting specimen with the intention to enhance knowledge about Africa. Thus, a variety of actors contribute to science, including adventurers, explorers, naturalists, scientists and experts, Africans and Europeans, from different fields and backgrounds. On a disciplinary level, the contributions to this volume are situated in the history of knowledge rather than merely in the history of science. They do not address the internal life of academic institutions and scientific disciplines but rather the activities of scientists beyond their laboratories, "in the field", and the use of scientific knowledge in non-academic contexts. This encompasses the ways in which knowledge takes on popular or "folk" expertise in non-academic contexts and comes to be marginalised, displaced or occluded by more established forms of scientific knowledge.[2] This understanding of knowledge and science, which underlies the book, is as polycentric as our notion of modernity.[3]

To understand the connected history between science, Africa and Europe, the volume combines three fields of research and their relevant states of the art: the history of Africa, imperial history (or the global history of colonial empires), and the history of knowledge, or of science and technology, with a focus on specific subfields like the history of medicine or the history of geography. We specifically aim at invigorating the history of Africa and imperial history by combining it with the perspectives of the history of science and the history of knowledge. In recent historiography, this combination between knowledge-based and space-based theoretical approaches proved highly innovative and original.[4] Thus, we investigate both, the process of creating truth by scientific actors and the mechanisms of connecting worlds through the knowledge produced. In the following paragraphs we will discuss some of major themes and debates in this field. In the latter part of the introduction, we will present the themes and structure of the book and the argument brought forward by the contributions.

Hybrid knowledge, cultural encounters and scientific pluralism: current historiography on science in Africa

Much of recent research in the history of science in Africa is driven by the theoretical concern to transcend the conceptual juxtaposition between Western science and non-Western indigenous knowledge. These categories, and the valuation implied by them, are themselves the product of a specific historic context. The case studies, presented in the following chapters, trace the complex mechanisms through which Europeans developed methods of scientific reasoning and how they came to consider them as a key to superior, privileged forms of knowledge. Scientific knowledge depended on a specific set of methodological rules, academic institutions, a process of specialisation along the lines of disciplines, and a growing professionalisation, patrolled and defined by scientists themselves. Western science also relegated African skills and expertise to a realm of magic and superstition held in place by communal beliefs and traditions.

The epistemological dichotomy between primitive and modern ways of thinking has a long history and was at the heart of modern anthropology since its early days. One of the founders, Lucien Lévy-Bruhl expressed this Eurocentric view in 1910 in *How Natives Think*, outlining the different ways in which "primitive" and "modern" people organised and structured reasoning. Later, evolutionist anthropologists challenged the fixity of Lévy-Bruhl's distinction between primitive and modern and stressed the possibility of transforming primitive cultures through a temporal development. The subsequent generation of functionalist anthropologists in turn questioned the racialised way in which evolutionist approaches developed a temporal hierarchy of knowledge. And Claude Lévi-Strauss, in his *La pensée sauvage* (1962), criticised the functionalists' view that knowledge was a mechanism in the struggle to survive. Instead, he stressed that people are "bricoleurs" who arrange and analyse data over time in logical ways. In all these approaches, the distinction between primitive and modern knowledge remained fundamental and linked – implicitly or explicitly – to the opposition of "European" and "non-European" knowledge.[5]

Since the 1960, historians and anthropologists increasingly questioned and criticised the superiority of Western science. As a philosopher and historian of science, Thomas Kuhn developed in the early 1960s the influential notion of knowledge paradigms and of scientific revolutions, thus preparing the ground for a social and historical constructivist understanding of scientific objectivity and of Western science.[6] Half a decade later, historian George Basalla outlined the process whereby Europeans had carried "Western science" into the corners of their expanding world, notably to Africa, giving way to a new field of research criticising the role of "colonial science" for the rise and expansion of Western imperialism since the late 19th century.[7]

However, over the past years, scholars in African history, African studies and science studies have begun to revise traditional narratives about the role of science in Colonial Africa and imperial development. A starting-point for these recent debates has been the critical assessment of concepts like "colonial science", "imperial science", or "colonial medicine". These concepts were brought forward in the 1990s, by Roy MacLeod and others who followed the work of George Basalla. Another influential figure was Edward Said, who offered with *Orientalism* a theoretical approach for highlighting the nexus between knowledge and power, in particular in colonial contexts.[8] All these approaches owed much to Michel Foucault's understanding of discourses as fields of asymmetrical power relations.

MacLeod and others assumed that in colonial contexts since the 18th century, modern science developed into specific forms serving the interests of colonial powers and their administrations – although which forms often remained unclear.[9] The notion of "colonial science" stressed the political role of science and was meant as a critique of a Mertonian understanding of science as a universalist, disinterested form of knowledge driven by the organised scepticism of the involved scientists. It was also critical of theories of modernisation that was also based on traditional understandings of science. Lay voices from rural Africa were seen as more authentic than the scientific perspectives of the Western metropoles.[10]

In a series of recent publications, concepts like "colonial science", together with the underlying conceptual dichotomies of expert vs. lay knowledge or metropoles vs. peripheries, were criticised as too linear and simplified. To begin with, there is no coherent, unified form of colonialism. Studies in the new imperial history rather pointed out a variety of colonialisms: settler colonialism and indirect forms of colonial rule; or British, French and American imperial traditions – each of them marked by national-specific sets of norms and values.[11] Moreover: Relying on science and technology for achieving political aims was no prerogative of empires. Post-independent governments too built their economic planning and modernisation policies on scientific knowledge and technical expertise.[12]

Another problem lies with the concept of science. Beinart et al. argue that science is often not as homogenous as assumed by the notion of "colonial science". The links between science and colonial authorities or capitalist corporations were neither clear nor direct. Reducing science to an instrumental role would underestimate the specific logic of scientific activities and inventions and the partially detached relation between scientists and colonial authorities.[13]

Patrick Harries's study on the scientific activities of missionaries is a case in point. The work of missionaries like the Swiss missionary Henri Alexandre Junod (1863–1934) oscillated between scientific endeavours, spiritual motivations and colonial settings. Junod understood himself as a missionary and a naturalist, a representative of a learned, civilised society, firmly anchored in the values of the enlightenment. He contributed to evangelisation, but also established schools and medical services. He distributed hymnbooks and the bible, but he also studied African languages and cultures as a linguist and anthropologist, not least to being able to preach the gospel in the language of the natives.[14]

Other authors too pointed at the variety and heterogeneity of scientific endeavours, shaped by local contexts and natural environments of the colonies, within which scientists were situated. In a survey of studies on the British empire, Mark Harison particularly stressed the various local forms of indigenous knowledge that shaped scientific activities and were integrated into academic forms of knowledge.[15] Studies on the use of science in French colonialism also point at the lacking coherence of the disciplines involved. Parts of geography were closely collaborating with the colonial administration, whereas historians seemed to be irrelevant, at least from the perspective of the government authorities. And colonial law existed as a term, but materialised eclectically in very disparate forms.[16] A lot of these experts acted, in the words of Roy McLeod, as scientists in the colonies rather than colonial scientists.[17] With David Livingstone, we can argue that all knowledge is local, as it is always produced and diffused in specific milieus. The context does not produce stable forms of knowledge. Information is rather continuously renewed and reinterpreted, its meaning reworked, as it moves through time and space.[18]

This notion of scientific knowledge as a hybrid resource is best illustrated with recent research on the history of medicine in Africa. Here too, medical knowledge is interpreted as the product of interactions between different, often hardly commensurable traditions of skills and information. As Anne Digby, Ernst Waltraud and Projit Muhkarji argued, medicine in Africa usually amalgamates Western medicine and indigenous forms of healing. They stress the effect of crossovers, connections, networks and circulations in the process of knowledge production – often on a transnational and global scale.[19] In these processes, brokers like African medical assistants, hospital clerks and midwives from families of healers play a decisive role.[20] Historian of medicine Walter Bruchhausen proposes the notion of *medical pluralism* in order to comprehend the locality and diversity of knowledge and expertise in the history of medicine in Africa.[21]

An important contribution to this debate was Helen Tilley's book on *Africa as a Living Laboratory*. Based on the analysis of the African Research Survey, a report of British scientists, published in 1938 and investigating Africa and its potential for modernisation, Tilley offers a comprehensive critique of the notion of "colonial sciences" for its 'theoretical fallacies and ambiguous dualisms'.[22] She argues that there was a colonial state but no colonial science, nor anything like any Western or European type of science. Tilley opposes any simplifying polarisation between scientific and indigenous knowledge in colonial settings. In sub-saharan Africa, Tilley sees no clear opposition between science and indigenous knowledge nor any comprehensive destruction of indigenous knowledge through colonialism.[23] Medicine is a case in point. African healers, when interacting with Western physicians, adopted certain foreign measures to their benefit. Indigenous medicine proved to be strong and widespread, also under colonial regimes. Moreover, colonial traditions of medicine in many ways depended on the collaboration with African research assistants, who operated to some degree in autonomy and had some control over how to collect information.[24] The role of indigenous savants as cultural brokers and as research collaborators was also stressed in other recent studies, such as Felix Driver's and Lowri Jones's work on the cooperation between explorers and local inhabitants in the formation of geographical knowledge in 19th-century Africa, or Dane Kennedy's analysis of the ambivalent relation between explorers and intermediaries (traders, cultural brokers, translators or research collaborators) in late 19th-century expeditions in sub-Saharan Africa.[25]

Tilley rather understands science as constituted by exchanges with other cultures, like with the Arab world. This also applies to science in Africa, where knowledge emerges in the field, based on local resources, and then travels to Europe where it is canonised as authentic and objective. 'Knowledge may be "situated", but it is designed to travel'. For Tilley, the circulation and travel between colonies and metropoles is more relevant than the context of origin. Roy MacLeod, in a case study on the British-Australian chemist Archibald Liversidge, similarly argued that the dynamics of cultural transmission go far beyond the dichotomy of the relationship between centres and peripheries.[26] Against this background, Tilly uses the notion of circulation in a broad sense, arguing that colonial and indigenous forms of knowledge often only loosely interact and co-exist in incommensurable fields.[27]

Thus, Tilley suggests a new understanding of the laboratory, the site, in which scientific findings are produced. For her, a laboratory is not an isolated box for experiments, detached from the outside world, but rather a hub that links actors from different fields, scientific and non-scientific ones.[28] Understood as a partly autonomous venture, scientific activities can even display subversive effects and undermine the power of regimes. As an illustration Beinart et al. mention the case of AIDS activism in post-apartheid South Africa, where scientific evidence was mobilised, after 2000, to criticise the misguided AIDS policies of the government.[29]

These shifts in understanding science in Africa raised the awareness for processes of mediation, communication and transfer. Such mechanisms were conceptualised, since the 1980s, by authors in social and cultural history with notions such as the "middle ground" (Richard White) or the "contact zone" (Howard Lamar, Leonhard Thompson, later also Mary Louise Pratt). These notions had in common that they transformed the idea of a border into a space for encounters, bringing together individuals from different cultures in often asymmetrical and hierarchical enganglements, initiating processes of transculturation.[30]

Thus, media and their impact have also become a focus of recent historiography in this field. Innes Keighren, Charles Withers and Bill Bell have analysed the role of the printing industry in the production of scientific knowledge in imperial Africa. They see travel writing and its publication as crucial for the making of imperialism and globalisation. Putting materiality and content in relation to each other, their analysis stretches from writing practices in the field, over producing printed manuscripts to the publishing and distribution process, and finally to the reading by a wider audience.[31] Related studies have stressed the role of cartography and maps as a media for representing and shaping knowledge about Africa.[32]

Mapping, counting, collecting: thematic structure of the book

Many themes and problems in current research on the history of science in Africa are relevant for structure of the book and the following chapters. The book is divided into three sections addressing three central themes – each is illustrated by a series of case studies.

The first section deals with the *history of scientific travels and expeditions* – cartographic, botanical, natural-historical – from the 18th to the early 20th centuries, with a particular focus on the production of maps and other spacial knowledge. The information collected in those expeditions and the knowledge produced on that basis was significant for the self-conception of the colonial empires. This section also investigates the ambiguous aspects of scientific knowledge between symbolic representation and political practice. Science provided, on the one hand, a privileged language for the representation of differences between Western and non-Western actors, while on the other hand offering an instrument for legitimating and sustaining – as well as questioning – social and political power relations in the colonial context. The spread of scientific knowledge had thus ambivalent effects. It legitimated the marginalised and under-privileged status of African colonies, but also offered, since the late 19th and early 20th centuries, a universalist language with which actors could criticise discriminatory colonial policies.

The chapters in this section also reflect how knowledge about African colonies was brought back to Europe where it proved to be a crucial factor in the constitution of modern disciplines in the natural sciences. This was due to the popularisation of scientific knowledge by learned societies, libraries, or zoological and botanical gardens in Europe. The cultural impact of colonial knowledge reached not only the colonial empires (among others: Britain, France, the Netherlands), but also other European nations not (or not yet) involved in a colonial expansion, such as Switzerland or Germany before 1850.

In their chapter on Peter Kolb, a German anthropologist and explorer of the Cape Colony in the early 18th century, *Nigel Penn* and *Adrien Delmas* analyse how knowledge on the Cape of Good Hope formed an important part of the transnational information networks between Europe, Asia and Africa. Naturalists, explorers, navigators, writers, missionaries, doctors, cartographers and linguists – they were all part of transnational networks and contributed to the emerging geographical and anthropological knowledge in Europe on Southern Africa. The case of Peter Kolb illustrates how some of this knowledge was not yet predetermined by codes of a European supremacy. When collaborating with local inhabitants, Kolb was aware of the limitations and distortions that his European perspective implied.

The chapter of *Sandra Näf-Gloor* focuses on Hinrich Lichtenstein, another exemplary figure in the history of 18th- and 19th-century naturalism. She argues that Lichtenstein, trained as a physician and venturing for extended travels to the Cape Colony in the early 19th century, continued to profit from his African networks after he returned to Berlin, where he became a professor of zoology and a founder of the Berlin Zoological Garden. Näf-Gloor points out the different levels upon which figures like Lichtenstein operated. He acted simultaneously as a scientist, a collector, a patron and a trader, accumulating material resources and symbolic prestige in his European contexts. Lichtenstein also plays a prominent role in the contribution of *Patrick Grogan*, who examines the collaboration between the German zoologist and another collector of German origin, Ludwig Krebs, who moved to Cape Town early in his career and spent the rest of his life in Southern Africa becoming one of the most important natural history collectors of his time. Grogan shows jow Krebs relied on a network of local collaborators, usually unnamed black servants, acting as informants and helpers.

The chapter of *Dag Henrichsen* traces the history of collecting museum specimens to the late 19th and early 20th centuries by examining the Swiss botanist Hans Schinz, whose travels led him, among other destinations, to southwestern Africa. Henrichsen argues that by collecting specimens and integrating them into a scientific taxonomy, Schinz detached local forms of knowledge from its context of origin and transformed them into a piece of universal science, to be integrated and exhibited in the sanctuaries of modernity, the museums, gardens and parks of European cities.

The second section of the book addresses the *role of the colonies*, since the late 19th century, as *laboratories for modern scientific research*. The section examines how interaction between colonial and local actors influenced the constitution of expert knowledge, academic institutions and government policies, for example in the field of public health. The chapters deal with both contexts, African and

European, with a particular focus on the repercussions that research practices in Africa had on the European metropoles. They also deal with the changing historical context in which epistemic interactions between Europe and Africa took place from colonisation to decolonisation.

The contribution of *Sonia Abun-Nasr* investigates the complex paths of diffusion and perception of a classic ethnographic publication of the early 19th century, Thomas Bowdich's *Mission from Cape Coast Castle to Ashantee*. The case study illustrates the ambivalent modes of knowledge-generating practices. Bowdich's work, as Abun-Nasr argues, cannot be reduced to a merely Eurocentric point of view. Rather, its author was also motivated by an ethic of accuracy and relied on local sources. The image he depicted of the Asante Empire rather corresponded with the self-understanding of the Asante elite and collided with the expectations of his European audience. The perception of Bowdich by later generations illustrates that his work could also be used for emancipatory purposes, criticising European colonial policies.

The chapters of *Heinrich Hartmann*, *Marcel Dreier* and *Lukas Meier* all treat aspects, in different periods, of the history of medicine and public health in Africa. Hartmann investigates discourses on public hygiene in the French and German colonial armies of the late 19th and early 20th centuries. He argues that the colonies acted as a testing ground for policies in public health. Different disciplines – from medicine over demography and geography to biology – developed notions of *tropical* pathologies, encoding the African environment as deficient, dangerous and outright morbid. The discourse on tropical pathologies also fed back on concerns in Germany and France about the physical fitness and the health of soldiers in general, contributing to a broad discourses on social degeneration of the *fin de siècle*.

Marcel Dreier investigates the entangled history of two disease categories in 20th-century Tanzania: the Western diagnosis of epilepsy and the local concept of *kifafa*, a Kiswahili term, usually translated into epilepsy. The chapter shows how kifafa became the object of a translocal and transcultural redefinition and circulation, with repercussions on the development of medical research and health services. Examining the local embeddedness of kifafa, Dreier argues that the medical concept was indeed a historical product, shaped by the colonial context of Tanzania and the emergence of an international medical discourse, not least marked by the World Health Organization (WHO).

Lukas Meier, in his contribution, examines the limits of the concept of a colonial "laboratory" and of transnational circulations of medical knowledge. His topic is the history of a failed medical innovation, the clinical trials of a malaria vaccine (SPf66) in the 1980s and '90s, which were stopped before the launch of a commercial product. Meier argues that the trials failed because they were embedded in local contexts. The tests delivered specific, locally shaped results that were hardly comparable on an international level. In this case, the standardisation of medical knowledge failed, because findings depended too much on local, heterogeneous "reality effects".

The third and final section investigates how the *relationship between Western and non-Western countries* increasingly came to be embedded, over the course of the 19th and 20th centuries, in *international discourses and transnational circulations of knowledge*. The contributions in this section show that global discourses on archaeological specimens, museum collections, human rights or development policies not just reflected the interests of Western countries, but also offered opportunities for African actors to participate. Already in the 19th century, European and African actors dealt with each other about the politics of archiving, exhibiting and remembering the past. Similarly, the emergence of reform policies in the late colonial period helped problematising global differences in the form of developmental stages and ultimately led to the post-war rise of developmental and health policies in African countries.

Tanja Hammel's contribution examines the history of archaeological knowledge in South Africa as an exemplary case for a multiple discovery and multi-site development of a scientific discipline. Investigating how the collections and related archaeological knowledge grew in different areas of the world, across different colonial regions (Cape Colony, Australia) and the European metropoles, Hammel points out how Eurocentrism and colonialism left an often-underestimated mark in contemporary South African archeology.

Similarly, the chapter of *Pascal Schmid* situates the development of health care delivery in a region of the Gold Coast (and independent Ghana) in an international context, involving British colonial authorities, missionaries and medical staff from Switzerland and experts from the WHO. The ways for treating diseases and organising hospitals were thus influenced by the medical state of the art, the agenda of WHO and UNICEF and the policies of the colonial and post-colonial authorities.

The chapter of *Daniel Speich Chassé*, finally, traces the development of macroeconomic theory, in particular approaches of development economic theory, in Africa from the late colonial to the post-colonial decades of the 20th century. Speich argues that the epistemic techniques of macroeconomic thinking, originally based in a Western context, were not helpful to understand African social problems. Economic models were based on a paradigm of modernisation, designating non-Western economic structures as deficient and thus fortifying the idea of a poor, underdeveloped Africa. Such epistemic perspectives also prevailed beyond the period of decolonisation.

Notes

1 Harries, *Butterflies & Barbarians*; Livingstone and Withers, *Geographies of Nineteenth Century Science*.
2 Tilley, 'Global Histories'.
3 AHR Roundtable, 'Historians and the Question of "Modernity" ', 631–7; Vogel, 'Von der Wissenschafts- zur Wissensgeschichte'; Vetter, *Knowing Global Environments*.
4 Cf. Livingstone, 'Landscapes of Knowledge'; Livingstone, *Putting Science in Its Place*.
5 Cf. Horton, *Patterns of Thought in Africa and the West*.
6 Kuhn, *The Structure of Scientific Revolutions*.
7 Basalla, 'The Spread of Western Science'.
8 Said, *Orientalism*.
9 MacLeod, 'Reading the Discourse of Colonial Science'; MacLeod, 'On Science and Colonialism'; Reingold and Rothenberg, *Scientific Colonialism*.

10 Beinart, Brown, Gilfoyle, 'Experts and Expertise in Colonial Africa Reconsidered', 416f.
11 Digby, Ernst and Muhkarji, 'Crossing Colonial Historiographies', ix. For the national-specific differences in colonial rule, see also: Stuchtey, *Science across the European Empires*.
12 Beinart, Brown and Gilfoyle, 'Experts and Expertise in Colonial Africa Reconsidered', 420.
13 Ibid., 418f., 424.
14 Harries, *Butterflies & Barbarians*; see also: Harries and Maxwell, *The Spiritual in the Secular*.
15 Harrison, 'Science and the British Empire'. See also: Chambers and Gillespie, 'Locality in the History of Science'.
16 For example: Singaravélou, *Professer l'Empire*.
17 MacLeod, 'Introduction'; Beinart, Brown and Gilfoyle, 'Experts and Expertise in Colonial Africa Reconsidered', 424.
18 Livingstone, *Putting Science in Its Place*.
19 Digby, Ernst, Muhkarji, *Crossing Colonial Historiographies*; Digby, Ernst and Muhkarji, 'Introduction', x–xii.
20 Digby, Ernst and Muhkarji, *Crossing Colonial Historiographies*; Digby, Ernst and Muhkarji, 'Introduction', xviii. See also: Hokkanen, 'Towards a Cultural History of Medicine(s)'.
21 Bruchhausen, 'Medical Pluralism as a Historical Phenomenon', 100.
22 Tilley, *Africa as a Living Laboratory*, 10f.
23 Ibid.
24 Tilley, 'Global Histories', 14f.; Tilley refers to Vaughan, *Curing Their Ills*.
25 Driver and Jones, *Hidden Histories of Exploration*; Kennedy, *The Last Blank Spaces*, 167–74; see also: Kennedy, 'Introduction: Reinterpreting Exploration', 7–9.
26 MacLeod, 'Imperial Science under the Southern Cross'.
27 Tilley, *Africa as a Living Laboratory*, 10f, 15f.
28 Ibid., 10.
29 Beinart, Brown and Gilfoyle, 'Experts and Expertise in Colonial Africa Reconsidered', 427.
30 Pratt, *Imperial Eyes*, 1–12; Lamar and Thompson, *The Frontier in History*; White, *The Middle Ground*.
31 Keighren, Withers and Bell, *Travels into Print. Exploration*, 2f.
32 Penn, 'Mapping the Cape'; Etherington, *Mapping Colonial Conquest*; Duminy, *Mapping South Africa*; Stone, *A Short History of the Cartography of Africa*.

Bibliography

AHR Roundtable, 'Historians and the Question of "Modernity", Introduction', *American Historical Review*, 116, no. 3 (2011): 631–37.

Basalla, George, 'The Spread of Western Science. A Three-Stage Model Describes the Introduction of Modern Science into Any Non-European Nation', *Science*, 156 (1967): 611–22.

Beinart, William, Karen Brown, and Daniel Gilfoyle, 'Experts and Expertise in Colonial Africa Reconsidered: Science and the Interpretation of Knowledge', *African Affairs*, 108, no. 432 (2009): 413–33.

Bruchhausen, Walter, 'Medical Pluralism as a Historical Phenomenon: A Regional and Multi-Level Approach to Health Care in German, British and Independent East Africa', in *Crossing Colonial Historiographies. Histories of Colonial and Indigenous Medicines in Transnational Perspective*, ed. Anne Digby; Waltraud Ernst, and Projit B. Muhkarji (Newcastle: Cambridge Scholars Publishing, 2010), 99–113.

Chambers, David Wade and Richard Gillespie, 'Locality in the History of Science: Colonial Science, Technoscience, and Indigenous Knowledge', *Osiris* 15 (2000): 221–40.

Digby, Anne, Waltraud Ernst, and Projit B. Muhkarji, eds., *Crossing Colonial Historiographies. Histories of Colonial and Indigenous Medicines in Transnational Perspective* (Newcastle: Cambridge Scholars Publishing, 2010).

———, 'Introduction: Crossing Historiographies, Connecting Histories and Their Historians', in *Crossing Colonial Historiographies. Histories of Colonial and Indigenous Medicines in Transnational Perspective*, ed. Anne Digby, Waltraud Ernst, and Projit B. Muhkarji (Newcastle: Cambridge Scholars Publishing, 2010), ix–xxii.

Driver, Felix and Lowri Jones, *Hidden Histories of Exploration* (London: Royal Holloway University of London, 2009).

Duminy, Andrew, *Mapping South Africa: A Historical Survey of South African Maps and Charts* (Auckland Park: Jacana, 2009).

Etherington, Norman, ed., *Mapping Colonial Conquest: Australia and Southern Africa* (Crawley: University of Western Australia Press, 2007).

Harries, Patrick, *Butterflies & Barbarians: Swiss Missionaries & Systems of Knowledge* (Oxford: James Currey, 2007).

Harries, Patrick and David Maxwell, eds., *The Spiritual in the Secular: Missionaries and Knowledge about Africa* (Cambridge: Eerdmans, 2012).

Harrison, Mark, 'Science and the British Empire', *Isis*, 96, no. 1 (2005): 56–63.

Hokkanen, Markku, 'Towards a Cultural History of Medicine(s) in Colonial Central Africa', in *Crossing Colonial Historiographies. Histories of Colonial and Indigenous Medicines in Transnational Perspective*, ed. Anne Digby, Waltraud Ernst, and Projit B. Muhkarji (Newcastle: Cambridge Scholars Publishing, 2010), 143–64.

Horton, Robin, *Patterns of Thought in Africa and the West: Essays on Magic, Religion and Science* (Cambridge: Cambridge University Press, 1993).

Keighren, Innes M., Charles W.J. Withers, and Bill Bell, *Travels Into Print. Expoloration, Writing, and Publishing with John Murray, 1773–1859* (Chicago: University of Chicago Press, 2015).

Kennedy, Dane, *The Last Blank Spaces. Exploring Africa and Australia* (Cambridge, MA: Harvard University Press, 2013).

———, 'Introduction: Reinterpreting Exploration', in *Reinterpreting Exploration: The West in the World*, ed. Dane Kennedy (Oxford. Oxford University Press, 2014), 1–20.

Kuhn, Thomas, *The Structure of Scientific Revolutions* (Chicago: Chicago University Press, 1962).

Lamar, Howard and Leonard Thompson, eds., *The Frontier in History: North America and Southern Africa Compared* (New Haven and London: Yale University Press, 1981).

Livingstone, David N., *Putting Science in Its Place: Geographies of Scientific Knowledge* (Chicago: University of Chicago Press, 2003).

———, 'Landscapes of Knowledge', in *Geographies of Science: Geographies of Scientific Knowledge*, ed. P. Meusburger, D. N. Livingstone, and Heike Jöns (London: Springer, 2010).

Livingstone, David N. and Charles W.J. Withers, eds., *Geographies of Nineteenth Century Science* (Chicago: University of Chicago Press, 2011).

MacLeod, Roy, 'Reading the Discourse of Colonial Science', in *Les Sciences Coloniales: Figures et Institutions*, ed. Patrick Petitjean (Paris: ORSTOM, 1996), 87–96.

———, 'On Science and Colonialism', in *Science and Society in Ireland: The Social Context of Science and Technology in Ireland, 1800–1950*, ed. Peter J. Bowler and Nicholas Whyte (Belfast: Queen's University, 1997), 1–17.

———, 'Introduction', *Osiris* 15 (2000): 1–13.

———, 'Imperial Science under the Southern Cross: Archibald Liversidge, FRS, and the Making of Anglo-Australian Science', in *Science Across the European Empires, 1800–1950*, ed. Benedikt Stuchtey (Oxford: Oxford University Press, 2005), 175–214.

Penn, Nigel, 'Mapping the Cape: John Barrow and the First British occupation of the Colony, 1795–1803', *Pretexts*, 4, no. 2 (1993): 20–43.

Pratt, Mary Louise, *Imperial Eyes. Travel Writing and Transculturation* (London: Routledge, 2008, second edition).

Reingold, Nathan and Marc Rothenberg, eds., *Scientific Colonialism* (Washington, DC: Smithsonian Press, 1987).

Said, Edward, *Orientalism* (London: Routledge, 1978).

Singaravélou, Pierre, *Professer l'Empire: Les "sciences coloniales" en France sous la IIIe République* (Paris: Publications de la Sorbonne, 2011).

Stone, Jeffrey C., *A Short History of the Cartography of Africa* (Dyfed, Wales: Edwin Mellon Press, 1995).

Stuchtey, Benedikt, *Science Across the European Empires, 1800–1950* (Oxford: Oxford University Press, 2005).

Tilley, Helen, 'Global Histories, Vernacular Science and African Genealogies; Or, Is the History of Science Ready for the World?', *Isis*, 101, no. 1 (2010): 110–19.

———, *Africa as a Living Laboratory. Empire, Development, and the Problem of Scientific Knowledge* (Chicago: University of Chicago Press, 2011).

Vaughan, Megan, *Curing Their Ills: Colonial Power and African Illness* (Stanford: Stanford University Press, 1991).

Vetter, Jeremy, ed., *Knowing Global Environments: New Historical Perspectives on the Field Sciences* (Piscataway, NJ: Rutgers University Press, 2010).

Vogel, Jakob, 'Von der Wissenschafts- zur Wissensgeschichte. Für eine Historisierung der "Wissensgesellschaft"', *Geschichte und Gesellschaft* 30 (2004): 639–60.

White, Richard, *The Middle Ground: Indians, Empires, and the Republics in the Great Lakes Region, 1650–1815* (Cambridge: Cambridge University Press, 1991).

Part I

Mapping and exploring

2 Peter Kolb and the circulation of knowledge about the Cape of Good Hope

Nigel Penn and Adrien Delmas

Introduction

In 1719 the German astronomer Peter Kolb published, in Nuremberg, a major work about the Cape entitled *Caput Bona Spei Hodiernum*, or *The Cape of Good Hope Today*.[1] This book, written in German, would subsequently be translated into Dutch, English and French and become one of the most influential sources of knowledge about the Cape in the 18th century. Though Kolb's reliability as a source of knowledge would be questioned by later writers, his book was regarded as an essential work on the early colonial Cape, the authority of which might be refuted or accepted but could certainly not be ignored. It was, indeed, the first work ever published to deal exclusively with the Cape. The history of the reception of Kolb's book (or books, if one takes into account the various editions) and the fluctuation of Kolb's reputation, serve as an interesting example of how, during the era of the rise of modern science, the status of authoritative knowledge was shaped by different processes involved in the circulation of knowledge. It is the intention of this chapter to consider the impact of some of these processes on the authority of Kolb's work and to end with a discussion of the reception of *Caput Bona Spei Hodiernum* at the Cape itself. In this way the circulatory route taken by Kolb's book will have been traced from its place of inception, to its place of reception and then back to its place of origin, during which journey the dynamic nature of knowledge, with its propensity to change its value according to its geographical location, will be revealed.

In order to place Kolb within the context of knowledge circulation in the 18th century it will be helpful to refer to Kapil Raj's ideas about the production of knowledge in the early modern period. In his book *Relocating Modern Science*, Raj emphasises 'the negotiated, contingent and situated nature of the propositions, skills and objects that constitute natural knowledge'.[2] Scientific research, in other words takes place in a particular place and is an outcome of many complicated interactions rather than the imposition of a pre-existing system of hegemonic knowledge. This is particularly true of those domains of natural knowledge that Raj calls the 'open air' sciences, branches of knowledge that were not dependent on laboratories for their observations, such as natural history, navigational astronomy or geographical exploration.[3] What Raj is concerned with here is to combat the idea that science in the modern period was like Edward Said's idea of Orientalism, that is that it was a discursive discourse associated with Western imperialism, that simultaneously subjugated and relegated the rest of the world to a type of Other. He is also concerned to emphasise the place where science took place, namely, not just in European laboratories, libraries or even coffee shops.[4] Knowledge creation often began in a non-European country or a colony and then circulated through the networks of the modern world, changing as it went. Though knowledge creation might have occurred in a specific geographical place, for instance the Cape, modern science also developed in a specific space, namely the globalised space of European expansion. This ensured that intercultural encounters as well as circulation would play a major role in 'the specialized knowledges that constituted science in this period'.[5]

All of these are important points to bear in mind when we consider the case of Peter Kolb who, in his own person, exemplified these characteristics and whose work was shaped by transnational movement and intercultural encounters.[6] Kolb was, by origin, German, born in Dorflas and trained in many branches of knowledge – philosophy, mathematics, physics, logic, metaphysics, oriental languages and theology – before being granted a doctorate in astronomy at the highly regarded University of Halle in 1700. He then accepted a job as tutor to the son of Baron Bernhard von Krosick before agreeing, at the latter's suggestion, to journey to the Cape in order to make astronomical observations. His trip to the Cape was enabled by the patronage of one of the directors of the Dutch East Indies Company (VOC), Nicolas Witsen, Mayor of Amsterdam, and a man renowned for his scientific curiosity and extensive contacts with men of learning.[7]

The VOC was, in 1705 (the year Kolb arrived at the Cape), the foremost international trading company of the day and the Cape the most visited spot on the oceanic trade route between Europe and Asia. The Cape was, as Kapil Raj stresses, one of 'the obligatory passage points' for trading ships in the Indian Ocean, particularly since it had been turned into a depot for supplying drugs and medicines to the crews of VOC vessels. These medicines were developed with the aid of plants, from all corners of the Indian Ocean or beyond, introduced into the Company Garden at the Cape and grown especially for their medicinal or commercial properties following the advice of botanical experts.[8] Kolb, therefore, found himself at one of the most important hubs, or nodes, of global circulation on earth and was well positioned to gather information and exchange ideas not just from passing travellers (who included some illustrious scientists) but also from the Cape's inhabitants.[9] These inhabitants encompassed not only the indigenous Khoikhoi but also the local settler community and their slaves. The East really did meet the West at the Cape which, besides, at that time, was the most important European entrepot into sub-Saharan Africa.

The role played by the VOC in the accumulation and circulation of knowledge has been receiving increasing attention from historians. Whilst some suggest that the institution itself was largely unsupportive of scientific investigation, others have stressed the key role that was played by curious or dynamic individuals within the system in accumulating knowledge despite the lack of Company support.[10] It is also evident that certain branches of knowledge, such as navigation and information about the properties of plants, were absolutely vital to the Company's commercial interests and could hardly be ignored.[11] As Harold J. Cook writes, 'employees of the company produced some of the most important works on early modern medicine and natural history of the New World, Africa, and Asia' despite the counter productive behaviour of the VOC in some instances.[12] Whatever the VOC's official attitude towards knowledge creation and processing may have been, however, it cannot be denied that the infrastructure of the VOC facilitated the rapid circulation of knowledge along its networks and that the Cape was a key point in these networks. Until recently the Cape's importance as a node of knowledge circulation has been somewhat overlooked by those historians who are interested in the connection between the VOC and the circulation of scientific knowledge, with emphasis falling on the collection of such knowledge in the east and/or its processing and dissemination in the west. There has been, indeed, a notable tendency of Dutch scholars (possibly influenced by the cultural boycott of the apartheid years) to ignore the Cape's place in the VOC's eastern empire altogether and to focus almost entirely on the East Indies alone. The Cape, when it was mentioned, was characterised as being a liminal space, between worlds, and possessing no qualities or significance of its own beyond its function as a refreshment station.[13] Though VOC period maps of the Cape sea route at least showed the Cape, albeit stressing its position as an outpost at the end of the world inhabited by savage "Hottentots", modern histories often had maps which showed the Asian sections of the VOC's domain alone and failed to show the Cape at all thereby ignoring the vital role it played at the southern end of the Indian Ocean trading system.[14] Post-apartheid scholarship is currently redressing this imbalance by re-inserting the Cape into its Indian Ocean context whilst emphasising its unique identity.[15] A reconsideration of Kolb's significance must be seen as an important part of this process.

Kolb had been sent to the Cape as an astronomer to map the stars of the southern sky but once he arrived he decided to describe the extraordinary world of the Cape of Good Hope instead. The reasons for his change of focus are obscure but probably unconnected with later charges of astronomical incompetence or habitual drunkenness. He was actively engaged in the project of collecting information about the Cape when the death of Baron von Krosick obliged him to earn his keep by entering VOC service as Secretary of the Stellenbosch district in 1711. In 1712 Kolb became almost completely blind and returned to Europe where, fortunately, his eyesight was partially restored by an operation. In 1718 he was appointed to the position of Rector at the gymnasium of Neustadt an der Aisch, in which position he wrote his Caput Bona Spei. He died in 1726.

From the above brief synopsis of Kolb's life it may be seen that a great deal of his initial intellectual authority stemmed from his position within the established networks of European knowledge and patronage. At one stage these networks nearly led him to the acceptance of the Chair of Mathematics at the University of Moscow which, had he taken it, would have removed him to the margins of European knowledge circulation. But even Russia was part of global networks.

Nicolas Witsen, the VOC Director, was an acknowledged expert on Russia having written a treatise on that country as well as having initiated Peter the Great into the mysteries of Dutch ship building.[16] Kolb, through Krosick, used Witsen's connections to open a more fruitful area of investigation for him. Somewhat against the more pragmatic wishes of the Directors of the VOC, whom Witsen accused of being more interested in money than in science, Witsen succeeded in getting Kolb the necessary documentation to undertake research at the Cape.[17] Happily for Kolb he eventually headed south, to 'the most famous Cape in the World',[18] and happily, too, he supplemented his mathematical and astronomical skills with those more in keeping with the new task that he set himself – a comprehensive description of the Cape.

It is no easy task to decide what sort of book Kolb set out to write, for in keeping with the fluid nature of circulatory knowledge it is, in many ways, a book in transition between genres. When the book was first published, in 1719, it had the appearance of a travel narrative, especially since it was written in the form of a first-person narrative addressed, in an epistolary form, to Kolb's sponsor, Baron von Krosick. In many respects it was similar to other travel narratives of the 16th and 17th centuries. Since travel was an activity inextricably bound to European global expansion travel writing was the characteristic genre of the era.[19] The genre was popular as it had the ability to contain a great deal of new or interesting material, supposedly observed by an individual during the course of a journey, in a loose and adaptable form. The presence of the narrator in the text as an eye-witness observer and the personal guarantor of factual veracity provided the book with its authority and plausibility, as did the fact that such books were often dedicated to respectable or trustworthy patrons.[20] As Joan-Pau Rubies has explained, such travel literature 'provided, in effect, the foundations for an empirical scientific discourse', as the genre initially focused on practical, empirical descriptions in everyday language but later developed 'more sophisticated ethnographic and naturalistic discourses and engaged in more theoretical, ideologically charged debates amongst the educated elites'.[21] Kolb certainly has an easy, colloquial style, full of anecdote and personal comments, but there are also passages in the book where he puts aside his persona as an entertaining raconteur and attempts to contribute to learned debates or to add to the stock of knowledge. Of great significance here is the fact that the Dutch edition of the work, published in 1727 in Amsterdam, dispensed with the epistolary style altogether and replaced each "letter" by a "chapter". In this we may see the shift in genre, within a decade, from an epistolary to an encyclopaedic form.

Another way in which the book is linked to travel narratives is through the introductory chapters, or letters, that describe the voyage out to the Cape. In the travel genre the voyage is an essential component of the book since it is the voyage that creates distance and traverses space. It is the voyage itself that provides the authenticity of travel and sets up the possibility of exotic encounters. The voyage is every bit as important as a detailed description of one's destination, for the "other" world begins at sea.[22] In travel narratives the voyage is often full of adventures – storms, shipwrecks, pirates – and hardships. There can be no ordinary journey. In this respect Kolb does not disappoint. He set sail on the VOC ship the *Unie* in 1704 and had a dreadful voyage. His narrative, written with one eye on other accounts by German-speaking VOC voyagers to the East, attempts to inform his land-locked readers of the strange world of oceanic travel.[23] Arguably, the transformative experience of the voyage in the prison-like confines of the ship, subject to strict discipline and the brutal company of European savages, namely, the sailors and soldiers of the VOC, predisposed him to view the Khoikhoi of the Cape more favourably once he arrived at his destination.[24]

Though the *Caput* begins with a description of Kolb's voyage out (pages 1–50) the voyage is not organised as a section on its own. It is, instead, in Part One of a three-part book. The 1719 edition is a large and very luxurious folio-book volume divided into three sections: the Erste Theil (from pages 1–316) constitutes a natural history or *physicalia* of the Cape; the Zweyte Theil (from pages 317–582) deals 'only with the Hottentots' and the Dritte Theil (from pages 583–846) deals with 'the European colonies in the Cape headland'. As noted above, Kolb wrote the book after his return to Europe in 1713 and based it upon notes that he had made at the Cape (some of which he lost) as well as reports given to him by, or which he had been allowed to copy from, various people at the Cape. These "borrowings" would later expose Kolb to charges of plagiarism, In fact, Kolb acknowledged his debts to others and took care throughout his book to cite his sources.[25] We should rather be inclined to see such borrowing as proof that there were men of learning at the Cape who were happy to exchange knowledge with a visiting scholar. Similarly, Kolb himself was able to share his knowledge with visiting scholars. In 1706, for instance, the missionaries Bartholomus Ziegenbalg and Heinrich Plutschau visited the Cape on their way to the Danish trading post at Tranquebar on the Malabar coast of India. These men were employed by the Halle mission and went on to write about the flora and fauna that they had seen on their travels in Africa and the East Indies.[26] Kolb met with these men from his old alma mater and very likely corresponded with them subsequently as he cites Ziegenenbalg's comments on the Khoikhoi language from the latter's account of his journeys as well as the Danish missionary Johannes Boving's *Curieuse Bescreibung und Nachricht von den Hottentoten*.[27] We must also assume that, back in Europe, Kolb consulted copies of his own personal correspondence to refresh his memory before beginning to write his book as there are, for instance, copies of his letters to Witsen in the archives of the Gymnasium at Neustadt.

Natural history

Kolb himself referred to his work as a history but his instincts are clearly encyclopaedic and parts of his book organise information in accordance with this venerable genre.[28] This is especially true of Part One, which deals with natural history. Although Kolb became most famous for what he wrote in the second part of the book, on the Khoikhoi, it is important not to neglect his observations on natural history. Significantly, these observations take place in Part One, signalling to us that Kolb believed that his readers would be most interested in this section. Since

Kolb's thoughts on natural history have not often received the consideration they deserve, one of the objectives of this chapter is to rectify this omission. Natural history enjoyed an enormous prestige at this time in Europe but especially in Holland where the most renowned natural historians of Asia happened to have been VOC officials: Hendrik Adriaan van Reede to Drakenstein (1636–91) who wrote about the Malabar Coast; Paulus Hermann (1646–95) who studied Ceylon and Georgius Rumphius (1628–1702) who wrote on Ambon and other places in the Indonesian archipelago.[29] The latter two men incidentally, were, like Kolb, Germans.[30]

Although one associates encyclopaedias with Diderot's Enlightenment project one should remember that Kolb is more typical of the pre-Enlightenment period, especially when it comes to classifying nature. Before the publication of Linnaeus's *Systema Naturaea* in 1735, in which that botanist attempted to systematise, classify and order the plant kingdom based on empirical observation and the precise measurement of observable differences, there was no agreed upon method of classifying plants. Linnaeus divided plants into classes based on structural variations in their sexual, or flowering, parts, but this was not a system available to Kolb.[31] Instead Kolb, like other scholars of his era (such as Conrad Gesner)[32] delighted in getting things down in ordered lists, the ordering principals of which are best sought for in the realms of resemblance and representation that Foucault sees as having been influential as a way of knowing in the 16th and 17th centuries.[33] Within these lists, things were organised according to the alphabetical order of their names, in German. Truth to be told, Kolb enjoyed listing for the sake of listing, and in this regard he is typical of an age, in Europe, where people delighted in collecting exotic facts, curiosities and rarities for their own sake. Benjamin Schmidt has noted the great outpouring of Dutch works celebrating the geographically exotic around about 1700 and Kolb's book should certainly be seen in this context.[34] It was the era, especially in the Netherlands, of the *rariteiten-verzamelingen* and the *naturalienkabinetten*. Witsen was famous for his *rariteiten-verzamelingen* and Kolb would certainly have seen it when he met the mayor in Amsterdam before he sailed for the Cape.[35] Kolb's book was, in many ways, a printed version of such a cabinet, a wonderful collection of extraordinary odds and ends crammed into one cover.

Once Kolb has described his voyage and arrival at the Cape, which takes place in letters I to IV, he turns to natural history. More specifically, he begins by describing the topographical features and agricultural productivity of the land. In some ways this is simply a review of the economic potential of the colony for such basic, pragmatic assessments of a region's wealth were a fundamental part of travel literature and uppermost in the thinking of colonial powers. The basic question was in what ways could the natural history of a region be turned to economic advantage?[36] How did the environment influence economics? It cannot be said that Kolb's survey of these matters in any way contributes to natural history since this was not his intention. What he is doing is describing the extent of the VOC's settlement at the Cape and its utilisation of natural resources. These are his opening remarks about the nature of Dutch colonialism at the Cape, a subject he will pursue at greater length in Part Three.

Once these introductory remarks have been made he turns to natural history proper, in letter XI, 'in which all the Animals are dealt with alphabetically, which are to be found at the Cape of Good Hope' and then deals with various other creatures through to the end of letter XIV. Altogether Kolb devotes 65 quarto pages to animals. The expert on the zoological exploration of southern Africa in this period, L.C. Rookmaker, is not impressed by Kolb's contribution and makes the following observations:

> Later workers who consulted the very voluminous natural history section in Kolb's book had few good words about it. De la Caille (1763) included a long section of criticism. Sparrman, in a letter to George Foster of 25 July 1777, called him the 'most impudent lying contradictory fellow'. Part of this criticism must stem from a general feeling of disappointment which we even sense today. . . . It is remarkable how little new or even useful information was presented. Kolb often excused himself, saying that he was not a zoologist. At the start of his chapter on birds, for instance, he mentioned having lost his notes which occasioned the short list of birds. When the identity of the animal is not clear from the name and more information is sought in the descriptions, these almost invariably are so general and meaningless that one cannot reach any conclusions. Kolb compared the African species with those in Europe and many of his notes refer to those European animals. This, of course, does not make the contribution of Kolb any worse than that of his contemporaries and we must call his book a brave attempt. However to read and study 65 pages without much value, does give the feeling of wasted time.[37]

Despite these harsh words it is possible for a non-zoologist to find a great deal to admire in Kolb's account. What Rookmaker does not seem to notice is that Kolb has gone to great pains to gather local knowledge about each animal and presented it with rich anecdotal detail, if not with accompanying accurate scientific observations. There are many ways, in fact, in which Kolb's description of the Cape's animals is similar to the natural histories of the Renaissance, such as Gesner's histories, which included information about the legends and folklore surrounding animals and their emblematic meaning in the cultures in which they were found.[38] Kolb's bestiary is a good guide to what Khoikhoi and colonists thought or believed about each creature. The section on the baboon, for instance, contains Kolb's personal observations and experiences of baboons as well as stories that he has been told about their behaviour by both European and Khoikhoi informants. If some of the stories told seem far-fetched they are, nonetheless, indicative of the folklore surrounding baboons in the colony and a good indication of how baboons were perceived by locals. He includes, for instance, a "joke" that will be familiar to all South Africans even today.[39]

In his willingness to entertain local knowledge Kolb is actually meeting one of the necessary criteria for doing "field science" in the colonies, for he is not trying to impose his own beliefs on an indigenous understanding. Kolb frequently names his informants, at least the European ones, and suggests that their observations are trustworthy. His writing is also humorous and lively whilst his attempt to compare African with European animals probably won him more readers than he lost when his book came to be published.[40] It is likely that Kolb consulted Hondius's *Klare Bescryving van Cabo de Bona Esperanca*, a book that was published in the year the Dutch settled at the Cape, 1652, and was almost certainly available at the Cape. At the beginning of the 18th century it was still the best and most detailed list of animals of the Cape, with their Dutch names, available.[41]

Although Kolb was no more a botanist than he was a zoologist he was better situated at the Cape to gather information about plants than he was about animals. This is because when he arrived at the Cape Governor Willem van der Stel provided him with accommodation in "a pleasant garden-house" in the Company gardens.[42] The gardens were already famous throughout Europe as one of the most remarkable features of the Cape. They had originated out of the VOC's need to grow fruit and vegetables as anti-scorbutics for the crews of passing ships but had developed into a testing ground to see which crops or plants from the rest of the world would do well at the Cape and might, perhaps, be profitably cultivated. The plants were mainly valued for their commercial or medicinal properties but, inevitably, such a varied collection of exotic species attracted the attention of visiting botanists as they travelled between Europe and the Dutch East Indies or other parts of Asia. In addition to exotic flora the Cape could also boast its own unique floral kingdom and it quickly became the Mecca of botanists and plant collectors throughout the world. Specimens of Cape plants, some living and some dead, found their way into collections and herbariums in many European countries but nowhere more so than in the Netherlands, due to the activities of the VOC. Largely thanks to the reach of its trading networks, which allowed it to gather both material objects and knowledge from the ends of the earth, Holland was, at the time, the pre-eminent centre for the study of natural history and medicine.[43]

One of the foremost scientific herbaria and gardens in the world was to be found in the University Garden at Leiden, in which 34 Cape specimens were growing as early as 1687. These plants had been collected by Paul Hermann, like Kolb a German in service with the VOC, who had visited the Cape in 1672 and, after his appointment as Director of the University Garden at Leiden in 1680, stocked it with plants from the East Indies, America and especially, thanks to his VOC contacts, from the Cape. Via Hermann, Cape plants reached collections in Denmark and England. One of the beneficiaries of his zeal was Linnaeus who spent three years in the Netherlands (1735–7) studying its famed botanical collections and who wrote thus of Hermann: 'O God. Hoe talrijk, hoe zeldzaam, en hoe wonderschoon waren de planten die zich aan Hermanns ogen voordeden. In enkele dagen ontdekte Hermann alleen meer Afrikaanse planten dan alle botanici voor hem'.[44] Hermann had described and collected 791 items before his death in 1695 and it is likely that copies of his work were available to the chief gardener of the Company Gardens at the Cape, Johannes Hertog, during Kolb's stay.[45]

Another botanical authority whose work on Cape flora pre-dated Kolb was Willem ten Rhyne, a physician in the VOC who left descriptions of Cape specimens and who, from his posting in Java, helped the VOC Commissioner-General and botanist Hendrik Adriaan Van Reede tot Drakenstein work on his famous *Hortus Malabaricus* (1678–1703).[46] Ten Rhyne also assisted Rumphius with his *Herbarium Amboinense* which eventually appeared between 1741 and 1755.[47] Both Van Reede tot Drakenstein and Rumphius had visited the Cape (the latter in 1652) and although their work was mainly on the natural history of the East Indies it serves to remind one that the Cape was intimately linked to the major scientists and botanists of the Dutch world. The Cape, in other words, was not a backwater. Rather, it was on the mainstream for the transfer of botanical and scientific knowledge between east and west. Kolb could not claim to be in the same league as these botanists, nor did he ever meet them. More modestly he acknowledged that what he knew about botany he owed to 'daily conversations with Johannes Hertog, in charge of this well-renowned garden of the said Company; and finally to the specimen-albums of the late Herr N. Olderland, which his stepsons Johann and Andreas Blum repeatedly allowed me to see'.[48]

Olderland, or, correctly, Heinrich Berhard Oldenland, was yet another German born physician in the service of the VOC. He had studied botany for three years at Leiden and began working for Governor Simon van der Stel as master-gardener and surveyor in about 1690. At the Company Gardens he formed a team with Jan Hartog, also German born and trained at the Leiden University Gardens. Oldenland was the more scientifically trained botanist and Hartog the more practical gardener. Oldenland worked on a "Kruid Boek" of dried and mounted plants, together with a catalogue of Latin descriptions and phrase names. He died in 1697 at the age of 34 but it is this book that his appropriately named stepsons – Blum – showed to Kolb.[49] The same book was examined and utilised by Francois Valentijn when he visited the Cape on his way to and from the East in 1685, 1695, 1705 and 1714. Valentijn's account of his travels appeared between 1724 and 1726 and included a section on the Cape.[50] From Valentijn's description it appears Oldenland's catalogue, or *Herbarius Vivus*, consisted of

> 13 to 14 volumes in folio, with a very fine description in Latin of each plant. I saw this work after his death, and often read it with great pleasure. There were some Englishmen who took a great interest in it, but the owner, Mr Donker, his heir, asked too much for it in their opinion. The specimens were unusually lovely, exceptionally well dried, and still so living in colour that it was a pleasure to see them.[51]

Valentijn's list of Cape plants, was, incidentally, almost identical to Kolb's list. This need not surprise us since Valentijn visited the Cape whilst Kolb was there, in 1705, and was friends with Governor Willem Adrian van der Stel.[52] Valentijn and Kolb may well have met each other. At the very least they were shown the same catalogue.

Another VOC official who had examined the book was Johannes Starrenburg, Landdrost of Stellenbosch who, a short while before Kolb arrived at the Cape, had been in correspondence with the English apothecary Petiver, offering him plant specimens from the Cape and stating that

> it would be a great pity, after the death of Dr Oldenland you should be destitute of collections, here to be made, his widow is marrying again to a man who will not trouble his head with these foelerys (as he calls them).[53]

It would seem that whilst Donker may not have valued botanical research, he was, at least, alert to its monetary worth for the collection eventually ended up in Holland in the possession of Johannes Burman, Professor of Botany at the University of Amsterdam, who published a list of the plants as an appendix to his *Thesarus Zeylanicus* in 1737.[54]

Kolb also mentioned having read the works of Georg Meister and Father Guy Tachard. Meister, another German in VOC employment, had travelled to the east as a gardener in the 1670s. From there he had helped to dispatch trees, such as tea, camphor, guava, banana and pineapple, to the Company's garden at the Cape. On his return to Holland in 1688 he stopped at the Cape, bearing more plants, and from there took 17 chests of soil and all sorts of plants, to various notables in Europe. He also took back with him plant bulbs and plant drawings. In 1692 he published an account of his travels, which Kolb read, called *Der Orientalisch-Indianische Kunst und Lustgartner*.

By mentioning Tachard, Kolb reveals that he must have been familiar with the work of Heinrich Claudius, a German apothecary in employment with the VOC who was one of the first to produce illustrations of Cape plants. Claudius began his work as the assistant of Andreas Cleyer in Batavia. Cleyer was a German physician studying medicinal plants in the east and he dispatched Claudius to the Cape to report on medicinal plants there. Claudius soon found employment with governor Simon van der Stel and accompanied several expeditions of exploration to Namaqualand in the northern Cape, one of them with the governor himself.[55] In March 1685 and 1686 Claudius met Tachard, a French Jesuit and scientist who visited the Cape on his way to and returning from Siam. Also present at the Cape at this time was Hendrik van Reede tot Drakenstein, the VOC Commissioner and author of *Hortus Malabaricus*. Tachard conversed with these distinguished botanists and in the first volume of his *Voyage de Siam* he refers to having been impressed by Claudius's detailed notes and his fine drawings and paintings of plants and animals. He went on to write:

> He has already completed two thick folio volumes of diverse plants, painted from nature, and he has collected specimens of all kinds which he has pasted into another volume. Doubtless the Baron van Reede who always keeps these books in his own apartment, and who it was allowed us to see them, intends publishing a *Hortus Africus*, after his *Hortus Malabarius*. Had these books been for sale we would have spared nothing to acquire them for the library

> of the King. As this learned Doctor has already made several journeys, to a distance of one hundred and twenty leagues North and East of the Cape, it is from him that we obtained all our knowledge of the country. He gave us a little map made by his own hand, and some drawings of the inhabitants, which I am inserting in my book.[56]

It was this acknowledgement from Tachard that got Claudius into trouble. On his next visit to the Cape in 1687 Tachard found that Claudius had been removed from his position and sent to Mauritius or Batavia. Governor van der Stel had decided that Claudius had shown too much familiarity to the French.

> What understanding he had with the Jesuits has been fully shown, to our great perturbation, by their book lately published regarding their Siamese voyage. In it his name is mentioned, and it is plainly stated that he communicated to them everything about the colony and our inland expeditions, and perhaps more besides than we know of.[57]

This incident alerts us to another aspect regarding the VOC's attitude towards the circulation of knowledge. When knowledge was deemed to be strategically or financially valuable the VOC could show a secretive, possessive and obstructive attitude towards those trying to obtain it. This was particularly the case if suspicious foreigners were involved and access to the Cape interior was closely regulated. Maps, in particular, were a closely guarded secret and the Dutch took great care to keep them from circulating.[58] It is perhaps for these reasons that Kolb was never allowed to travel much further than the hot springs at present-day Caledon. Similar suspicions may also help to explain Kolb's subsequent fall from grace at the Cape.

From this brief account of the sources Kolb consulted for the creation of his plant list it is clear that, as far as botanical science was concerned, the Cape was a most important node of knowledge circulation. It is also clear that even though Kolb was not a botanist he was no worse qualified than many other men of letters who wished to contribute to the circulation of knowledge in that he had consulted the best available authorities on Cape flora at the time, namely, Oldenland and Hartog and had seen Witsen's collection before he left the Netherlands. Kolb gave the plant names in German and Latin. He also attempted to take an inventory of both indigenous and exotic plants, taking great care to describe which foreign plants, or plants of commercial potential, did well in the Cape. What makes his contribution exceptional, however, are the same virtues that he brought to his list of animals. Kolb took care to try to provide both Khoikhoi and local Dutch names for the plants he described. He also tried to explain what uses indigenous people made of the plants, describing how the Khoikhoi had not only a knowledge of edible plants but were also able to produce a variety of salves, ointments, powders and infusions from plants that they used with great efficacy to heal people, stating that: 'They know how to seek out and name the herbs' . . . 'nor lack healing herbs . . . of which they also use some very skilfully'.[59] Kolb specifically mentions *buchu* and *dagga* as useful in healing wounds but adds that the Khoikhoi were very secretive about their other healing herbs. He does, however, mention aloes, wild sallow leaves, wild garlic, marsh-mallow, wild fennel and wild figs as being medicinal plants.[60]

Like his descriptions of animals, Kolb's chapters on plants (letters XVI–XVII) are not so much botanical descriptions as a cultural history of plant knowledge at the Cape. Consider, for example, his description of wild asparagus or asparagus *syvestris aculeatus*:

> With very sharp thorns on the branches. This plant is found in the open country everywhere around this Cape, and also near water or in other damp places, and is so prickly that one's stockings are often torn when passing near or through it. The stems are grass-green, and nevertheless so soft, if cut before they bloom and bear berries, as any cultivated asparagus could be. Their taste is also very pleasant, as I often found from curiosity, although the cultivated asparagus grows in the greatest abundance. The slaves and Hottentots bring it to sell for very insignificant trifles, so that one can indulge one's taste for it before the cultivated asparagus becomes available, or when the season is over.[61]

Here may be seen Kolb at his best. He does not only describe the plant but he describes his personal experience of it in vivid terms. He also describes how the plant had been assimilated into the lives of the Khoikhoi, slaves and Europeans in the context of the colonial Cape. Once again, this was an approach that was perhaps more in keeping with the emblematic natural history of the Renaissance than with the post-Linnaean classificatory systems of the Enlightenment but this merely serves to remind us that, in matters of natural history, Kolb was of his time and not ahead of it. The same cannot be said of his description of the Khoikhoi.

The Khoikhoi

Part Two of the *Caput Bona Spei* deals exclusively with the "Hottentots" and was the section of the work that contributed most to the book's success. It is not the intention of this chapter to describe the contents of Part Two in detail but rather to describe how Kolb's writings and ideas about the Khoikhoi were circulated and received in both Europe and the Cape. A major aspect of Kolb's work was that it was seen, at the time, as an attempt to rehabilitate the image of "the Hottentot", a people who had been portrayed in travel literature as 'the most vile and horrible people of the creation' or the most pitiful savages ever discovered.[62] Before Kolb, European visitors to the Cape, and, indeed, many who had never been there, portrayed the Khoikhoi as being a lazy, ugly, rude, stinking and shameless people. Previous publications had competed with each other in terms of deprication, depicting the Khoikhoi as savages endowed with nothing save bestiality and stupidity.[63] In his preface Kolb spoke of having to find the necessary courage to write his book in the face of accounts of authors 'who wrote useless, deceitful and false things on the Hottentots and their country'.[64] He presented his rehabilitative undertaking as being the result of prolonged, direct and objective observations and not motivated by any laudatory intentions. The transformation that Kolb was able to bring about in the way that the Khoikhoi were represented by Europeans – from ignoble to noble savage – has been the subject of several recent publications which agree on the success of this rehabilitation.[65] So successful was Kolb's argument that, on the strength of his observations on the Khoikhoi, the encyclopaedic *Caput* became the recognised authority on anything to do with the Cape until the end of the century.

What evidence is there for this success? The success can partially be measured by the book's publishing history and partially by the attention that it received. The first edition, published in German in Nuremberg in 1719, was a large and extremely luxurious folio-book volume divided into three sections.[66] A Dutch version, that dispensed with the epistolary form, was published in Amsterdam in 1727. It contained a list of subscribers – several of them VOC officials – at the front and had new, improved illustrations. (For instance, the armour-plated rhinoceros of the German edition is replaced by a more realistic representation.) It is significant that this publication, bearing signs of approval by Dutch authorities and the VOC, should have followed shortly after the very well-received publication of Valentijn's *Oud en Nieuw Oost Indien* between 1724 and 1726.[67] As Schmidt informs us,

> Six hundred readers signed on for Francois Valentijn's Oud en Nieuw Oost-Indien at a cost of 38 guilders per five volume set, netting the publisher a phenomenal gross of 20,000 guilders – this at a time when the average VOC sailor earned 120 guilders per annum.[68]

Before this date the VOC had been very secretive about its affairs, declining even to publish the official history it had commissioned from Pieter van Dam.[69] But Valentijn's work, based on his experience as a VOC minister of religion in Ambon and drawing on access to voluminous Company records, seems to have lifted the veil of secrecy somewhat, making the climate more favourable in the Netherlands to the reception of works containing information about the VOC. A cheap, two-volume English version, much abridged, in a rather loose translation by Guido Medley, saw the light of day in London in 1731.[70] A French version, *Description du Cap de Bonne Espérance*, was published by Jean Catuffe in 1741. It was an abridged version of the German text of 1719 but contained the illustrations from the Dutch volume of 1727. It retained the three-part structure and was published in three duodecimo volumes.[71] Volumes 2 and 3, the parts about the Khoikhoi and the Dutch colony, were reprinted in 1742 and 1743.[72] A second, abridged German version, based on the French translation, appeared in 1741 with illustrations from the first edition.[73] The French version, the *Description*, was reviewed in the *Journal des Savants*, the official organ of the Republic of Letters, in 1741. The *Journal* immediately conferred upon the *Description* and its author a certain authority: 'This detailed description, as is, is far more curious and instructive that those found in other travelogues'.[74] Parts of the book, or translations from it, soon found their way into various travel collections that were blossoming in those days.[75]

The fictional abuse of Kolb's depiction of the "Hottentots" is further evidence of the wide reception and reuse of the *Caput* in Europe. The London tabloid, the *Grub Street Journal* was quick to make use of the English version of 1731. In issue number 59 of 18 February 1731, for instance, the author of a satire made reference to a Khoikhoi rites of passage ceremony, described by Kolb, where the initiates were urinated upon by their elders. Several weeks later the *Gentleman's Magazine* outdid the *Grub Street Journal* in its reference to this custom. *The Connoisseur* ridiculed both Dutch and Khoikhoi in article of 1754 called 'Tquassouw and Knonmquaiha, A Hottentot Story' whilst other saw Kolb's influence writ large in parts of *Gulliver's Travels*.[76]

Yet, without doubt, the reading, interpretation and reuse of the *Description* by Jean-Jacque Rousseau are the most obvious marks of success of Kolb's rehabilitation of the Khoikhoi in Europe. The "Hottentot" as introduced by Kolb, became, if not the main character, at least the hero of the footnotes and, for this reason, the ultimate justification and guarantee of the veracity of the *Discours sur l'origine et les fondements de l'inegalie parmi les homes*, published in 1755 in Amsterdam. Rousseau said: 'I will mention just one confirmed example which admirers of European civility can examine. . . . All the endeavours of Dutch missionaries at the Cape of Good Hope have never lead to the conversion of a single Hottentot'.[77] These remarks were based on Kolb's story of Pegu, a Hottentot who 'had been taught the rudiments of Christian civilization but who chose to return to his savage life and renounce all things European after returning to his country'.[78] The story of Pegu seems to have impressed Rousseau enough for him to have it illustrated and to make the illustration the frontispiece of his book. Other famous authors of the 18th century also had a copy of Kolb's book in their libraries, such as Voltaire and Diderot, who quotes it explicitly, as does Volume 8 of the *Encyclopedie*. Created within the framework of the Republic of Letters Kolb's "Hottentot" had become a figure of the Enlightenment. But what did the colonial inhabitants of the Cape make of Kolb's attempt to rehabilitate the reputation of the Khoikhoi?

Amongst the most enlightened members of late 18th-century Cape society was the commander of the Cape garrison, Colonel Robert Jacob Gordon. Gordon, of Scottish descent, was a correspondent of Professor Allemand, professor of natural history in Leiden, and supplied the latter with information and specimens of fauna and flora. Whilst visiting Allemand in 1774 Gordon met the famous *philosophe*, Denis Diderot, and answered some of his many questions about the Khoikhoi. Gordon had travelled widely in the Cape interior and was able to assure Diderot that the Khoikhoi 'were not as stupid as was believed'. He went on to describe many of their customs, but remarked that 'Kolb, upon whom Monsieur de Buffon relied, knew nothing and was a liar, but he praised the truthfulness and accuracy of Monsieur l'Abbe de Lacaille, who he added was not often mistaken'.[79] Despite his favourable disposition towards the Khoikhoi, Gordon refused to see a kindred spirit in Kolb. Indeed, a few years later he wrote to Rousseau (unaware that the philosopher had just passed away) suggesting that the Frenchman journeyed to the southern tip of Africa to see for himself what the Khoikhoi were really like. 'I am quite certain that Rousseau, notwithstanding his attachment to how he feels (i.e. that primitive people are happier people), would totally change his tune should he find himself travelling among them'.[80] Once again, one feels, it is Kolb that Gordon wishes to discredit, or, more precisely, Kolb's book. For though his book was a source of inspiration in Europe it was being ridiculed in the Cape, reminding us that a book does not cross the ocean without changing its meaning.

The first reference to the *Caput* in the Cape of Good Hope was made by Johan Wolfgang Heydt, a German employee of the VOC. In his comments on the Cape, which he visited for a few weeks on the way to the East Indies in 1734, and on his way back from there in 1740, he claims to have read the monumental work of Peter Kolb in its 1727 Dutch edition.[81] Both in his comments and in his references to Kolb Heydt spoke very highly of the book, recommending it as an accurate guide to travellers wishing to visit the southern tip of the African continent. '[A]ny who desires to know more . . . let him read Kolb's description, who, as I was assured in Africa, has described everything truthfully, so that from it he will reach a good end'.[82] Such praise, however, was exceptional, and if the *Caput* once enjoyed an early reputation for excellence at the Cape this quickly deteriorated.

The leader of the attack on Kolb's reputation was the Abbe de la Caille, who, like Kolb, had first arrived at the Cape as an astronomer but left it as with much broader scientific and literary ambitions. De la Caille was at the Cape between 1751 and 1753 and wrote an account of his experiences ten years later. In them, he writes in the third person, that:

> The description of the Cape of Good Hope by Pierre Kolb made famous the name of this writer in the Republic of Letters. After the Abbe de la Caille had purchased Kolb's publication as a guide of great accuracy, before leaving for the Cape, he was very surprised, when arriving on site, to see that Kolb's account was inaccurate in almost every part of the book, and that instead of an exact description, Kolb's three volumes were just a novel full of tales.[83]

This is vicious criticism indeed. What was it that inspired such hostility? It is not unreasonable to suspect that De la Caille was jealous of Kolb's reputation and sought to advance his own book on the Cape at Kolb's expense. If he could demonstrate that Kolb's work was full of falsehood, readers would favour his own. De la Caille took pains to denigrate Kolb, referring to him as a man who spent most of his time at the Cape either in idleness, smoking and drinking to excess. He deemed him to have been a complete failure as an astronomer and declared that he had rightly earned the suspicion and mistrust of the governor, Willem Adriaan van der Stel, by taking sides in political disputes that did not concern him. De la Caille, in fact, suggests that Kolb acted the part of conduit for the colonists' complaints against the governor, writing down their complaints so that he could publish them later in Europe.

> On several occasions the residents tried, always unsuccessfully, to send their grievances against the bad administration of the Cape Colony to Holland.

> The reports they sent to the States of Holland in this regard were all intercepted and the colony continued to suffer under the oppression of the colonial administration. They imagined they could make a description of the Cape to Kolb, and in order to make it look more interesting, gathered people's opinions and even went as far as entreating Kolb, who did not know the country, with an infinity of marvelous descriptions which they took delight in imagining. Kolb's revelations were also based on the compilation of Mr. Grevenbroek of whom he spoke already; they included memorandums on the government of the Cape, in which they exposed the Government's injustice and indicated means to remedy this. Such matters being exposed by a foreigner could become public without compromising anyone. Kolb, enchanted by the service rendered to him, left the Cape with this work. He had it printed in Holland, as a translation from German. Surprisingly, the work was read avidly. The edition went out of stock in a short time and the Government of Holland, shocked by what was being said in it on the affairs of the Cape, made enquiries which turned out to be in conformity with Kolb's report. The main officers of the Cape Colony were recalled and dealt with ruthlessly on their return from Africa.[84]

What this passage reveals, in fact, is that De la Caille was far worse informed about Cape affairs than was Kolb. Though Kolb was indeed a hostile witness to Willem van der Stel's governorship he was not the man who conveyed the colonists' grievances to Holland. The colonists themselves did so. Van der Stel was recalled, in disgrace for corruption, in 1707, whilst Kolb was still at the Cape. Though Kolb was unpopular with certain VOC officials he was respected enough to be employed as secretary to the *landdrost* of Stellenbosch between March 1711 and January 1713 at a time when he needed employment due to the death of his patron, Von Krosick. Only when Kolb suddenly went blind, in 1713, did one of Van der Stel's surviving henchmen, the *secunde*, Helot, insist on Kolb being dismissed from his post and being returned to Europe on the first available ship.[85] Kolb's book was eventually published long after Van der Stel's dismissal and much too late to have influenced it.

De la Caille also attempted to damage the reputation of Kolb and his book by stating that Kolb had plagiarised Grevenbroek's notes:

> M. Grevenbroek, Secretary of the Court of Justice in the Cape at the beginning of this century and an extraordinary man, had conducted research on the customs of the Hottentots: after his death his papers were handed over to Kolb who compiled them without distinction or judgment, according to what the most sensible people from here are saying, namely the Governor, M. Grand-Pre and M. Dessin.[86]

But Kolb himself had acknowledged his debt to Grevenbroek and had certainly never pretended to know more than the local experts he relied upon.[87] If De la Caille's remarks about Kolb reveal anything other than the Abbe's jealousy it is probably this: that by the 1740s Kolb's book and its reputation were well known at the Cape and some of "the most sensible people" there believed that they knew better. Gerald Groenewald has written about the circulation of books at the Cape during the VOC period with certain elite individuals owning large collections. The above mentioned Joachim von Dessin, for instance, owned a large library of 3856 volumes with the Dutch edition of Kolb in it. Many of these individuals, like the governors and religious ministers, seemed to be rather conservative in their reading choices and could not allow that an outsider knew more about the Cape than they did.[88]

Another indication we have concerning the local reception of Kolb's book comes from Otto Mentzel, a German who had worked at the Cape as a VOC soldier in the 1730s. Mentzel unexpectedly returned to Europe in 1740 and, years later, in 1785, published an account of his years at the Cape. To some extent, like De la Caille, Mentzel would have seen Kolb as a rival whose work had to be denigrated and supplanted in order to make room for his own. Mentzel's memory of the 1730s was not always accurate but he asserted quite clearly that: 'It is quite true that, even today, Kolb did not leave a good impression in the Cape'.[89]

What emerges from these and other criticisms of Kolb is that there was a great difference between the reception of a book published in Europe about life outside of Europe and the reception of that same book in the place the book was about. This draws attention to the difference in written and cultural practices in the colonies and the metropolitan powers. European commentators, like Kolb, evoked an information network in the Cape that seemed composed of secrets, rumours, hearsay and tall tales, which the publication of a book in Europe was going to shatter or clarify. Cape residents felt that their information monopoly was being taken away from them by passing strangers. Through his practice and discourse Kolb often ran down those who told tales about African people at the Cape, or, at least, distanced himself from them. 'I am therefore persuaded that this author is wrong', he concluded concerning a religious ceremony reported by Tachard, 'and that he heard this story not from actual Hottentots but from a few Europeans'.[90] Kolb often spoke about preferring a direct relationship between the observer and the person being observed, doing away with intermediaries and mocking authors who found in *topoi* and other "rumours" conveyed by Cape settlers, subject matters to write their books. Conversely, for Kolb, distance was likely to authenticate the truth of a matter – "the truth is out there" – could have been one of his maxims. In commenting on previous authorities, he remarked that: 'This author could never have seen any if, as I believe, he did not go past the mountain of the Lion, which is not far from the Cape'.[91] Ironically, Kolb's honesty would expose him to similar criticisms from his rivals, who accused him of not having travelled far enough himself. Kolb, who had travelled as far east as the Caledon hot springs admitted the limitations of his travels:

> While I only saw and consulted [the Hottentots] who live with the Dutch or in their neighbourhood, not only did I not know what to make of the situation, but I also believed I could never see the end of travelling in these deep mysterious lands.[92]

These modest disclaimers were, however, ignored by his critics and local informants who scoffed at his portrayal of the Khoikhoi and resented the fact that Kolb should acquire literary glory in Europe at their expense. His criticisms of knowledge based on local hearsay, instead of personal observation, wounded parochial pride and confirmed Kolb's bad reputation in the colony.

History

Yet Kolb's poor reputation at the Cape could not have been solely due to local pettiness and literary rivalry. Independently of any literary issue Kolb's book declined in reputation over the years because of wider developments in colonial society. What happened, essentially, is that time continued to advance, rendering some of Kolb's observations, especially those dealing with the colony's history, obsolete. Kolb thought of his book, as a whole, as being a "History", though it is really the third part of his work, that deals with the 'European colonies in the Cape headland', that merits this description most truly.[93] History, nonetheless, left its mark on all sections of the book. When Kolb boarded the ship that returned him to Europe in 1713 he left in his wake a colony that was being transformed by a terrible smallpox epidemic. Particularly hard hit were the Khoikhoi, whose societies in the southwestern Cape were almost completely eradicated. The destruction of Khoikhoi societies by disease was augmented by the expansion of the colonial frontier after 1713 as land-hungry settlers moved into the Cape interior, benefiting from a new system of land allocation – the loan farm system – granted by the VOC.[94] The world that the *Caput* had described had vanished. Travellers who hoped to use Kolb's book as a guide to the Cape would have noticed a huge disparity between the situation he described and the reality they now observed.

Ultimately, then, Kolb's work and its reception reflect characteristics widely found in writing practices originating in travel literature. Firstly there is the theoretical argument that consists in authors differentiating themselves, sometimes very artificially, from their predecessors who dealt with the same subject. Both Kolb and those writers who came after him who wrote about the "Hottentots" sought to produce a discourse that was new, or at least different, from what was already available on the subject. Kolb insistently differentiates himself from his predecessors, just as De la Caille did from him. This practice is not only the preserve of travel writing since it seems to have been widespread throughout the Republic of Letters.[95] But it is particularly flagrant in the specific case of attempting to write colonial history since there was always a gap between what was happening on the ground in the colony and what was being represented in literature in Europe. It is almost as though such writing suffered from a built in obsolescence that necessitated the creation of new discourses. The new "truth" could, as in Kolb's case, enjoy a temporary literary and scientific success in Europe but be found wanting back on the other side of the world.

Later travellers who took Kolb's book with them into the interior could not help but be struck by such disparities. Heinrich Lichtenstein, for instance, who travelled inland between 1803 and 1806, wrote explicitly of the gap between European reader and African traveller in his introduction:

> Besides in compiling my work, it was not to German, or even to European readers alone that I wished to address myself; I had equally in view the rendering of my labours useful to citizens of Africa, and to future travelers in the southern part of that vast peninsula. I was the most desirous of this, from finding the little attention that had been paid to such objects by my precursors both in the route I travelled, and in the reports given of their travels. Their sole object in their publications seems to have been to make them entertaining to their own country-men, or, at the most, to their contemporaries in general; – they seem never to have thought of rendering them useful to the travelling part of the community.[96]

The provisional nature of observations by travel writers – which were not thought of as being provisional at the time – caused authors to criticise their predecessors rather than to accept that their own observations would also become obsolete. As Lichtenstein observed:

> Thence it happens that each one in succession has found great fault with the immediate precursor, and indeed too often not undeservedly. La Caille and Mentzel are severe upon Kolbe, Sparrman criticizes La Caille and Menzel, and Le Vaillant comes under the censures of Barrow.[97]

Here again Lichtenstein is perceptive, but the comments of Father Jose de Acosta, more than two centuries before, show that not only the gap between Europeans and travel writers, but also the awareness of such a gap, goes back to the first decades of European expansion overseas:

> This explains why our times do not have great regard for prior authors who wrote on the Indies: there is a noticeable gap with the current situation. Consequently, we need to assume that those who, by publishing today will be topical, will no longer be so soon afterwards.[98]

The inherent tension that exists between written observations and historical development was thus brought to light, in the case of the Cape, by the circulation of Kolb's book between the continents. Like all books that are taken to places about which they are supposed to speak, but where reality remains elusive, this led to paradoxical and even contradictory readings. The endless comments to establish whether one author or another was more or less "trustworthy" could not transcend the conditions of writing and book circulation. Nor could European and African readers be reconciled by mere proclamations of "a love of truth", such as that issued by Kolb in his introduction. Printed books, as a special form of writing, found limits to their representativeness when they travelled to other continents.

The subject of Kolb as an historian, perhaps the first historian of the Cape has, as yet, received little attention and its full treatment is beyond the scope of this chapter. We do not plan to discuss here the ways in which Kolb's statements about the Cape's history either conform, or do not conform, to what modern historians deem to be historically true. Instead, what we hope to achieve here is to show how Kolb dealt with the problem of writing a history of a colony that included two very different societies – colonists and Khoikhoi – and what the precedents behind his historical methodology were. As we hope to demonstrate, Kolb's solution was to separate "European" or "settler" history into one section and the description of the Khoikhoi into another. Despite the fact that Kolb announced that his section on the Khoikhoi was a history[99] he evolved a style of history, when writing about the Khoikhoi, that presaged the methodologies of anthropology.

When writing of the Khoikhoi, Kolb could state, almost like a good anthropologist before the fact, that 'it is the custom of the Hottentots' and 'that they have never acted otherwise' when he dealt successively with their names, languages, appearances, government, religion, music, food, habitat, clothing, etc. Kolb was less concerned with describing events, or history, and more concerned with providing an example – 'a few examples will introduce the capacity of these people' – to establish timeless behaviour. Throughout the Caput the Khoikhoi are depicted as a people who themselves say it has "always" been so. Thus, 'any reply I got from them was that it was a law established among them from times immemorial'.[100] The lack of historicity in Kolb's ethnographic account is all the more flagrant since it coincides with that which is truly historical – the establishment of a Dutch colony at the Cape. Even in the history section, historicity is exclusively reserved for Europeans and not for the Khoikhoi. In other words, within the same book, two paradigms of events are being opposed, depending on whether Europeans or Khoikhoi are being discussed.

A consequence of this disjuncture is that Kolb finds it difficult, if not impossible, to write of encounters between Europeans and "Khoikhoi", almost as though they cannot co-exist in the same time frame. For example, he writes that the "Hottentots" are mad about European food:

> they find our savoury and refined dishes delicious; they are fond of them and there are no dishes they do not look on with desire. But these meals are harmful to them. They develop stomach aches and are often subject to fevers after eating this type of food, and all among them in particular who, being in service with the Dutch, are forced to eat like Europeans, soon become weak, and in the long term become prone to various illnesses and die younger than the others who stick to unrefined and plain food of the country and their ancestors.[101]

Many similar examples of such mis-encounters may be found. Kolb delights in describing moments of encounter that seem to suggest that the two peoples are about to connect only to refute such a possibility in the end. If by chance a European wanted to enter a "Hottentot" house, he writes, he would find that 'the smoke escapes through the door. No European could stay in this oven' and he would have to exit immediately.[102]

Writing the history of the young colony presented a real challenge, but one that had to be met if Kolb was going to fulfil the promise of the title of the German edition – a "vollstandige Beschreibung". We cannot say that Kolb's solution – to use two different paradigms for two different peoples – was entirely innovative. On the contrary, he was pushing to the limit a reasoning that had already been widely rooted in travelogues. Such writing contained moments where the writer stood at a cross roads, between the daily account of navigation and the decontextualised moment of landscape description. A temporal gap opened up between the historicity of the ship (the journey, the crossing, the trip) and the historicity of what was external to it.

This temporal gap is visible, in the case of the *Caput*, from a narrative point of view but is prefigured in earlier travel writers. Vespucci, for instance, in the 16th century appealed to his readers: 'Let us dwell here, since the opportunity arises, on their morals, as they appeared to us'.[103] This switch in the narrative style is similar to the point of departure of many digressions in many other travelogues. Once the digression was over the author, like Vespucci, could stop himself from going further: 'I am now coming back to our first trip from which I wandered'.[104] An author like Peter Martyr, whose writings introduced Columbus's discoveries to well-read Europeans, also used digression when compiling his information in Rome. 'We now need to pick up where we left off', he wrote, with a view to closing an ethnographic description of the inhabitants of Hispaniola and resuming the chronological narrative of Columbus's voyage.[105] A consequence of this switch is the double present. That of the journey through time that is part of the voyage and that of the unchangeable present of the landscape, geography, natural history and peoples described.

Such opposition between European and extra-European historicity becomes even more visible when it is part of the theme of the story. In his preface, Kolb admitted that he had drawn inspiration from Robert Knox's *Description of Ceylon*, a book that was published in 1681 and that was definitely to be found in Cape Town in the 18th century.[106] Knox's *Description* is also to be found in three parts. The general description of the island combines geography and natural history in a first part and is followed by an ethnography of the inhabitants in the second and third part. Only in a fourth part does the author chronologically relate his travels, adventures and misadventures, in which he and fifteen other seamen were kept prisoner in Ceylon until their "miraculous escape" in 1679.

There was, in fact, something about the Cape's position half way between east and west that encouraged travellers to digress. The first Dutch ship sailing to the East Indies under the command of Houtman arrived at the Cape in August 1596. The narrator of this voyage recorded that 'On the 2nd of August . . . we saw the Cape of Good Hope . . . on the 7th we reached the shore . . . on the 9th the natives were waiting for us on shore'. The narration is in the past tense. It then continues with a few lines narrated in a referent-free present –'these people are small, of a brown-reddish colour . . . they go about completely naked' – before resuming the subject of navigation: 'on the 11th of August it was decided to navigate'. Such a split in narration is to be found in all travelogues dealing with the Cape, which was a paradigmatic stop over place on the long sea voyage and a temporal gap in a the narration of such a voyage. Whilst the crew went ashore to replenish supplies the story could digress into dissertations and other comments on its flora, fauna, topography and, most especially, its inhabitants. Although Kolb lived for several years at the Cape he remained true to this narrative convention, opening and closing his account of his stay at the Cape with the story of his voyage, both there and back. Thus we may see that the pattern of contrasting written to oral culture, or history to ethnography, was already in place before Kolb and his way of representing history was in no way unique.

In the end during the 18th century (and Kolb's *Caput* contributed to this result) the description of morals or manners would become ahistorical. Ethnographic description ceased to be a tool of chronological ambition. As Kolb's work demonstrates, an ahistorical account of the Khoikhoi was the best solution to the problem of writing the Cape's history for it enabled the Khoikhoi, as extra-Europeans, to be placed outside of the human chronology in which the Dutch colonists took their place.

Conclusion

From the above, brief overview of Kolb's work it may be seen that his book, though germinated or conceived at one of the ends of the earth, nonetheless played a central role in the networks of European thought. Though the Cape was a remote and seemingly uncivilised place relative to Europe it was these very qualities that made it so important in European consciousness. Its peoples, flora and fauna were thought to contain vital information about the order of the natural world and the nature of humanity. Closer inspection, in fact, revealed that the Cape was not only remote but also crucially situated on a major information network, the sea route between East and West. This meant that it served, to some extent, as a repository of knowledge from other quarters of the globe and a stop over point for travellers and knowledge seekers of all descriptions: botanists, curiosity collectors, explorers, navigators, writers, doctors, linguists, priests, artists, cartographers and astronomers all passed by and contributed to the circulation of global information. Not only were learned people to be found at the Cape but books too were owned and read by the settler inhabitants, providing Kolb with a means to keep abreast of current debates.

Kolb was thus well placed, in Cape Town, to gather and assess information that he judged to be valued in Europe. Nowhere does he speak of his desire to make a commercial success of his book but we may assume that one of his goals was that the book would sell well. The publication and republication of the *Caput*, in different editions and in different translations, suggests that the book was indeed a commercial success and found a wide audience in Europe. Part of its success lay in the fact that it appealed to a wide market. It did so by adopting the tried and true form of a travelogue, featuring a personal narrator who experiences adventures.

But it was also a success because it identified and catered for crucial areas of interest amongst its potential readership.

The first of these areas, corresponding to Part One of his book, was natural history. The prominence given to this subject attests to its popularity in Europe at the beginning of the 18th century. Though Kolb was neither a botanist nor a zoologist by training it is evident that his training had equipped him to observe and describe natural phenomena. He was as well read on the subject as many specialists in the topic and, at the Cape, had access to local knowledge, and local collections of plants and animals, which were as advanced as any to be found in Europe. Though he did not contribute to the creation of any new knowledge in the field of natural history, to be fair, he did not seek to do so. His work was a valuable compilation of existing knowledge about the Cape's natural history and was accepted as such until superseded by the Linnaean revolution.

The second area of interest in his book, and that which created the most interest in Europe, was his section on the Khoikhoi, Part Two in the German edition but Part One in the French. It may be seen, from the way that Kolb approached this subject, that he had read the existing literature on extra-European people and employed methodologies of comparison and description in an attempt to place them in the world. This rather ahistorical approach encouraged a timeless or static view of Khoikhoi customs and traditions and presaged the future human science of anthropology. Notwithstanding these methodological problems, there is no doubt that Kolb based his observations on the best available local knowledge and made a real contribution to an understanding of the Khoikhoi. Kolb's essential sympathetic view of the Khoikhoi went a long way towards rehabilitating their reputation in Europe, so much so that they became exemplars of the new Enlightenment paradigm of "the noble savage". Through Kolb's book, or through popularised versions and re-editions of the book, Kolb's Khoikhoi became familiar to both *philosophes* and ordinary readers in Europe. What is more, his book also reached the Cape. Despite an initially favourable reception there it soon became evident that local readers were critical of certain parts of the "Full Description". It was Kolb's depiction of the Khoikhoi that irritated readers at the Cape the most and the reception of his book there is evidence of the continuous interplay and revision of knowledge between Europe and its colonies. Though the locals were, at first, flattered to be the objects of curious scrutiny, they soon objected to what they perceived to be misrepresentation. They also felt, in the case of the Khoikhoi, that they knew these people better than any foreign visitor could ever hope to know. These criticisms were received in Europe and did help to discredit Kolb's reputation somewhat without, however, demolishing his essentially benign view of the Koikhoi.

The reception of Kolb's book at the Cape also provides evidence that there was bound to be a feeling of dissatisfaction upon reading a book which set out to be both a travelogue and a history. In such a book there was bound to be a temporal gap between what was described at a particular moment in time and what the reality was at a later date. History moved very fast, especially in an expanding colony, and its affect on the Khoikhoi was particularly brutal. Kolb's response to this was to separate the history of the Khoikhoi from that of the colonists, to suggest that the former lived in a world where chronology was either absent or unknowable. In his defence, however, he cannot have foreseen the devastating affect of the smallpox epidemic of 1713 that completely transformed the world he had described.

In conclusion, we should remark that Kolb himself was not unaware of the problems of representing, or misrepresenting, other societies. Apart from acknowledging the inspiration of Robert Knox's work on his own, Kolb also mentioned that of Lahontan, whose book, *Dialogues avec un sauvage dans l'Amerique*, was in the early library collection at Cape Town and probably read by Kolb whilst he was there. The book contains a scene in which an American Indian has read to him – because savages obviously cannot read – a description of his country from a book written by a Jesuit priest. His response is: 'I have been read books which the Jesuits wrote about our country. Those who read them to me explained them in my language, and I identified twenty lies one after the other'. Kolb uses this story to comment on the possibility that one day a Khoikhoi reader would be similarly unimpressed by works about his people:

> Should one of them come one day in our beloved Germany, and learn perfectly our language, to the point of being able to read and understand our written books, he would certainly have given a disparaging judgment to our nation.

Later on in Kolb's narrative he actually allows a Khoikhoi informant, when confronted with evidence of misrepresentation, to shout out in frustration: 'We are not as we are being depicted!' For all his faults Kolb, at least, allowed that local opinion counted.

Notes

1 Kolb, *Caput*. We have used an unpublished manuscript of the late R. Raven-Hart's English translation of the first German edition as our primary version of Kolb's book. We are busy preparing a new English edition of the *Caput* for publication by the Van Riebeeck Press.
2 Raj, *Relocating Modern Science*, 8.
3 Ibid., 14. Raj borrows this concept from Michel Callon.
4 In this instance he challenges Peter Burke's argument that knowledge from the peripheries was processed in European centres or cities of influence. See Burke, *A Social History of Knowledge*, 53–80.
5 Raj, *Relocating Modern Science*, 10.
6 Biographical information about Kolb may be obtained from the *Allgemeine Deutsch Biographie*, vol. 16, and from Moritz, *Die Deutschen am Kap*, 69–71. See also Good, 'Primitive Man'. There is also an autobiographical essay in the Kolb papers at the Gymnasium at Neustadt an der Aisch. Thanks to Professor David Wardel for translating this from the Latin and to Professor Hans Raum for accompanying Nigel Penn there.
7 For Witsen see Rietbergen, 'Witsen's World', 121–34 and Parker Brienen, 'Nicolas Witsen', 439–49.
8 Raj, 'Surgeons, Fakirs, Merchants and Craftspeople', 252–69; Dear, 'Space, Revolution and Science', 37, 41–2.

9 See Ward, *Networks of Empire: Forced Migration in the Dutch East India Company*, 10–11 for a discussion of how Cape Town was a "node" in the networks of the VOC empire.
10 See for instance van Schoor, 'Handel en wetenschap' and van Berkel, 'Een onwillige mecenas?'. See also Blusse and Ooms, Kennis en Compagnie.
11 See for instance Davids, 'Navigeren in Azie' and Baas, 'De VOC in Flora's Lusthoven'.
12 Cook, 'Global Economies and Local Knowledge', 100–18.
13 For a discussion of this view of the Cape see Jamal, 'Africa's Appendix', 155–69.
14 See for instance the map in Femme S. Gaastra's De geschiedenis van de VOC, 42. For an interesting discussion of early Dutch maps of the Cape see Brotton, 'Printing the Map', 137–59.
15 See for instance Worden, 'VOC Cape Town', 142–62 as well as the essays in Worden, *Cape Town, Between East and West.*
16 Witsen, Noord en oost Tartarye and Witsen, *Aeloude en hedendaegsche scheepbouw en bestier*; Marion Peters, *De Wiyze Koopman: Het Wereldwiyde Onderzoek van Nicolaes Witsen (1641-1717), Burgermeester en VOC-Bewindhebber van Amsterdam.*
17 In 1712 Witsen wrote to his friend, the Mayor of Deventer in a depressed mood: 'What do your eminences expect from gaining knowledge in the Indies? No, sir, our people are only interested in money, not knowledge, an appalling fact'. See Blusse and Ooms, *Kennis en Compagnie*, 8 and Rietbergen, 'Witsen's world', 121–34.
18 The words are those of J. Hondius in Hondius, *Klare Beschrvinghe*, 9.
19 For travel writing in this period see Hulme and Youngs, *The Cambridge Companion.*
20 In the forward to the first edition Kolb states explicitly that the fact that the letters are addressed to Baron von Krosick confirms 'the authenticity of what he says'. See Kolb, Caput, Vorrede.
21 Rubies, 'Travel Writing as a Genre', 5, 28.
22 The space of the voyage is something that greatly interested Michel de Certeau. See Girard, 'Epilogue' and Certeau, 'Travel Narratives of the French to Brazil', 313–28.
23 For an account of these voyages see van Gelder, *Het Oost-Indische Avontuur.*
24 Penn, 'The Voyage Out'.
25 A point made by Raum, 'Reflections on Reading Peter Kolb', 31.
26 Spohr, 'The First Danish-German Missionaries'.
27 Kolb, *Caput*, Part 2, First Letter.
28 Encyclopaedias date back at least as far as the Middle Ages. See Hamesse, 'The Scholastic Model of Reading', 109.
29 Cook, 'Global Economies and Local Knowledge', 101–2'.
30 For Hermann, Rumpius and Van Rheede see p. 12 below.
31 There is a useful summary of Linneaus's system in Bowler, *The Fontana History*, chapter titled 'Nature and the Enlightenment', 137–99.
32 For Gesner (1516–65) see Ashwoth, 'Emblematic Natural History of the Renaissance'.
33 Foucault, *The Order of Things*, 117–77.
34 Schmidt, 'Inventing Exoticism', 347–69. Schmidt refers to the decennium mirabilius of exotic natural history as being 1695–1705 and includes Maria Merian's 'glorious Metamorphosis insectorum Surinamensium as amongst the productions of these years,' see 350.
35 See van Gelder, 'De Wereld Binnen Handbereik' and van Berkel, 'Citaten uit het boek der natuur'.
36 As Lisbet Koerner shows, this was an important element in Linnaeus's study of natural history. Koerner 'Carl Linnaeus in His Time and Place'.
37 Rookmaker, *Zoological Exploration*, 29.
38 Ashworth, 'Emblematic Natural History of the Renaissance', 17–37.
39 'Since now the outward form of the baboons so much resembles that of men, the Hottentots have an unusual idea concerning them, in that they believe that they are men, who

refuse to speak since otherwise they would have to work like themselves (although what they do can hardly be called "work", as will be said later)'. Kolb, *Caput*, letter XI, 3.

40 Kolb's style of writing here is very similar to that of Jacobus Bontius's descriptions of the animals of the East Indies in the 17th century. Cook writes that 'Bontius' descriptions of animals are full of interesting anecdotal as well as morphological information'. Cook, 'Global Economies and Local Knowledge', 106.

41 Rookmaker, *Zoological Exploration*, 12.

42 Kolb, *Caput*, letter XVI, 147.

43 See Cook, *Matters of Exchange* for an account of the inter-connections between science, medicine, botany and commerce.

44 'O God. How rich, unique and wonderful were the plants that revealed themselves to Hermann's eyes. In a single day Hermann discovered more African plants than any other botanists before him'. In Baas, 'De VOC in Flora's Lusthoven', 128.

45 On the Company Gardens see Karsten, *The Old Company's Garden*. For early botanical research at the Cape see Gunn and Codd, *Botanical Exploration* and Mike and Liz Fraser, *The Smallest Kingdom*.

46 Heniger, *Hendrik Adriaan Van Reede Tot Drakenstein*.

47 Beekman, *The Ambonese Curiosity Cabinet*; Buize, 'De natuurhistoricus Georg Everhard Rumphius', 12–19.

48 Kolb, *Caput*, letter XVI, 148.

49 Gunn and Codd, *Botanical Exploration*, 178–9 and 265–6. Oldenland married Margaretha van Otteren, widow of Johann Blum in 1693. When Oldenland died in 1697 his widow married Hendrik Donker.

50 Valentiyn, *Oud en Nieuw Oost-Indien*.

51 Gunn and Codd, *Botanical Exploration*, 265–6.

52 Habiboe, *Tot Verheffing Van Mijne Natie*, 62, 69.

53 Gunn and Codd, *Botanical Exploration*, 332–3.

54 Ibid., 47, 265–6.

55 Claudius's illustrations eventually ended up in the collection of Nicolas Witsen as part of the Codex Witsenii. The drawings may have been copies made of Claudius's work. Witsen states that they were 'made for me at the Cape' in 1692. Witsen had a collection of over 1,700 plant drawings. See Wilson, van Hove-Exalto, and van Rijssen, *Codex Witsenii*.

56 Gunn and Codd, *Botanical Exploration*, 117.

57 Ibid., 118.

58 See Zandvliet, *Mapping for Money*.

59 Kolb, *Caput*, Part 2, letter IV, XIII and XXI.

60 Ibid., Part 2, letter XXI.

61 Ibid., letter XVI, 157–8.

62 Schouten, *Oost Indische Voyagie*, cited by Raven-Hart, *Cape of Good Hope*, 79.

63 See for instance Herbert, *Some Years Travel into Africa and Asia*, 16–17.

64 'Hernach hat mich dazu veranlasset und bewongen/weil bey einigen Scribenten/viele nichts nuszige/irrige und ganze falsche Sachen/von den Hottentoten und ihrem Land/ habe aufgeschrieben gefunden'. Kolb, *Caput, Vorrede*.

65 Pratt, *Imperial Eyes*, 41–9; Merians, *Envisioning the Worst*, chapter 5; Fauvelle-Aymer, *L'invention du Hottentot*; Goode, 'The Construction of an Authoritative Text'; Huigen, *Knowledge and Colonialism*, chapter 2; Penn, 'Written Culture and the Khoikhoi'.

66 Siegfried Huigen provides a list of the different editions of Kolb's *Caput* in an appendix at the back of his book *Knowledge and Colonialism*, 241–4.

67 For Valentijn and his work see Habiboe, *Tot Verheffing Van Mijne Natie*, 93–118.

68 Schmidt, 'Inventing Exoticism', 362.

69 Gaastra, *De geschiedenis van de VOC*, 106, 180. Van Dam's work, *De Beschryvinge van de Oostindische Compagnie*, completed around 1700, was finally published in 1927.

70 Kolb, *Naaukeurige en uitvoerige beschryving van de Kaap de Gooed Hoop* and Kolb, *The Present State of the Cape of Good Hope*.
71 Kolb, *Description du Cap de Bonne Espérance*.
72 Huigen, *Knowledge and Colonialism*, 244.
73 Ibid.
74 *Journal des Savants*, Paris, Juillet 1741. Cited by Good, 'Primitive Man', 397.
75 Such as Astley and Green's *New General Collection of Voyages and Travels*, which was published in London between 1745 and 1747, Arkstee and Merkus's *Allgemeine Histoire der Reisen zu Wasser und Lande*, published in Leipzig from 1747, and the *Histoire Generale des Voyages* compiled by the Abbe Prevost from 1748.
76 Merians, *Envisioning the Worst*, chapter 5; Good, 'Primitive Man', 381; Fauvelle-Aymar, *L'invention du Hottentot*, 259.
77 Rousseau was citing Kolb in a footnote on Prevost's *Voyages*. See Fauvelle-Aymer, *L'invention du Hottentot*, 275.
78 Rubies, 'Ethnography, Philosophy and the Rise of Natural Man 1500–1750', 103. The authors thank Professor Rubies for an advanced copy of this chapter.
79 Cullinan, *Robert Jacob Gordon*, 23.
80 Smith and Pheiffer, 'Col. Robert Jacob Gordon's note on the Khoikhoi', 34.
81 Heydt, *Scenes of the Cape of Good Hope*, 20.
82 Ibid., 21.
83 de La Caille, *Journal historique*.
84 Ibid., 317–9.
85 Penn, 'Notes towards a Rereading of Peter Kolb', 45.
86 de la Caille, *Journal historique*, 157.
87 Grevenbroek's observations on the Khoikhoi may be found in Schapera, *The Early Cape Hottentots*.
88 J.N. von Dessin was secretary to the Orphan Chamber and owner of the largest library in the colony. See Groenewald, 'On Not Spreading the Word', 302–23. Schoeman, *Cape Lives of the Eighteenth Century*, 224–5.
89 Mentzel, *Vollstandige und zuverlassige geographische und topographische beschreibung des beruhmten und in aller betrachtung merwurdigen afrkanischen vorgeburges der Guten Hoffnung*, 21–2.
90 Kolb, *Description*, I, 214.
91 Ibid., I, 248.
92 Ibid., I, 172.
93 The layout of the three sections differed according to the different editions. And is related to the readers' expectations depending on the country of publication. Thus natural history came first in German and English editions because it was highly regarded in those countries. Colonial history came first in Holland and the "Hottentot" section came first in France.
94 See Penn, *The Forgotten Frontier* for an account of frontier expansion at this time.
95 An author like Pierre Bayle, the father of the footnote according to Anthony Grafton, was also very found of practising distinction, which was widely spread at this time. See Grafton, *The Footnote*.
96 Lichtenstein, *Travels in Southern Africa*, xi–xii.
97 Ibid., xii.
98 Acosta, *De Procuranda Indorum Salute*, 55.
99 In the Description he wrote 'Je travaillais a l'Histoire des Hottentots', *Description*, II, 72.
100 Ibid., I, 277.
101 Ibid., I, 252.
102 Ibid., I, 290.
103 Vespucci, *Le Nouveau Monde*, 89.

104 'Since everything [in my book] is dealt with in particular and in more detail, and since at this stage I am simply endeavouring to give an overall view, I am now coming back to the development of our first trip, from which I digressed a little. At the beginning of the trip'. Ibid., 95.

105 d'Anghiera, *Le Noveau Monde*, 35–9, 55.

106 Knox, *An Historical Relation of the Island of Ceylon*. The book is in the Cape Town Library and there is every reason to believe that Kolb read this book during his stay in the Cape.

Bibliography

de Acosta, J., *De Procuranda Indorum Salute* (Madrid: Consejo Superior de Investigaciones Cientificas, 1984).

Ashwoth, William B., 'Emblematic Natural History of the Renaissance', in *Cultures of Natural History*, ed. N. Jardine, J.A. Secord and E.C. Spary (Cambridge: Cambridge University Press, 1996), 17–37.

Baas, Pieter, 'De VOC in Flora's Lusthoven', in *Kennis en Compagnie: De Verenigde Oost-Indische Compagnie en de moderne Wetenschap*, ed. Leonard Blusse and Ilonke Ooms (Leiden: Balans Uitgeverij, 2002), 124–37.

Beekman, E.M., *The Ambonese Curiosity Cabinet: Georg Everhardus Rumphius* (New Haven: Yale University Press, 1999).

Blusse, Leonard and Ilonke Ooms, eds., *Kennis en Compagnie: De Verenigde Oost-Indische Compagnie en de moderne Wetenschap* (Leiden: Balans Uitgeverij, 2002).

Bowler, Peter J., *The Fontana History of the Environmental Sciences* (London: Fontana Press, 1992).

Brotton, Jerry, 'Printing the Map, Making a Difference: Mapping the Cape of Good Hope, 1488–1652', in *Geography and Revolution*, ed. David N. Livingston and Charles W.J. Withers (Chicago: University of Chicago Press, 2005), 137–59.

Buize, Wim, 'De natuurhistoricus Georg Everhard Rumphius en ziijn bezoek aan Tafelbaai, 1653', *Quarterly Bulletin of the National Library of South Africa* 56, no. 1 (2001): 12–19.

Burke, Peter, *A Social History of Knowledge from Gutenberg to Diderot* (Cambridge: Polity, 2000).

Cook, Harold J., 'Global Economies and Local Knowledge in the East Indies: Jacobus Bontius Learns the Facts of Nature', in *Colonial Botany: Science, Commerce, and Politics in the Early Modern World*, ed. Londa Schiebinger and Claudia Swan (Philadelphia: Pennsylvania Press, 2005), 100–18.

———, *Matters of Exchange: Commerce, Medicine, and Science in the Dutch Golden Age* (New Haven, CT: Yale University Press, 2007).

Cullinan, Patrick, *Robert Jacob Gordon, 1743–1795: The Man and his Travels at the Cape* (Cape Town: Struik Winchester, 1992).

d'Anghiera, P. Martyr, *Le Noveau Monde* (Paris: Les Belles lettres, 1992).

Davids, C.A., 'Navigeren in Azie. De uitwisseling van kennis tussen Aziaten en navigatiepersoneel bij de voorcompagnieen en de VOC, 1596–1795', in *VOC en Cultuur: Wetenschappelijke en culturele relaties tussen Europa en Azie ten tijde van de Verenigde Oostindische Compagnie*, ed. J. Bethlehem and A.C. Meijer (Amsterdam: Schiphouwer en Brinkman, 1993), 17–37.

de Certeau, M. 'Travel Narratives of the French to Brazil: Sixteenth to Eighteenth Centuries' in *New World Encounters*, ed. S. Greenblatt (Berkley: University of California Press, 1993), 313–28.

de La Caille, Nicolas-Louis, *Journal historique du voyage fait au Cap de Bonne Esperance* (Paris: Chez Guillyn, Quai des Augustins, 1763).

Dear, Peter, 'Space, Revolution and Science', in *Geography and Revolution*, ed. David N. Livingstone and Charles W.J. Withers (Chicago: University of Chicago Press, 2005), 27–42.

Fauvelle-Aymer, F.X., *L'invention du Hottentot, Histoire du regard occidental sur les Khoisan (XVe-XIXe siècle)* (Paris: Publications de la Sorbonne, 2002).

Foucault, Michel, *The Order of Things: An Archaeology of the Human Sciences* (London: Tavistock, 1977).

Mike and Liz Fraser, *The Smallest Kingdom: Plants and Plant Collectors at the Cape of Good Hope* (Kew: Royal Botanic Gardens, 2012).

Gaastra, Femme S., *De geschiedenis van de VOC* (Zutphen: Walburg Pers, 2002).

Girard, L., 'Epilogue: Michel de Certeau's Heterology and the New World', in *New World Encounters*, ed. S. Greenblatt (Berkley: University of California Press, 1993), 313–28.

Good, Anne, 'Primitive Man and the Enlightened Observer: Peter Kolb Among the Khoikhoi' (PhD diss., University of Minnesota, 2005).

———, 'The Construction of an Authoritative Text: Peter Kolb's Description of the Khoikhoi at the Cape of Good Hope in the Eighteenth Century', *Journal of Early Modern History* 10, no. 1 (2006): 61–94.

Grafton, Anthony, *The Footnote: A Curious History* (Cambridge, MA: Harvard University Press, 1999).

Groenewald, Gerald, 'On Not Spreading the Word: Ministers of Religion, and Written Culture at the Cape of Good Hope in the 18th Century', in *Written Culture in a Colonial Context*, ed. A. Delmas and N. Penn (Cape Town: University of Cape Town Press, 2011), 302–23.

Gunn, Mary and L.E. Codd, *Botanical Exploration of Southern Africa* (Cape Town: A.A. Balkema, 1981).

Habiboe, R.R.F., *Tot Verheffing Van Mijne Natie: Het leven en werk van Francois Valentijn (1666–1727)* (Franeker: Van Wijnen, 2004).

Hamesse, Jacquelin, 'The Scholastic Model of Reading', in *A History of Reading in the West*, ed. Guglielmo Cavallo and Roger Chartier (Amherst, MA: University of Massachusetts Press, 2003), 103–19.

Heniger, J., *Hendrik Adriaan Van Reede Tot Drakenstein (1636–1691) And Hortus Malabaricus: A Contribution To The History Of Dutch Colonial Botany* (Rotterdam, Boston: A.A. Balkema, 1986).

Herbert, T., *Some Years travel into Africa and Asia* (London: Jacob Blome and Richard Bishop, 1638).

Heydt, J.W., *Scenes of the Cape of Good Hope in 1741 as drawn by Johann Wolffgang Heydt* (Cape Town: Struik, 1967).

Hondius, J., *Klare Beschrvinghe van Cabo de Bona Esperanca* (Cape Town: Van Riebeeck Society reprint, 1952 [1652]).

Huigen, Siegfried, *Knowledge and Colonialism: Eighteenth-Century Travelers in South Africa* (Leiden: Brill, 2009).

Hulme, Peter and Tim Youngs, eds., *The Cambridge Companion to Travel Writing* (New York: Cambridge University Press, 2002).

Jamal, Ashraf, '"Africa's Appendix": Distortion, Forgery and Superfluity on a Southern Littoral', in *Eyes Across the Water: Navigating the Indian Ocean*, ed. Pamila Gupta, Isabel Hofmeyer, and Michael Pearson (Pretoria: Unisa Press, 2010), 155–69.

Karsten, M.C., *The Old Company's Garden at the Cape and Its Superintendents* (Cape Town, 1951).

Knox, Robert, *An Historical Relation of the Island of Ceylon, in the East-Indies. Together with an Account of the Detaining in Captivity of the Author and Diverse Other Englishmen Now Living There, and of the Author's Miraculous Escape* (London: Printed by Richard Chiswell, 1681).

Koerner, Lisbet, 'Carl Linnaeus in His Time and Place', in *Cultures of Natural History* and *Linnaeus: Nature and Nation*, ed. N. Jardine, J.A. Secord and E.C. Spary (London: University of Cambridge Press, 1999), 145–62.

Kolb, Peter, *Caput Bonae Spei Hodiernum. Das ist, vollstandige Beschreibung des Afrikanischen Vorgeburges der Guten Hofnung* (Nuremberg: P.C. Monath, 1719).

———, *Naaukeurige en uitvoerige beschryving van de Kaap de Gooed Hoop* (Amsterdam: Balthasa Lakeman, 1727).

———, *The Present State of the Cape of Good Hope* (London: William Innys, 1731).

———, *Description du Cap de Bonne Espérance ou l'on trouve tout ce qui concerne l'histoire naturelle du pays, la religion, les moeurs et les usages des hottentots, et l'etablissement des Hollandois* (Amsterdam: Jean Catuffe, 1741).

Lichtenstein, H., *Travels in Southern Africa* (1811–1812; reprint Cape Town: The Van Riebeeck Society, 1928).

Mentzel, O., *Vollstandige und zuverlassige geographische und topographische beschreibung des beruhmten und in aller betrachtung merwurdigen afrkanischen vorgeburges der Guten Hoffnung* (Glogau, 1785–1786; reprinted as *A Geographical and Topographical Description of the Cape of Good Hope*, Cape Town: The Van Riebeeck Society, 1921).

Merians, Linda E., *Envisioning the Worst: Representations of "Hottentots" in Early-Modern England* (London: University of Delaware Press, 2001).

Moritz, E., *Die Deutschen am Kap unter Hollandischer Herrschaft, 1652–1806* (Weimar: Böhlams, 1938).

Parker Brienen, Rebecca, 'Nicolas Witsen and His Circle: Globalization, Art Patronage, and Collecting in Amsterdam, circa 1700', in *Contingent Lives: Social Identity and Material Culture in the VOC World*, ed. Nigel Worden (Cape Town: University of Cape Town Historical Studies Department, 2007), 439–49.

Penn, Nigel, 'Notes Towards A Rereading Of Peter Kolb', *Kronos: Journal of Cape History* 24 (1997): 41–5.

———, *The Forgotten Frontier: Colonist and Khoisan on the Cape's Northern Frontier in the 18th Century* (Cape Town: Double Story Books; Athens: University of Ohio Press, 2005).

———, 'The Voyage Out: Peter Kolb and VOC Voyages to the Cape', in *Many Middle Passages: Forced Migration and the Making of the Modern World*, ed. Emma Christopher, Cassandra Pybus, and Marcus Rediker (Berkley: University of California Press, 2007).

———, 'Written Culture and the Khoikhoi: From Travel Writing to Full Description', in *Written Culture in a Colonial Context*, ed. A. Delmas and N. Penn (Cape Town: University of Cape Town Press, 2011), 166–88.

Marion Peters, *De Wiyze Koopman: Het Wereldwiyde Onderzoek van Nicolaes Witsen (1641-1717), Burgermeester en VOC-Bewindhebber van Amsterdam* (Amsterdam, Bert Bakker, 2010).

Pratt, Mary Louise, *Imperial Eyes: Travel Writing and Transculturation* (London: Routledge, 1992).

Raj, Kapil, 'Surgeons, Fakirs, Merchants and Craftspeople: Making L'Empereur's Jardin in Early Modern South Asia', in *Colonial Botany: Science, Commerce, and Politics in the Early Modern World*, ed. Londa Schiebinger and Claudia Swan (Philadelphia: University of Pennsylvania Press, 2005), 252–69.

———, *Relocating Modern Science: Circulation and the Construction of Knowledge in South Asia and Europe, 1650–1900* (London: Palgrave Macmillan, 2007).

Raum, Johannes W., 'Reflections on Re-Reading Peter Kolb with Regard to the Cultural Heritage of the Khoisan', *Kronos: Journal of Cape History* 24 (1997): 30–40.

Raven-Hart, Rowland, *Cape of Good Hope 1652–1702: The First Fifty Years of Dutch Colonization as Seen By Callers* (Cape Town: A.A. Balkema, 1971).

Rietbergen, P.J.A.N., 'Witsen's World: Nicolaas Witsen (1641–1717) Between the Dutch East India Company and the Republic of Letters', *Itinerario*, 9, no. 2 (1985): 121–34.

Rookmaker, L.C., *The Zoological Exploration Of Southern Africa 1650–1790* (Rotterdam, Brookfield: Balkema, 1989).

Rubies, Joan-Pau, 'Travel Writing as a Genre: Facts, Fictions and the Invention of a Scientific Discourse in Early Modern Europe', *Journeys*, 1, no. 1/2 (2000): 5–35.

———, 'Ethnography, Philosophy and the Rise of Natural Man 1500–1750', in *Europe's New Worlds. Travel Writing and the Origins of the Enlightenment*, ed. Joan-Pau Rubies (forthcoming).

Schapera, Isaac, ed., *The Early Cape Hottentots, Described in the Writings of Dapper, Ten Rhyne, and Grevenbroek* (Cape Town: Van Riebeeck Society, 1933).

Schmidt, Benjamin, 'Inventing Exoticism: The Project of Dutch Geography and the Marketing of the World, circa 1700', in *Merchants and Marvels: Commerce, Science and Art in Early Modern Europe*, ed. Pamela H. Smith and Paula Findlen (New York: Routledge, 2002), 347–69.

Schoeman, Karel, *Cape Lives of the Eighteenth Century* (Pretoria: Protea Book House, 2011).

Schouten, Wouter, *Oost Indische Voyagie* (Amsterdam: Jacob Meurs op de Keysers-Graft, 1676).

Smith, A.B. and R.H. Pheiffer, 'Col. Robert Jacob Gordon's Note on the Khoikhoi 1779–80', *Annals of the South African Cultural Museum*, 5, no. 1 (1992): 1–56.

Spohr, O.H., 'The First Danish-German Missionaries at the Cape of Good Hope, 1706', *Quarterly Bulletin of the South African Library* 21, no. 4 (1967).

Valentiyn, Francois, *Oud en Nieuw Oost-Indien*, 2 vols (Amsterdam: Dordrecht, 1724–1726).

van Berkel, K., 'Een onwillige mecenas? De rol van de VOC bij het natuurwetenschappelijk onderzoek in de zeventiende eeuw', in *VOC en Cultuur: Wetenschappelijke en culturele relaties tussen Europa en Azie ten tijde van de Verenigde Oostindische Compagnie*, ed. J. Bethlehem and A.C. Meijer (Amsterdam: Schiphouwer and Brinkman, 1993).

———, 'Citaten uit het boek der natuur: Zeventiende-eeuwse Nederlandse naturalienkabinetten en de ontwikkeling van de natuurwetenschap', in *De Wereld Binnen Handbereik*, ed. E. Bervelt and R. Kistemaker (Amsterdam: Museum, 1992).

van Gelder, Roelof, 'De Wereld Binnen Handbereik: Nederlands Kunst-en Rariteitenverzamelingen, 1585–1735', in *De Wereld Binnen Handbereik*, ed. E. Bervelt and R. Kistemaker (Amsterdam: Museum, 1992).

———, *Het Oost-Indische Avontuur: Duitsers in Dienst van der Niederlandischen West-und-Ost-Indischen Kompagnien, 1602–1797* (Nijmegen: SUN, 1997).

van Schoor, Joseph, 'Handel en wetenschap', in *VOC en Cultuur: Wetenschappelijke en culturele relaties tussen Europa en Azie ten tijde van de Verenigde Oostindische Compagnie*, ed. J. Bethlehem and A.C. Meijer (Amsterdam: Schiphouwer and Brinkman, 1993).

Vespucci, A., *Le Nouveau Monde* (Paris: Chandeigne, 2005).

Ward, Kerry, *Networks of Empire: Forced Migration in the Dutch East India Company* (Cambridge: Cambridge University Press, 2009).

Wilson, M.L. Th. Toussaint van Hove-Exalto, and W.J.J. van Rijssen, eds, *Codex Witsenii* (Cape Town, Amsterdam: Iziko Museums of Cape Town, Davidii Media, 2002).

Witsen, N., *Aeloude en hedendaegsche scheepbouw en bestier* (Amsterdam: Constantijn Huygens, 1690).

———, *Noord en oost Tartarye ofte donding ontwerp van eenige dier lande en volken* (Amsterdam: Francois Halma, 1705).

Worden, Nigel, 'VOC Cape Town as an Indian Ocean Port', in *Cross Currents and Community Networks: The History of the Indian Ocean World*, ed. Himanshu Ray and Edward Alpers (New Delhi, New York: Oxford University Press, 2007), 142–62.

Zandvliet, Kees, *Mapping for Money: Maps, Plans and Topographic Paintings and Their Role in Dutch Overseas Expansion During the 16th and 17th Centuries* (Amsterdam: Batavian Lion International, 1998).

3 A naturalist's career

Hinrich Lichtenstein (1780–1857)

Sandra Näf-Gloor

The decision on South Africa

Martin Hinrich Carl Lichtenstein was born on 10 January 1780 into an academic family living in Hamburg.[1] He serves in this case study as an exemplary figure for the analysis of the networks of colonial powers, naturalists and European museums. Through his travels in South Africa and the resultant papers he published, he not only offers insights into academic Germany in the 19th century but also played a major role in establishing scientific disciplines and academic institutions.

His life as a transnational scientist starts with his academic career, which at the beginning ran parallel with his father's. Anton August Heinrich Lichtenstein (1753–1816) held the position as headmaster at the "Johanneum" in Hamburg, where his son completed his schooling. The "Johanneum" was the oldest and most established *Gelehrtenschule* for higher education in Hamburg. August Lichtenstein held a chair for eastern languages and was temporarily responsible for the school's library. His interest in books was not only professional, as he owned an eminent reference collection of bibliographical works, including many maps and geographical volumes kept at home.

Due to his father's passion, Lichtenstein's education was marked by the development of a robust interest in natural history, which was even more stimulated by his contacts with Count Johann Centurius von Hoffmannsegg. Lichtenstein came across Hoffmannsegg at Hamburg in 1797. Count von Hoffmannsegg was not only an explorer who went to Hungary, Italy and Portugal but also the owner of both a huge collection of insects and a sizeable collection of birds. He was a renowned botanist and entomologist of considerable means, and Lichtenstein worked since 1797 for the Count's collections, becoming increasingly familiar with natural history.[2]

At that time, career opportunities for naturalists were scarce and professional openings were often linked to the patronage of wealthy figures, providing an opportunity for eager students to get on with their work.[3] In contrast, the medical education promised a regular occupation. Many students therefore chose a medical degree while keeping the option of a later transfer to the natural sciences in mind. The life of Carl Linneaus, who first studied medicine and later became the most renowned botanist of his time, is just the most prominent example.

On 26 April 1802, Lichtenstein qualified for his doctorate in medicine. But instead of accepting an employment as a doctor, he saw his wish come true. This was 'an ardent desire to be acquainted with a country upon which, even in my boyish years, my imagination had eagerly dwelt, and which since my arrival at a mature age I had always had an unbounded curiosity to explore'.[4] He spoke of travelling as 'an unconquerable inclination to try my powers amid the vicissitudes and toils of wandering through new climes and under a different heaven'. This idea of broadening the mind was a common practice at his time. Essner claims that many young scientists were motivated to travel in order to quicken the mind. The diversity of experiences gained during a voyage would stimulate ideas and improve intellectual productivity.[5] The image of travelling as an element of education had just one particular disadvantage: it cost money.

Therefore, no better opportunity, certainly no more agreeable, could have offered itself to the young medical doctor Lichtenstein than employment at the Cape. The possibility appeared in the person of the Cape's newly appointed Dutch Governor, General Jan Willem Janssens who required a German tutor for his son, Henry.

Lichtenstein had only two months time to prepare himself for the new situation in South Africa. During these weeks, he caught up on reading books written by earlier travellers such as Peter Kolbe, Anders Sparrman, Carl Peter Thunberg, François Le Vaillant and John Barrow.

Additionally, he met Count von Hoffmannsegg (1766–1849), who was in Braunschweig since 1801 because of his engagement to order his collections. He was supported in his work by the co-proprietor of the country's biggest collection of insects, the court counsellor and well-known entomologist Johann Hellwig (1743–1831), and his gifted assistance Johann Karl Wilhelm Illiger (1775–1813). All three combined to give Lichstenstein instruction and guidance concerning observation, collection and preservation of African flora and fauna.[6]

Travelling to South Africa's inland

When Lichtenstein left Germany, it was at a time of crisis and upheaval at the Cape of Good Hope. The colony had twice changed hands in the previous decade. When troops of the French Revolution conquered the Republic of United Netherlands in 1795 and proclaimed the Batavian Republic, the supposed heir to the Dutch throne, Wilhelm V of Orange, escaped to England. The Batavian Republic became a satellite regime of revolutionary France and a British force seized the Cape in 1795 and put an end to almost 150 years of Dutch rule.[7] Although the British had – with the help of the Prince of Orange – occupied the Cape with very little resistance, this first British occupation only lasted eight years.[8] Following the Peace of Amiens of 1802, the Cape was restored to the Batavian Republic the following year.

Lichtenstein arrived in Cape Town on 23 December 1802, together with Commissary-General de Mist and Governor Janssens. When de Mist planned a six months' tour in the Cape Colony for October 1803, Lichtenstein was given permission to join him. This was possible because his pupil joined the excursion as well and because de Mist hoped to derive some advantages 'from the observations it would induce me to make, and the objects which it was my purpose to collect'. Consequently, Lichtenstein was also permitted to join the group due to the 'little knowledge I had in natural history'.[9] The venture became the first of altogether five journeys that Lichtenstein undertook in South Africa.

On this first tour, Lichtenstein set out together with a party of 50 people through the western, northern and eastern parts of the Colony. The party performed the journey on horseback and was accompanied by wagons. They travelled up north along the west coast to Saldanha and further north through the Nardouw Mountain until the District of Hantam. There, they reached for the first time the border of the colony and turned south over the Roggeveld Mountains and through the Great Karroo, touching at the first mission station of the Herrnhuter at Baviaanskloof, nowadays Genadendal.[10] They passed Swellendam, Mossel Bay and Plettenberg Bay until they arrived at Algoa Bay on 1 January 1804. Having visited Bethelsdorp, the mission station of the London Mission Society, they then turned northeast until they crossed the Fish River. According to Lichtenstein's directory, they then travelled along the 'borders of the Caffre country to Graaf Reynett', where they stayed for a week and then returned through the Karroo and the Hex River Pass via Paarl, Franschehoek and Stellenbosch to Cape Town. According to Lichtenstein, this journey was

> unlike any which has perhaps ever fallen under the reader's observation, since no travels into the interior of Africa resemble it in any way . . . and through the novelty of the objects to which we were obliged to attend.[11]

They arrived in Cape Town eventually on 23 March 1804, after 167 days, respectively 900 hours of travel.[12]

Due to the unpredictable condition, de Mist und Janssens had afterwards to remain at the fort in Cape Town. For the planned journey to Swellendam and its adjacent countries in September 1804, Lichtenstein was asked to join together with Henry Janssens but without either de Mist or Janssens.[13] The mission was conducted by Captain Paravinci di Capelli and three further officers. Meanwhile, Lichtenstein acquired the confidence of both the Governor and the Commissary-General through his talent, skills and his honest behaviour.[14] Therefore, Lichtenstein went on a short journey with his principal object being a more accurate examination of the wooded chasm in the mountains of Swellendam. The party set off on 2 September 1804 in an easterly direction, passed by the mission of Baviaanskloof again and headed then to Swellendam. They went back via Rotterdam and came back to Cape Town on 20 September 1804.

That the Batavian Government allowed Lichtenstein to be part of the journey, was only one way of demonstrating their support. They even fostered his natural scientific researches, as the following quote highlights:

> General Janssens, who was always most obligingly anxious to promote, in every possible way, my desire of becoming more intimately acquainted with the natural history of the country.[15]

As Saul Dubow states, colonial administration always had its own interest in sciences. He claims that the colonial enterprise always utilised science not only to observe, measure and control the "native other" but also to proclaim and shape the self-image of the colonisers themselves.[16]

In this regard, the Dutch administration attempted to accomplish a more complete and correct map of the colony. Maps held an important function within the colonial system in order to serve as a sign of (re-)organisation, control and conquest of non-European territories by Europeans. The land could be overwritten and the new colonial power distanced itself from its precursor by correcting its maps.[17] Governor Janssens and Captain Paravinci di Capelli were instrumental in this undertaking. They employed experienced officers who revised the material already in existence, with the intention of correcting the errors it contained. The southern coast for example was carefully examined in several voyages from Table Bay to Algoa Bay by 'some very intelligent and experienced sea-officers'.[18] According to Lichtenstein, they especially directed their attention to correcting the errors in Barrow's map and therefore to doubt the English competence. Lichtenstein himself was also involved in this process, since the observations he made in his different journeys further promoted the maps' development.

The Batavian Government encouraged everyone in the colony to study nature with the support of available knowledge from other nations. As England was the declared enemy, their scientific results were doubted and the colony's population was invited to support the advancement of science. Lichtenstein added in his book the example of the request to find a unicorn and therefore to prove Barrow's claim of its existence:

> The Commissary-general repeated here the engagement made by the governor both at this and many other places in his journey, to all its appurtenances (worth all together about five hundred dollars) as a reward to any one who should bring a complete skin of this animal, with the horn and skull-bone, to the Cape Town.[19]

This quote illustrates that the Dutch were familiar with the available literature. They knew about the accounts of scientific expeditions and were interested in advancing their own findings and publish their discoveries under their name.

Lichtenstein's principal journey led him beyond the official boundaries of the colony to the unknown 'country of the Bosjesmans, the Corans,[20] and the Beetjuans'.[21] Again, Lichtenstein was invited by Governor Janssens to join the expedition, in order to examine the nature of the country on the other side of the Orange River as well as the situation of the inhabitants. As a basis, he handed him over the account of the furthest inland expedition so far, which was undertaken by Petrus Johannes Truter and William Sommerville in 1801. Barrow had published the journal of Truter, who was his father-in-law, and described therein the Bechuana. Lichtenstein did not hesitate to accept the offer:

> Long as I had wished to become more acquainted with these remote countries, nothing could be more accordant with my wishes than such a proposal.[22]

The main objects of this journey were the negotiations with the Bushmen and the settlement of a matter with the missionaries. Both should be handled by the Landdrost Van de Graaff but Lichtenstein served as a Government Commissioner in this enterprise. Lichtenstein was furthermore supported in completing of the colony's maps and was therefore – for the first time – furnished with astronomical instruments.[23] His personal interests belonged to the Bechuana: he made many observations and inquiries about their language and their way of life.

Lichtenstein and Van de Graaf met in April 1805 in Cape Town where they had discussed, on several occasions, the preparations needed for the expedition. The governor allowed them to be furnished with everything they desired and let them choose their own guides. Lichtenstein left Cape Town without any companion on April 24th, towards Tyger Mountain where he visited Governor Janssens and where 'this excellent man explained me more fully the objects he had on view in the present expedition'.[24] Lichtenstein received a written paper of instruction and headed towards Tulbagh where he stayed for ten days with the Landdrost. On May 7th, they set off enjoying the confidence of the colony's population during the whole journey because they were employed by the government.

Throughout the journey, Lichtenstein increasingly took over the leadership and commissioned the other members of the group. They visited villages of the "Hottentot republic", mainly formed under the government of missionaries, where they obtained a list of all the heads of the families with their places of birth, their former modes of life and the number of their children.[25]

They continued their way to the Corans, whom Lichtenstein referred to the oldest original inhabitants of the country.[26] There, they visited a mission station again and sent a report of the situation to Cape Town. After a strenuous and dangerous journey, the troop finally arrived at the Betchuana country. They first interviewed some herdsmen and afterwards arrived at the residence of the King Mulihawang, bringing him presents. After several meetings with the king, Lichtenstein went back to Tulbagh with his troop, where he arrived on 15 August 1805 and from where he started his next journey, only joined by one Khoikhoi.[27]

When he was not travelling and therefore in Cape Town, he also devoted his whole spare time to botanical and zoological studies. He used every occasion to collect botanical, zoological and geological specimens. Lichtenstein, thrilled by the impression of this unknown nature was increasingly passionate about it: 'I should conceive that a man could scarcely explore this country without almost involuntarily becoming a naturalist'.[28]

He built up an herbarium and a detailed description of all his collected plants. A collection of insects, seeds, minerals and even skulls was also among his findings. Depending on the specimens, he either kept them to take back to Germany or was 'collecting seeds for the benefit of the governmental botanical garden at the Cape Town, and of my friends in Europe'.[29]

After the colony went again into the hands of England in 1806, Lichtenstein left the Cape and went back to Holland and later to Germany.[30]

Lichtenstein's findings in the hands of renowned scientists

Back in Germany, Lichtenstein met with many contemporary intellectuals, among them Johann Wolfgang von Goethe. He encountered him in Weimar in 1808 and talked to him about his notes from South Africa.[31] He also went to Count von Hoffmannsegg with his records, and even handed him his herbarium. As a patron of Lichtenstein's training in natural sciences, Hoffmannsegg introduced him to Carl Ludwig Willdenow (1765–1812) who was in charge of Hoffmannsegg's collection of insects. In 1804, Willdenow published the book *Anleitung zum Selbststudium der Botanik*, according to which Lichtenstein ordered his collected plants.

A letter written in 1812 mentions that Lichtenstein gave most of his dried plants to Willdenow after his return. Willdenow was considered to be one of the most important taxonomists of his time and became the intellectual centre for young natural scientists of whom he mentored many. Alexander von Humboldt, too, let him examine the plants he collected in South America.[32] Von Hoffmansegg and Willdenow recognised in Lichtenstein's collection several new plant species. Under the name of *Lichtensteinia undulate*, Willdenow gave a description and engraving for the annual publication of the *Gesellschaft naturforschender Freunde zu Berlin* (GNF) for the year 1808.[33]

Lichtenstein planned to publish a book with some of his botanical findings:

> Professor Willdenow, who . . . undertook to examine my collection of plants, gave to these the name of gladiolus papillonaceus. It is my intention hereafter to describe them with a hundred and eighty other sorts of plants, which are in my herbarium, more accurately: this will be done in a little work under the title of Spicilegium Floroe Capensis, which is already more than half prepared for publication.[34]

Unfortunately, this book was never published.

When Lichtenstein returned from South Africa, he also held drawings and notes about insects and animals and some zoological items. Concerning the aspects of animals, he contacted Karl Illiger. Illiger thought that about half of the 600–700 species in Lichtenstein's insect collection were new.

Lichtenstein also brought back from South Africa findings that belonged neither to the discipline of botany, zoology or mineralogy. Among them was for example a skull of a Bushman, including the skin of his face. With these human remains, he went to Johann Friedrich Blumenbach (1752–1840), a brilliant Professor of Medicine at the famous University of Goettingen.[35] Blumenbach was considered an excellent anatomist, anthropologist and zoologist who believed that human beings should be classified in the same way as other living things.[36] He divided human species into distinct races based on physical characters such as the shape of skulls – for which he maintained his famous collection of skulls from all over the world.[37]

During Lichtenstein's encounters in South Africa with different tribes – which he distinguished as *Caffers* (Bantu), *Hottentots* (Khoikhoi), *Bosjesmans* (Bushmen) and *Corans* – Lichtenstein carefully observed their way of life and made note of the various languages. He discussed these examinations with experts to whom he only referred to as 'persons skilled in languages'.[38] Spohr speculates that with the help of his linguistically well-versed father, Lichtenstein was able to find a connection between Arabic, Hebrew and other North African words and South African native words. According to different scholars, this seemed to be the first effort of its kind. Lichtenstein himself learned the Bechuana language on his journey and preferred his own method of describing the African languages. However, he was challenged in finding a way to transcribe them:

> For various reasons, indeed, it is almost impossible to give, in writing, information that shall be intelligible upon languages of such a nature, or to treat of them after the manner in which we should treat of a language reduced to a regular system.[39]

He listed each of the words or sentences that he had collected as spoken by the Corans, Bushmen, Xhosa and Bechuana, and developed a vocabulary for each of these languages.

The importance of learned societies and scientific journals

At this time, the *Akademie der Wissenschaften zu Berlin* belonged to the most important academies in Germany. Together with the *Royal Society of London* and the *Académie Royale des Sciences* in Paris it was one of the three leading societies in Europe.[40] All of them were founded within 40 years; the *Akademie der Wissenschaften* in 1700 by Gottfried Wilhelm Leibniz who was also a member of the other two and became the first president of the Berlin academy. Scientists saw advantages in associating with like-minded people. Universities at that time had no adequate place for the exchange of ideas among equals (there were neither faculty clubs nor professorial offices). In the halls of the academy, controversies could be aired, alliances forged and criticism vetted.[41] Learned societies not only provided an avenue for communicating results of scientific investigations, they also increasingly formalised the system of communication due to their regular meetings and the dissemination of their results in journals. The medium of the society had also been used by scientists in their attempts to communicate with interested or potentially interested outside parties. The motives for these networks were often explicitly self-serving. As science had become ever more expensive, patrons outside the academic world had to be convinced of its value, in order to provide the necessary resources. The source of the patronage had changed, from royalty and the nobility in the 17th century to governments, industry and the general public, but the need to "sell" science remained the same. These societies had other functions besides mere communication, though. Very often they served as gate keepers by defining who was accepted to the scientific community and who excluded.[42] Membership of a formal society could easily be regulated in order to keep figures with unwanted views outside.[43]

However, the learned societies provided a new professionalism to the study of nature. Scientists began to organise themselves both into specialised societies – such as the GNF – for the promotion of individual disciplines, and into general organisations designed to promote the role of science in society.[44] They stimulated research and provided for the diffusion of findings through their publications.[45] The *Akademie der Wissenschaften zu Berlin* was explicitly established to foster and generate sciences and thereby top promote the natural sciences. It also housed books in its library, displayed specimens and collected instruments in its cabinet. All these services assisted the investigations of individual members. The structure of the academy consisted of permanent members who lived in Berlin, and external members which participated only as corresponding and eventually honorary members.[46]

Lichtenstein knew: a scientist could only be acknowledged if people got to know about his discoveries. In order to earn credits, scientific achievements had to be communicated to other scientists and to the academic world.[47] By the 19th century, scientific publishing was a well-established field, with authors aiming their works at a variety of audiences.[48] In Germany, 718 new titles of scientific journals appeared in the 1780s alone. This indicates how rapidly the learned journal became a prime mover of the scientific enterprise and supplied scientists with a quick, accessible medium for publishing their findings. To the readers, it also brought information that otherwise might escape their attention, whether issued in foreign languages or hardly available sources.

Lichtenstein wanted his studies of South African natural history to be placed on a scientific basis and to spread the knowledge gained in the fields of geography, anthropology, linguistics and medicine, through scientific journals to the learned societies in Germany. Probably inherent to the aim of publishing new scientific knowledge was the intention to spread his own name and make the scientific society aware of his personal achievements. Through his studies – achieved only thanks to his travels in a little-known area on the scientific landscape in Germany – he could now establish himself among the country's community of naturalists. Since he did not take up an employment as a doctor, it can be assumed that Lichtenstein hoped to qualify for a career in the sciences. He pursued this ambition this idea by publishing articles – and later a book – about his journey (see Figure 3.1).

Position at the Berlin University

Although Lichtenstein lived mainly in Helmstedt near Hamburg, he visited Berlin several times to visit his academic friends living there. The political situation in Berlin was unstable, since Prussia lost the War of the Fourth Coalition and Napoleon had occupied the capital. Following defeat during the Napoleonic wars, Prussia saw itself not only dispossessed of parts of its territory but also of its most important university, Halle an der Saale.[49] After this military and political disaster,

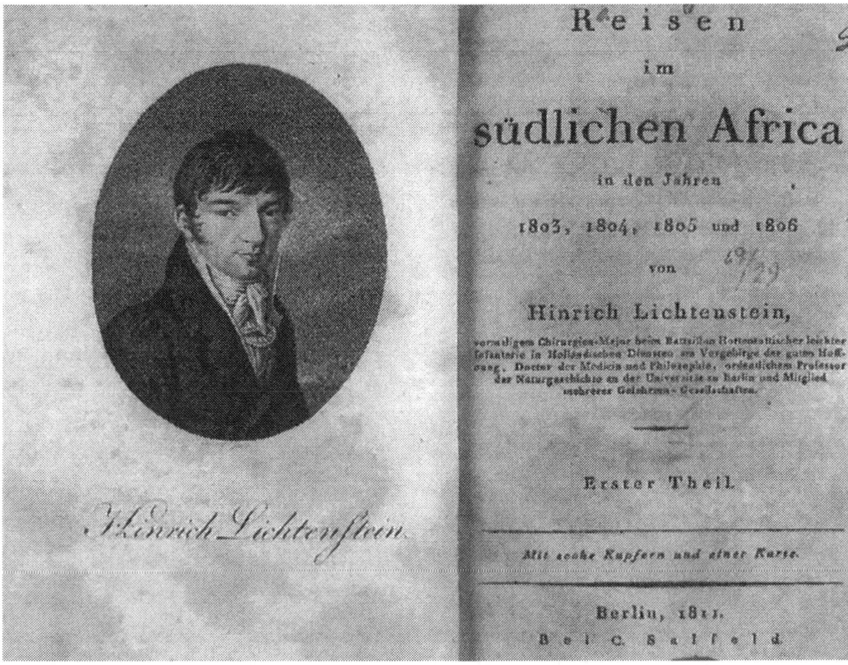

Figure 3.1 Front page of the first volume of Lichtenstein's travelogue

Source: Lichtenstein, Hinrich: Reisen im südlichen Afrika in den Jahren 1803, 1804, 1805 und 1806. Berlin 1811.

Prussia undertook a reformation of the state, which also included an extensive educational reform including planning for a university in Berlin. In this university, professors were expected to do research and train their graduate students to become researchers in their own right.[50] Hence, professors were now expected not only to contribute new knowledge but also to teach existing knowledge. Under the guidance of the privy counsellor and historian Wilhelm von Humboldt (1767–1835), brother of the naturalist Alexander von Humboldt, the notion of pure learning was laid at the centre of the organisation of this new university.[51] It became one of the centres of 19th-century German learning and thus one of the principal settings for the meteoric rise of German science.[52] It was here that the modern research university was created and the specification of disciplines arose.

Before the foundation of the university, Berlin had two scientific institutions: the *Collegium medico-chirurgicum* and the *Theatrum anatomicum*[53] and was therefore equipped with faculty-like medical institutions.[54] The former had already transcended the borders of a mere medical education towards an institution that included the entire natural history and which chair was assigned by Willdenow. The subjects taught included chemistry, physics, pure and applied mathematics, botany, zoology and even philosophy.[55]

Humboldt conceived of a new form of education, which not only envisioned the alliance of research and teaching, but also the particular significance of the humanities. In this system, the philosophical faculty served as the common fundament for the inherent independent disciplines. Since priority went to the philosophy faculty, the formerly "higher" faculties of law, theology and medicine were of less relevance.[56] In the new philosophical faculty, the subjects of zoology, botany and mineralogy were also included; the faculty was therefore detached from the medical faculty.[57] This was an important step towards the development of independent disciplines and gave new importance to the natural sciences. This constituted a new approach in the university landscape of Germany.[58] According to Bowler, the establishment of new disciplines can be attributed to the competition between the different German universities.[59] As Germany was consisted of a number of states, each sought to rival its neighbours by gaining a better reputation. Prussia in particular wanted to put itself on the academic map by founding an outstanding university. Sciences were no longer considered "minors" for students of medicine, but students could now specialise in natural sciences. In fact, a considerable number of students chose to do so. Already since the early 19th century, it was possible to study the subject of zoology and to earn a respective doctor's degree at the Berlin University. Consequently, zoology – as well as botany and mineralogy – was no longer only a subject in which to do research but also one being taught and studied.[60]

Other already existing institutions fostered the foundation of autonomous chairs within natural history at the new Berlin University. Important for botany was the already existing Botanical Garden which had been maintained by the *Akademie der Wissenschaft* since 1716.[61] The Botanical Garden with its systematically ordered plants and greenhouses constituted an important facility for botanical studies. The Botanical Gardens emerged throughout the 18th century and functioned essentially as an encyclopaedia.[62] Like the pages of a book, plants of each genus were subdivided into species and placed according to a predetermined plan for reference purposes. Unlike the dried specimens in an herbarium, plants in a real garden could be read in their living form and a splendid diversity. Similar to museums, gardens sought to bring the earth's lush flora into a smaller compass, where it could be appreciated, enjoyed and savoured. Botanical Gardens had originated in the medieval universities from physic gardens as a place for training doctors to recognise the plants from which many medicines were derived. From the beginning, physic gardens were associated with the medical faculties of universities, where one professor usually taught both botany and medicine.[63] In the 18th century, the gardens emerged as leading institutions for research in natural history as new plants from all over the world were brought to Europe.[64] In the plans of the university administration, the Botanical Garden was soon linked to a chair of botany. Hence, it became integrated in the body of the university. Willdenow was its successful director since 1802; he also became the first Professor of Botany at the newly established Berlin University in 1810. He managed to broaden the holdings of plants from 1,200 species in 1804 to 7,700 species in 1812. The achievement is remarkable, as Prussia was at war with Napoleon these years.[65] According to Lenz, the garden could even rank with the much vaunted *Jardin des plantes* in Paris.[66]

The *Akademie der Wissenschaften* also kept a cabinet of specimens, which could be used by the students of the medical institutions as well as twice a week by the public.[67] It contained objects of all three kingdoms of nature and consisted therefore of plants, animals and minerals. But it was more than just a place where specimens were exhibited. It was characterised by being at the same time a meeting place, a place of communication and a system of sharing. It must therefore not only be understood in the classical sense, namely as a collection of material objects, but also as a centre for socialising and gaining power.[68]

As important as the Botanical Garden was for the emergence of botany as an independent subject, so too was a museum for the discipline of zoology. Count von Hoffmannsegg – who was consulted by Wilhelm von Humboldt in order to fill the university's chair – knew the *Muséum nationale d'histoire naturelle* of Paris, and believed that Berlin should now exemplify in a similar way – and therefore for the first time in Germany – a complete collection of fauna for the naturalists as well as for the students and the laymen.[69]

Wilhelm von Humboldt heavily counted on the former professors of the by then closed University of Halle to assign the chairs at the Berlin University. When Halle unexpectedly re-opened at the end of 1808, many designated professors for the Berlin University withdrew their applications and an active quest started to find appropriate scholars for the planned institution in Berlin. The king signed the deed of foundation on 16 August 1809, and a committee tried to persuade several professors outside Prussia to accept the appointment to one of the four faculties which were to open in 1810.[70] The division of the German-speaking region into a number of states created a situation in which each sought to rival its neighbours in the acquisition of scientific and scholarly talent. Due to the formation of the disciplines, suddenly more professors were needed than available. But for many chairs at the new university there was no professor of reputation available and the committee therefore had to turn to young scholars, mainly from Berlin itself. Illiger was proposed by Hoffmannsegg for the curatorship of the Zoological Museum and also both, for a Professorship of Zoology and a membership of the *Akademie der Wissenschaften*.[71] When Illiger refused to teach zoology, Lichtenstein was approached. The accounts of his journeys in Southern Africa had made him famous, and he had influential friends in Berlin supporting him. Even before his book was published, he was considered one of the best experts in Natural History and an outstanding collector and observer of South African plants and nature throughout Europe. Yet for all this, he had neither any zoological research credits nor any education in Zoology. Kocka argues that this was no disadvantage, as the Berlin University often employed new scholars due to their general reputation, acquired by publications[72] – especially it was in desperate need of academic staff.

Lichtenstein accepted without hesitation and was already prepared to lecture in autumn 1810. He lectured about mammals and amphibians and was one of the first professors at a German university to lecture about birds. Since Lichtenstein had studied medicine, the university conferred upon him, at the suggestion of the dean, an honorary degree of Doctor of Philosophy in February 1811 – probably to fulfil the requirements – and one month later, he was appointed Professor of Zoology.[73]

Socialising and working in Berlin

In the early 19th century, Berlin became the intellectual centre of Prussia, and Lichtenstein was part of this development.[74] The upper intelligentsia met at different places in Berlin. Lichtenstein's apartment at the university became one such place. He joined the famous *Akademie der Wissenschaften* in 1814. All members of the academy were authorised to teach at the university – although most of its members first taught and then became members, as in Lichtenstein's case.[75] The academy was structured into classes, integrating the natural sciences as well as the humanities.[76] The members of the faculties of law, theology and medicine were only incorporated as far as they touched common interests, which in fact excluded them from the academy. The previously most important faculties were now seen as preparatory for jobs and not as pure sciences anymore.[77]

In the following years, Lichtenstein joined two other societies. In 1816, he was elected to the exclusive *Montagsklub*, a learned society and the intellectual heart of Berlin. The members – not more than 30 and all drawn from the *Bildungsbürgertum* – discussed philosophical and literary questions every Monday in a public house. The club was originally founded in 1749 by men of the enlightenment for men of similar backgrounds and tastes 'with the sole purpose of pleasant and stimulating conversation and free discussion over food and wine'. The philosophy of the Enlightenment, with its emphasis on the cultivation of the intellect was an integral part of each member's outlook.[78] Other university professors and members of the GNF, including Willdenow, belonged to the society as well.[79] As Lichtenstein had a preference for historical studies,[80] he compiled historical notes for the *Montagsklub*, published after his death in 1879.[81]

The second society he joined, in 1818, was the *Kaiserlich Leopoldinisch-Carolinische Akademie der Naturforscher*, briefly called the *Leopoldina*. This society existed since 1652 and elected an unlimited number of naturalists as its members. Although many members belonged to both the *Akademie der Wissenschaften* and the *Leopoldina*, the societies were organised differently. Members of the *Leopoldina* communicated mainly through letters, individual encounters and through the spreading of the member's results by publishing them. Moreover, the *Leopoldina* did not distinguish between permanent and external members.[82]

Director of the Zoological Museum

After Illiger's death, Lichtenstein became the director of the Zoological Museum in 1815. When it opened in 1814, the museum was one of the first zoological museums in Europe. Lichtenstein provided that the objects were not only systematically ordered and labelled but also colour-coded according to the object's country of origin.[83] It was the first time that this kind of geographical identification was used in a museum; previously, objects were labelled according to their donators or donors. Inherent to this new approach was the expectation that the collection would be useful for students and for learned societies. Lichtenstein foresaw that it was not enough to purchase collections but also necessary to support explorations. The king and the minister agreed and Lichtenstein was now authorised to employ collectors and to send them all over the world. Berlin could now follow in the footsteps of France, England, Holland and Russia, which all sent keen young botanists and naturalists to collect new specimens for their museums.[84] The explorations started in 1816 and were paid, in most cases, by the King.[85] Naturalists were sent to Brazil (from where Freyreiss, Feldner, Beseke and Sellow sent 262 mammals, 5,457 birds and almost 10,000 insects to Berlin between 1817 and 1831), to Mexico and California (where Schiede and Deppe collected 958 birds of 315 species of which, according to Stresemann, many were new for Berlin scientists), to North America (Becker, Zimmermann, Koch and Bachmann), Siberia (Eversmann), Madagascar (Goudot), Australia (Sieber and Lothsky) and Colombia (Haeberlin).[86] Lichtenstein used the travellers' notes – previous and contemporary ones – to compose articles.

The support he gave to travellers to and in South Africa is perhaps not surprising. Lichtenstein was highly interested in any Cape naturalist. Already in 1815, he sent the gardener Louis Maire and the pharmacist Leopold Mund via St. Helena to the Cape – they were probably the first collectors engaged by him. Later he engaged Hesse, who stayed for 17 years at the Cape and was paid for the items collected and their delivery to Germany, and Bergius, a pharmacist assistant working for Lichtenstein's friend Polemann in Cape Town. Carl Bergius had been sent to Cape Town in 1815 by Lichtenstein on condition that he would collect botanical and entomological specimens for the Berlin Museum in his spare time. However, Bergius broke the contract before it expired due to bad relations with the pharmacy's owner.[87]

He was soon replaced by the 25-year-old Ludwig Krebs, who arrived at the Cape in 1817. At the end, Krebs worked for Lichtenstein and the Zoological Museum for 25 years and sent 15 shipments with a total of 3,245 birds, thousands of insects as well as many other important species (as for instance a quagga), some of which were still new to German science (for an extended account of this relationship, see the chapter of Patrick Grogan in this book).[88]

While Krebs could probably be considered responsible for one of the most important zoological collections of South Africa in a European museum, Carl August Ehrenberg's and Friedrich Willhelm Hemprich's expedition to the Orient were possibly the most important and successful expeditions of these days. Ehrenberg, a former student of Lichtenstein, and Hemprich, who ordered the amphibians in the museum for Lichtenstein and replaced him during his journeys, travelled to Arabia, Egypt, Nubia, Syria and Abyssinia in the years between 1820 and 1826.

For every collector abroad, Lichtenstein acted as an important facilitator: he was the person who brought forward the motion for money. He whole-heartedly sought to acquire subsidies for younger scholars willing to explore the unknown as he had done years before.

In his collecting, Lichtenstein followed the "Noah's Ark" principle, and wanted a pair of every species in his museum and focused on collecting a variety of specimens. He compared specimens newly received to the ones he already possessed. If the new ones were similar to any in the collection, they were sold as duplicates.[89] The first auction took place at the museum in 1818. The amount of money raised by the auction was so high that it became common practice to trade duplicates. Lichtenstein wrote that the success justified the sale. The auction served in his view both the purchaser and the museum. Directors and owners of collections found a good opportunity to enrich their collections with rare animals, which were already correctly classified and named. And the museum received financial means to fill gaps and continue its explorations.[90] Lichtenstein documented every auction and carefully classified the sold specimens according to Linnaeus. Where necessary, he described the animals as well and provided every species with a price. Individual specimens in 1823 sold for between 1 and 150 Thaler, with the highest bid offered for an African ostrich. Many of the objects sent by Ehrenberg and Hemprich were consequently sold or traded with other museums.[91] The auction in 1823 included among the 1,481 species some of the "Nubian rarities" the two travellers had sent to Berlin.[92] Lichtenstein's auctions insulted some contemporaries including Ehrenberg, who was frustrated to find, after coming back to Germany, that many of the items he had collected had already been sold. Scientists of later generations, such as Lichtenstein's worst critic Professor Stresemann, the museum's director from 1946 to 1959, were also critical. Stresemann blamed Lichtenstein retrospectively for having sold important objects and risking the museum's good reputation.[93]

In order to grow the museum's collection and to sell and buy objects, Lichtenstein travelled throughout Europe almost every semester break, visiting professors, private scholars, collectors, animal traders or directors of zoological museums. From 1819 to 1857, he undertook some 18 major journeys in Europe. Due to his untiring efforts, Lichtenstein achieved an incredible acquisition of objects for the museum in a short time. Jahn states that in 1850, the museum had such a rich collection that it could compete with the biggest museums in Europe, as the ones in London, Paris, Vienna and Leiden.[94]

Conclusion

Considering his many years of active academic life, Lichtenstein's scientific output was not immense. Two former assistants wrote to Hemprich and Ehrenberg: 'Unfortunately, Lichtenstein is overloaded with unpleasant affairs'[95] and 'he is not able to pursue his research because of the administrative work'.[96] Not only administrative details but also financial worries about the museum occupied a great deal of his time and mind and barred him from writing books or academic essays.

He additionally contributed to founding the Zoological Garden in Berlin in 1843 and acted as its first scientific director. Lichtenstein saw the potential of the zoo as a living museum and an opportunity to gain additional material for descriptive zoology, since it provided the only source for observing living animals outside of their natural habitats.

At the university, he offered crucial support to young scholars who were eager to explore the world and its specific nature. He spared no efforts to gather the money they needed. His strength was not only in motivating these young people but also in convincing the broader society of the necessity of such explorations. Lichtenstein's role in Berlin was rather that of an organiser and patron of knowledge than that of a researcher. Furthermore, he acted as an important contemporary witness and his papers offer insight into the networks and institutions in a time when modern disciplines in the natural sciences were constituted and the modern research university was arisen.

Additionally, he seemed to pave the way for a new career pattern in which explorations amounted to a career step. It became common that young men who had received a doctor's degree (mostly in medicine or natural sciences) first went on extended travels before settling for a position. This pattern guaranteed them a better advancement in their career after their return. The pattern emerged, according to Essner, in the early 19th century and only ended in the last quarter of the 19th century.[97]

Can Lichtenstein therefore be seen as a pioneer? He was innovative indeed, in several ways: It was him who ushered in a new age of Southern African travel accounts. And he fearlessly and passionately engaged in two intellectual adventures: first, the travel to South Africa, and second, the university chair of zoology. His stay in Africa certainly accompanied him throughout his life and fostered his reputation as an energetic scientist who dedicated his life to the services of natural science. Since he was responsible for raising funds, acquiring collections, buying specimens, mounting expeditions and sending collectors to the Cape and elsewhere, Lichtenstein shaped the scientific landscape of Prussia probably more than is evident at first glance.

Notes

1 Volkmann, 'Biographisches und Bibliographisches der Helmstedter Familie Lichtenstein', in *Zoologisches Institut Berlin, Historische Bild- und Schriftgutsammlung*, Bestand Zool. Mus; S I, Lichtenstein I-X, 13–19.
2 Gebhardt, *Die Ornithologen Mitteleuropas*, 216.
3 Bowler, *Fontana History*, 140.
4 Lichtenstein, *Travels in Southern Africa*, vol. I, 3.
5 Essner, *Deutsche Afrikareisende*, 74.
6 Stresemann, 'Hinrich Lichtenstein', 74.
7 If not differently indicated, all historical details are taken from: Freund, 'The Cape', 324–50.
8 Grünewald, *Geschichte der Deutschen in Südafrika*, 47.
9 Lichtenstein, *Travels in Southern Africa*, vol. I, 11.
10 Society of United Brethren.
11 Lichtenstein, *Travels in Southern Africa*, vol. I, 10.
12 Ibid., 14.
13 Lichtenstein, *Travels in Southern Africa*, vol. II, 137.
14 Lichtenstein, 'Nachrichten von Teneriffa', 37.
15 Lichtenstein, *Travels in Southern Africa*, vol. II, 138.
16 Dubow, *Science and Society*, 3.

17 Fiedler, *Zwischen Abenteuer, Wissenschaft und Kolonialismus*, 155.
18 Lichtenstein, *Travels in Southern Africa*, vol. I, xix.
19 Ibid., 206.
20 A Khoi group settled on the north bank of the Orange River.
21 Lichtenstein, *Travels in Southern Africa*, vol. II, 188.
22 Ibid., 191.
23 Lichtenstein, *Travels in Southern Africa*, vol. I, xx.
24 Lichtenstein, *Travels in Southern Africa*, vol. II, 193.
25 Ibid., 314.
26 Ibid., 316.
27 Ibid., 439.
28 Lichtenstein, *Travels in Southern Africa*, vol. I, 39.
29 Ibid., 201.
30 Ibid., 23.
31 Staatsbibliothek zu Berlin – Preussischer Kulturbesitz, Signatur: Slg. Darmst. Afrika 1804, Martin Karl Heinrich Lichtenstein, 79.
32 König, 'Willdenow, Karl Ludwig', 252–4.
33 The generic character he indicated as *hexandria tryginia*. Lichtenstein found another species at another place to which the name of Lichtensteinia loevigate was given, in: Lichtenstein, *Travels in Southern Africa*, vol. II, 208.
34 Ibid., 141.
35 Ibid., 453.
36 Rupke, *Göttingen and the Development of Natural Sciences*, 93.
37 Bowler and Morus, *Modern Science*, 421.
38 Lichtenstein, *Travels in Southern Africa*, vol. I, 298.
39 Lichtenstein, *Travels in Southern Africa*, vol. II, 464.
40 Kocka, *Preussische Akademie der Wissenschaften*, 65.
41 Pyenson and Sheets-Pyenson, *Servants of Nature*, 75–84.
42 Bowler and Morus, *Modern Science*, 319–20.
43 Withers, *Placing the Enlightenment*, 70.
44 Bowler, *Fontana History*, 195.
45 Pyenson and Sheets-Pyenson, *Servants of Nature*, 90.
46 Kocka, *Preussische Akademie der Wissenschaften*, 8–9.
47 Bowler and Morus, *Modern Science*, 319.
48 Ibid., 381.
49 Schröder, '150 Jahre Humboldt-Universität', 2.
50 Bowler and Morus, *Modern Science*, 330.
51 Olesko, *Science in Germany*, 29.
52 Ibid., 6.
53 Both surgical clinics.
54 Schröder, '150 Jahre Humboldt-Universität', 1.
55 Lenz, *Geschichte der Königlichen Friedrich-Wilhelms-Universität*, 40.
56 Pyenson and Sheets-Pyenson, *Servants of Nature*, 54.
57 Jahn and Sucker, 'Zur Geschichte der Botanik', 190.
58 Jahn, 'Zur Vertretung der Zoologie', 261.
59 Bowler and Morus, *Modern Science*, 183.
60 Jahn, 'Zur Vertretung der Zoologie', 262.
61 Ibid., 262.
62 Olesko, *Science in Germany*, 67.
63 Pyenson and Sheets-Pyenson, *Servants of Nature*, 152–5.
64 Bowler and Morus, *Modern Science*, 328.
65 Jahn and Sucker, 'Zur Geschichte der Botanik', 190.
66 Lenz, *Geschichte der Königlichen Friedrich-Wilhelms-Universität*, 248.

67 Jahn, 'Zur Vertretung der Zoologie', 260.
68 Te Heesen and Spary, *Sammeln als Wissen*, 62.
69 Hamel, Knobloch, and Pieper, *Alexander von Humboldt*, 137.
70 Faculty of law, medicine, theology and philosophy.
71 Lenz, *Geschichte der Königlichen Friedrich-Wilhelms-Universität*, 204.
72 Kocka, *Preussische Akademie der Wissenschaften*, 79.
73 Mauersberger, 'Gründer des Berliner Zoologischen Gartens', 5.
74 Ibid., 13.
75 Kocka, *Preussische Akademie der Wissenschaften*, 17.
76 Ibid., 12.
77 Ibid., 42.
78 Keeton, 'The Berliner Montags Klub', 148.
79 Lenz, *Geschichte der Königlichen Friedrich-Wilhelms-Universität*, 549.
80 In 1835 he also composed a genealogical tablet about his family. It started in the year 1690 and was named 'Die Stammtafel der bürgerlichen Familie Lichtenstein nebst hist. Nachrichten über einige Glieder derselben'.
81 Notizen aus den Urkunden des Montags-Klubs.
82 Kocka, *Preussische Akademie der Wissenschaften*, 48.
83 Lenz, *Geschichte der Königlichen Friedrich-Wilhelms-Universität*, 374.
84 Stresemann, *Reisen zweier naturforschender Freunde*, 1.
85 Lenz, *Geschichte der Königlichen Friedrich-Wilhelms-Universität*, 375.
86 All the names and their destinations are to be found in: Ibid., 376.
87 Ffolliott and Liversidge, *Ludwig Krebs*, 13.
88 Ibid., 1.
89 Ibid., 25.
90 Lichtenstein, *Verzeichnis der Doubletten*, III.
91 Stresemann, *Reisen zweier naturforschender Freunde*, 170.
92 Lichtenstein, *Verzeichnis der Doubletten*, X.
93 Lenz, *Geschichte der Königlichen Friedrich-Wilhelms-Universität*, 378.
94 Jahn, 'Zur Vertretung der Zoologie', 266.
95 Stresemann, *Reisen zweier naturforschender Freunde*, 31.
96 Ibid., 136.
97 Essner, *Deutsche Afrikareisende*, 121.

Bibliography

Printed sources

Bowler, Peter J., *The Fontana History of The Environmental Sciences* (London: Fontana Press, 1992).

——— and Iwan Rhys Morus, *Making Modern Science. A Historical Survey* (Chicago, London: University of Chicago Press, 2005).

Dubow, Saul, ed., *Science and Society in Southern Africa* (Manchester: Manchester University Press, 2000).

Essner, Cornelia, *Deutsche Afrikareisende im neunzehnten Jahrhundert. Zur Sozialgeschichte des Reisens* (Stuttgart: Steiner, 1985).

Ffolliott, Pamela and Richard Liversidge, *Ludwig Krebs. Cape Naturalist to the King of Prussia 1792–1844* (Cape Town: A. A. Balkema, 1971).

Fiedler, Matthias, *Zwischen Abenteuer, Wissenschaft und Kolonialismus. Der deutsche Afrikadiskurs im 18. und 19. Jahrhundert* (Köln: Böhlau Verlag Köln Weimar, 2005).

Freund, William M., 'The Cape Under the Transitional Governments, 1795–1814', in *The Shaping of South African Society, 1652–1840*, ed. R. Elphick and H. Giliomee (Cape Town: Maskew Miller Longman, 1989), 324–57.

Gebhardt, Ludwig, *Die Ornithologen Mitteleuropas. Ein Nachschlagewerk*, vol. II (Möggingen: Verlag der Deutschen Ornithologen-Gesellschaft, 1970).

Grünewald, Hildemarie, *Die Geschichte der Deutschen in Südafrika* (Cape Town: Ulrich Naumann Verlag Kapstadt, 1998).

Hamel, Jürgen, Eberhard Knobloch, and Herbert Pieper, eds., *Alexander von Humboldt in Berlin. Sein Einfluss auf die Entwicklung der Wissenschaften* (Augsburg: Rauner, 2003).

Jahn, Ilse, 'Zur Vertretung der Zoologie und zur Entwicklung ihrer institutionellen Grundlagen an der Berliner Universität von ihrer Gründung bis 1920', *Wiss. Zeitschrift der Humboldt-Universität zu Berlin, Math.-Nat. Reihe* 34 (1985): 260–80.

——— and Ulrich Sucker, 'Zur Geschichte der Botanik an der Berliner Universität von 1810 bis 1945', *Wiss. Zeitschrift der Humboldt-Universität zu Berlin, Math.-Nat. Reihe* 34 (1985): 189–202.

Keeton, Kenneth, 'The Berliner Montags Klub – A Center of German Enlightenment', *Germanic Review* 36, no. 2 (1961): 148–53.

Kocka, Jürgen, ed., *Die Königlich Preussische Akademie der Wissenschaften zu Berlin im Kaiserreich* (Berlin: Akademie Verlag, 1999).

König, Clemens, 'Willdenow, Karl Ludwig', in *Allgemeine Deutsche Biographie (ADB)*, vol. 43 (Leipzig: Duncker & Humblot, 1898), 252–4.

Lenz, Max, *Geschichte der Königlichen Friedrich-Wilhelms-Universität zu Berlin*, vols. 1–4 (Halle: Verlag der Buchhandlung des Waisenhauses, 1910).

Lichtenstein, Martin Hinrich Carl, 'Nachrichten von Teneriffa, mit einem Vorbericht von dem Hrn. Hofrath Brun's in Helmstädt', in *Allgemeine Geographische Ephemeriden*, vol. 19 (Weimar: Verlag des Landes-Industrie-Comptoirs, 1806) 37–58.

———, *Verzeichnis der Doubletten des zool. Museums der königl. Universität zu Berlin nebst Beschreibung vieler bisher unbekannter Arten von Säugethieren, Vögeln, Amphibien und Fischen* (Berlin: Zoologisches Museum Berlin, 1823).

———, *Travels in Southern Africa in the Years 1803, 1804, 1805 and 1806*, vol. I and II (Cape Town: Van Riebeeck Society, 1928 and 1930).

Mauersberger, Gottfried, 'Der Gründer des Berliner Zoologischen Gartens, Martin Hinrich Lichtenstein (1780–1857). Eine biographische Skizze', *Bongo* 23 (1994): 3–34.

Olesko, Kathryn M., ed., *Science in Germany. The Intersection of Institutional and Intellectual issues* (Philadelphia: University of Chicago Press, 1989).

Pyenson, Lewis and Susan Sheets-Pyenson, *Servants of Nature. A History of Scientific Institutions, Enterprises and Sensibilities* (London: HarperCollins, 1999).

Rupke, Nicolaas, ed., *Göttingen and the Development of Natural Sciences* (Göttingen: Wallstein Verlag, 2002).

Schröder, Kurt, '150 Jahre Humboldt-Universität zu Berlin. Das Werden einer jungen Universität', in *Forschen und Wirken. Festschrift zur 150-Jahr-Feier der Humboldt-Universität zu Berlin 1810–1960*, vol. 1 (Berlin: Deutscher Verl. der Wiss., 1960), 1–13.

Stresemann, Erwin, ed., *Reisen zweier naturforschender Freunde im Orient. Geschildert in ihren Briefen aus den Jahren 1819–1826* (Berlin: Akademie Verlag, 1954).

———, 'Hinrich Lichtenstein. Lebensbild des ersten Zoologen der Berliner Universität', in *Forschen und Wirken. Festschrift zur 150-Jahr-Feier der Humboldt-Universität zu Berlin 1810–1960*, vol. 1 (Berlin: Deutscher Verlag der Wissenschaften, 1960), 73–96.

Te Heesen, Anke and E.C. Spary, eds., *Sammeln als Wissen. Das Sammeln und seine wissenschaftsgeschichtliche Bedeutung* (Göttingen: Wallstein Verlag, 2001).

Volkmann, 'Biographisches und Bibliographisches der Helmstedter Familie Lichtenstein', in *Zoologisches Institut Berlin, Historische Bild- und Schriftgutsammlung, Bestand Zool. Mus; S I, Lichtenstein I–X*, 13–19.

Withers, Charles W.J., *Placing the Enlightenment. Thinking Geographically About the Age of Reason* (Chicago and London: University of Chicago Press, 2007).

Unpublished documents

Staatsbibliothek zu Berlin – Preussischer Kulturbesitz:
Signatur: Slg. Darmst. Afrika 1804, Martin Karl Heinrich Lichtenstein, 79.

4 'Nothing but love for natural history and my desire to help your Museum'?

Ludwig Krebs's transcontinental collecting partnership with Hinrich Lichtenstein

Patrick Grogan

Introduction: collecting at the Cape

With the rise of the natural history museum as an institution across Europe from the mid-18th century, the transcontinental 'commercial trade in natural history specimens', as Bregman observes, became 'an important part of the [contemporary] scientific world'.[1] This trend was driven by the continent's growing access to hitherto unseen specimens originating from the distant lands being explored and/or conquered by its maritime and imperial powers. For many naturalists, these new finds awoke a fervent desire to fit organic life, in all its formerly unimagined diversity, into an all-embracing and systematic web of life.[2] For European powers such as Britain and France, meanwhile, the burgeoning trade in collections of natural specimens represented an opportunity to establish natural history museums, the prestige projects of the era, which were soon engaged in a 'Wettstreit um Besonderheiten'.[3]

As Bregman shows in his innovative article on 'Botanists and the Commercial Trade in Botanical Specimens at the Cape' between 1750 and 1850, the natural history trade also flourished in the Cape Colony, which, especially after Britain took permanent control of the territory in 1806, became an accessible and favoured destination for collectors of natural history specimens from all over Europe.[4] The Cape was an attractive land for aspirant collectors for a number of reasons. As early as the late 18th century, the colony was well-known for its biological diversity, with the writings of 18th-century traveller-naturalists such as the Swedes Anders Sparrman and Carl Thunberg as well as the Frenchman François Levaillant – all of which were enthusiastically read in Europe – having vividly recounted the colony's natural beauty and organic richness.[5] Furthermore, the colony offered an unparalleled gateway to the interior of Africa, a continent that had hitherto appeared almost inaccessible to Europeans due to its extreme climate, the threat of disease, its forbidding landscapes, lack of infrastructure for travel and "hostile political structures" which, until the late 19th century, were yet to be subjugated by European powers.[6] In contrast, the Cape promised the European traveller a temperate climate, a favourable geographic position on the shipping route to Asia, established colonial settlements, a basic road infrastructure and sufficient equipment and supplies to make long journeys into the interior feasible, rendering it a popular destination for a cosmopolitan range of traders, adventurers, traveller-diarists and naturalist-collectors alike.[7] The latter group sought to take advantage of these favourable conditions to make a living from gathering specimens of natural history, a career which promised an attractive combination of autonomous travel and adventure. Meanwhile, in distant Europe, the directors of natural history museums and other scientific institutions saw in this demographic of young naturalist-adventurers at the Cape, and their like-minded counterparts elsewhere, the opportunity to obtain specimens from areas previously unstudied by European science.

"Salaried" collecting,[8] one type of partnership between sedentary naturalists in Europe and travelling collectors further afield, is the focus of this chapter, in which I explore the relationship between the director of the Berlin Zoological Museum from 1813 until 1857, Hinrich Lichtenstein (1780–1857) – himself an enthusiastic collector at the Batavian-ruled Cape during the first decade of the 19th century – and his long-time Cape collector, Ludwig Krebs (1792–1844). After introducing both men, an examination of Lichtenstein's correspondence with Krebs follows. After an analysis of Krebs's collecting practices in the field, it will be argued that collecting as an activity cannot be understood separately from the socio-political and social- cultural dynamics of the society in which it occurred as well as the personal, interpersonal, commercial and institutional factors which shaped its extent and form. In so doing, I posit that the Lichtenstein-Krebs partnership was driven by inherently commercial imperatives which manifested themselves in the destructive practices of uncontrolled hunting and the claiming of human remains.

Introducing Hinrich Lichtenstein

As Näf-Gloor describes in this volume, Hinrich Lichtenstein spent a productive four-year period in the Cape Colony from 1803 to 1806, expanding his knowledge of local geography, linguistics, ethnology and colonial politics while working as a tutor to the son of the Cape's Batavian Governor, Jan Willem Janssens, and later as the governor's personal physician, a report writer on official expeditions, and as a military physician for the so-called Hottentot Light Infantry. However, it was natural history and collecting which awoke within Lichtenstein the deepest passion, with his curiosity towards the natural environment and enthusiasm for collecting springing from the pages of his well-received account of his travels in southern Africa.[9] When, as Näf-Gloor describes, Lichtenstein was forced to return to Europe with his employers immediately after the Cape Batavian Republic's capitulation to Britain in January 1806, he took with him a large personal collection of natural history specimens and a trove of information and data which he had gathered in the field, all of which he made use of in publishing several scientific articles as well as his account of his travels. From these publications, Lichtenstein, who had only trained as a physician, built a reputation for himself as a naturalist and, by 1811, had been appointed as Professor of Zoology at the newly established University of Berlin.[10] Two years later, he was also appointed director of the university's Zoological Museum, a position he would hold until his death in 1857 and from which he would commission collectors to operate throughout the world gathering specimens, including at the Cape.[11]

Lichtenstein, Harries argues, 'marks the bridge between old and new ways of collecting and ordering nature':[12] an eager traveller who, through his experiences at the Cape, identified opportunities for a more systematised commercial collecting regime to improve on the more general personal collections of amateur naturalist-collectors like himself. Under Lichtenstein's directorship, the museum expanded rapidly, both in terms of the overall number of specimens which it held and in size.[13] For example, while in 1814 the museum had possessed three rooms, by 1817, when it was opened to the public, it had nine rooms; by 1822 it held so much material that there was no longer sufficient space to store and display it all.[14] Lichtenstein, however, continued to meet the challenge of securing more space, even giving up his adjoining apartment for museum use.[15] Furthermore, as Stresemann argues, Lichtenstein aimed to make the museum attractive to the lay public, extending opening hours, hosting school tours and tailoring displays accordingly, including introducing coloured labelling to indicate the geographical origin of animals.[16] Nevertheless, Lichtenstein's sudden rise in the field of zoology had left him unprepared for the demands of this rapidly evolving science. Consequently, he shied away from participating in scientific debate as his lack of zoological training made him wary of contributing his own theories. As a result, he did not seek to match the cutting-edge research of many of his contemporaries in comparative anatomy, choosing instead to stick to superficial external descriptions of the specimens in the museum's possession.[17]

As museum director, Lichtenstein thus emphasised commercial imperatives above scientific research. In so doing, he would earn the Berlin Zoological Museum a reputation as a clearing house for zoological specimens by instructing its collectors to seek not only rare or hitherto undescribed specimens, but also to provide duplicate specimens of species already represented in the museum's collections.[18] These Lichtenstein would organise to be auctioned off or sold to other institutions. With financial support from the Prussian state gradually decreasing, the sale of duplicates ultimately became the museum's chief source of funds. However, this had negative implications for its scientific reputation: Stresemann, for example, claims that Lichtenstein sold many rare specimens – falsely identified as duplicates of other species already in the museum's possession, including a number of soon-to-be-extinct southern African animals – to foreign museums.[19] Consequently, Lichtenstein enjoys an ambiguous legacy: a consummate fundraiser for scientific causes and a populariser of science, but a poor scientist who was out of his depth as a zoologist.

Introducing Ludwig Krebs

Ludwig Krebs, an apothecary from Hannover, arrived at the Cape in 1817 to work at a pharmacy co-owned by the Alsatian doctor Dietrich Pallas (1768–1840) and the Hamburg chemist Pieter Heinrich Polemann (1779–1839). Pallas & Polemann, as the shop was known, was the first chemist's store in Cape Town and would become familiar in German naturalist circles for recruiting aspiring collectors to work as apothecaries' assistants. During his stay at the Cape, Lichtenstein had developed a lasting friendship with Pallas, whom he accompanied on numerous collecting excursions. This connection later allowed him, in his role as museum director, to organise an assistant's position at Pallas & Polemann for Carl Bergius, a promising young chemist and naturalist from Berlin. Lichtenstein had hoped that this arrangement would allow Bergius to collect for the Berlin Zoological Museum in his spare time. Although this initial agreement would turn sour as Bergius became involved in a protracted dispute with Polemann – possibly resulting from the latter's perception that Bergius was devoting too much of his time to collecting in neglect of his day job[20] – the shop continued to attract aspirant German collectors, including Krebs.[21] There they could benefit from a position which allowed them to establish themselves in the colony – both socially and financially – and which provided a base from which they could explore the local area and acclimatise themselves to the region's unique flora and fauna. Indeed, although he left no written description of his time at Pallas & Polemann, participating in regular Sunday botanising excursions to Table Mountain, Devil's Peak or further afield in the company of the enthusiastic hobby-naturalist Polemann and other acquaintances may very well have influenced Krebs's ultimate decision to devote himself to a career in collecting, as his replacement at the pharmacy, Carl Drège, attested to in his own case.[22]

By the time his five-year contract at the shop was nearing its end, Krebs had grown eager to escape the hard and monotonous work and to fulfil his desire to explore the colony.[23] As a result, he wrote to Lichtenstein to offer his services as a collector. Lichtenstein, grateful for a renewed opportunity to commission a regular supply of Cape specimens after hitherto ill-fated arrangements in the colony,[24] responded enthusiastically, turning to the Prussian culture minister, Karl vom Stein zum Altenstein (1770–1840), to pass on Krebs's request for the

> authority to call himself a collector in the service of our country and for a letter from Your Excellency, by which he could identify himself with the English authorities, so that he could obtain permission to travel into the Interior of the Colony, and to hunt everywhere freely.[25]

An influential figure, Altenstein is best remembered as a cameralist; also a self-professed amateur natural historian who, as Brüsch describes, 'had always had affection for botany, horticulture, and gardening', Altenstein unsurprisingly took a keen interest in surveying the potential benefits for Prussia which might result from ready access to foreign plants.[26] He was thus willing to grant Krebs an official letter of recommendation appointing him to the grandly titled position of 'Cape Naturalist to the King of Prussia' as well as to provide him with a contract to assemble 12 collections of natural history specimens. Krebs, for whom the only contractual condition was to collect for the exclusive benefit of the Berlin Zoological Museum, was to be paid per specimen collected according to prices drawn up by Lichtenstein.

Krebs followed Lichtenstein's advice to establish his base on the then less frequently explored eastern edge of the Cape Colony, where he took up residence, first in Uitenhage, then on a farm on the Baviaans River near present-day Bedford. In 1828 he bought the farm, settling there permanently and renaming it "Lichtenstein" in honour of his patron. By 1829, Krebs had sent his twelfth collection to Berlin, the majority of his specimens having originated from the Uitenhage and Albany districts as well as from beyond the colonial frontier in territories controlled by the amaRharhabe or abaThembu peoples across the Kei River or near present-day Queenstown respectively. After the automatic expiry of his contract following his twelfth collection, a renegotiated arrangement saw Krebs continue to collect for Lichtenstein, who paid him out of the profits accrued from auctioning off his specimens in Berlin. However, Krebs's ill-health – he suffered from severe rheumatism – and the outbreak of the Sixth Frontier War (1834–5) hampered his collecting endeavours and resulted in him sending only one more collection to Berlin by 1837. Nevertheless, this enforced break and his resulting frustrations appear to have driven Krebs to prepare for what his biographers, Ffolliott and Liversidge, term his 'Great Expedition',[27] which saw his party cross the Vaal River, reaching as far north as present-day North West Province, far beyond the Cape's northern frontier, in 1839. But, as a result of his continuously deteriorating health, a further expedition which Krebs had planned to explore what would in 1843 become the Colony of Natal never occurred. Soon thereafter, in May 1844, Krebs died on his farm "Lichtenstein".[28]

Despite the *Graham's Town Journal* remembering Krebs in a glowing obituary as 'much esteemed for his high intelligence [. . . and] scientific attainments',[29] – he soon became and remained a largely forgotten figure in South Africa until a local settler historian, Pamela Ffolliott, chanced upon his grave and wondered 'what manner of man he had been'.[30] As her interest in Krebs waxed, Ffolliott decided to partner with the South African ornithologist Richard Liversidge in writing a biography of the collector,[31] feeling it their 'great sense of responsibility' to reintroduce Krebs's 'great contribution to Natural History'. In their book, Krebs is presented as 'stand[ing] out as a pioneer of mean calibre, who was recognised by his contemporaries at the Cape and by his neighbours on the Eastern Frontier, who were themselves ou tstanding settlers, as an intrepid traveller and ardent collector'.[32] Collectors, they argue on the first page of their introduction, had a crucial role to play in 'the opening up of South Africa' by 'contribut[ing] something to the further knowledge of the country's fauna and flora and [extending] the tenuous routes reaching north and east from Cape Town'.[33] Krebs, they felt, had embodied this quest to perfection.

In the decades since the publication of this biography, the study of both colonial science and the accounts of colonial travellers have undergone a paradigm shift. Ffolliott and Liversidge had followed what, in a slightly different context, Huigen has termed the "positivistic" approach to the critique of southern African colonial travel accounts, in which historians primarily seek to recapitulate routes travelled – and, in the case of collectors, list specimens gathered.[34] For Huigen, this perspective is underpinned by a belief that colonial travellers 'heralded the coming of European civilization', which is exactly what 'made them so admirable in the view of these historians'.[35] The positivistic approach, Huigen argues, has since been superseded by a post-colonial perspective in which pivotal authors such as Pratt[36] "accuse" colonial travellers of 'attempt[ing] to intellectually annex non-European history and to pave the way for colonial expansion'.[37] Meanwhile, in South African historiography, as Dubow notes, there has been a turn away from colonial history to a 'history "from below"' and, in particular, an emphasis on 'recovering the voices of the forgotten and the dispossessed' through a critical rereading of archival sources.[38] However, as crucial as it is to embrace the fresh perspectives which these new approaches offer, Dubow has warned that '[i]mportant dimensions of South African history risk being occluded or lost if the role of whites is viewed too narrowly in terms of settler colonialism and exploitation'.[39] Huigen, Beinart[40] and Jacobs[41] also criticise the post-colonial perspective for failing to recognise that not all aspects of colonial knowledge production were explicitly aimed to further "colonial expansion".[42] A re-examination of Krebs's collecting career and partnership with Lichtenstein from a third perspective – through the wider prism of the social history of science – offers an alternative yet critical approach which may reveal important insights which exclusively laudatory or accusatory accounts may overlook or neglect to examine. As we shall see in the following sections, these include the influence of local knowledge on collecting practice and the significance of commercial, interpersonal and other non-scientific and apolitical imperatives in determining collecting strategies.

Lichtenstein and Krebs: a partnership in correspondence

With both men grateful for the opportunities provided by the other, Krebs's relationship with Lichtenstein was marked by a warmth apparent throughout their correspondence. Lichtenstein, as we have seen, had successfully petitioned Altenstein to grant Krebs a promise of financial and institutional support, and in 1823, by which time Krebs had already supplied Berlin with eight collections, he made a further successful request for the establishment of an advance credit for Krebs at the Cape. In his accompanying letter to Altenstein, Lichtenstein emphasised not only the opportunity for the Berlin Zoological Museum to acquire 'African objects' hitherto lacking 'in our Museum, as well as in other European collections', but also Krebs's 'honesty, punctuality and faithfulness'. Lichtenstein continued, praising 'the industrious young man who indeed, for comparatively small wages, has given us his whole time, without having been left with more than the bare necessities for his travelling and living expenses', on whom it would be very hard to

> break off suddenly an enterprise which could be of great value to science, and with his limited resources, would have to find the almost impossible expenses for his return to Europe, where, after an absence of 6 years, he would have difficulty in obtaining a position as a pharmacy assistant.[43]

In return for this support, Krebs held a deep respect for and gratitude towards Lichtenstein, after whom, as we have seen, he even renamed his farm. Indeed, some two decades into their partnership, Krebs would still declare to Lichtenstein that his travels and collections were undertaken, despite financial losses,[44] 'only on account of my esteem for you and your Institute of Natural History'.[45]

Despite Krebs's reverence for Lichtenstein and the latter's reciprocated support, Krebs was often left frustrated by the Prussian state's weak or delayed financial backing for his endeavours. As he awaited payment for the first of the collections which he had shipped to Berlin, Krebs reported to Lichtenstein how he had been forced to spend 'more than 8000 Cape Thalers [on equipment], from which you may conclude that nothing but love for Natural History and my desire to help your Museums and institutes could have induced me to make these sacrifices'.[46] In earlier letters, he had unsuccessfully requested advance payments to cover the 'enormous . . . transport charges and packing costs'[47] as well as the high price of equipment for collecting itself, including 'instruments for catching', such as 'forceps' and 'insect needles', and 'good brandy' for preserving fish.[48] In another letter, he included a list of further expenses:

> Apart from a tent, I have now most of the necessities indispensable for a journey. Their prices were higher than I expected. However, without them I would not be able to make any decent collection. It might not be out of the way if I give you some of the prices so that you may see that I do not ask for advances without reason.
>
> A wagon – 360 thalers. A span of oxen – 300 thalers. . . . Two horses, 250 thalers, which is indispensable on longer journeys for carrying foodstuffs. Saddle and Harness 70 thalers. Six guns, of which most cost 200 thalers each, and are especially good. Mattresses with accessories – 70 thalers. Table and chair – 20 thalers. Pots and other cooking utensils and several other things also come to a fair amount, plus the salary for Strydom and one hottentot.[49]

Krebs, it seems, envied the 'good support' enjoyed by his French counterpart at the Cape, Pierre Delalande, which 'has enabled him to send such large amounts of treasures of nature objects to Paris, for which he now enjoys such well earned fame'.[50] Perhaps as a result of this grievance, Krebs made a conscious point of emphasising in his correspondence the lengths to which he was prepared to go to collect for museum and country – not only financially, but also in physical terms. As Krebs promised Lichtenstein after describing the collection of a human skull near Cape Town in 1821 (see also section "Collecting human remains"):

> Referring to your remark with regard to the price of the kaffir skull, I have to say that at the time I did not so much consider the monetary gain, but I wanted to prove *how much I like to enrich your Museum with objects which are very difficult to get for the Museum*, the maintenance of which costs such a lot of money and trouble.
>
> In future *you will see yourself with what zeal I will go for the rare and the new*, for every new object will give fresh interest among the collectors [my emphasis].[51]

In his letter to Lichtenstein describing his longest journey, the 1839 expedition to the Vaal River and beyond, Krebs emphasised the 'very unfavourable circumstances' under which it took place, including his own rheumatism and the 'unrest' caused by the Great Trek, which made the journey 'very dangerous'.[52] However, Krebs had always made it clear to Lichtenstein from the beginning of their partnership that

> I enjoy suffering trouble and danger because I am convinced that you, dear Professor, will contribute towards my advancement, with all your power and will, as far as I merit it, giving favourable reports to his Excellency the State Minister von Altenstein.[53]

Krebs thus left no doubt in Lichtenstein's mind that collecting was a painful and dangerous pursuit, the sufferings and hazards of which he was nevertheless eager to endure for their mutual benefit. Moreover, as we shall see in sections "Collecting practice" and "Collecting human remains", Krebs was also prepared to transcend ethical boundaries which proved to be inconvenient to or incompatible with his collecting ambitions. Next, however, I shall consider the dynamics of life in the field which Krebs encountered as he conducted his quest for specimens.

Navigating through the field

Jacobs observes that the historiography of science in colonial Africa has been preoccupied with the role of the applied sciences in 'European domination', thereby neglecting the 'politics of science[s] . . . conducted quietly in small arenas', namely the '[n]on-instrumentalist sciences such as ornithology [and] non-economic botany [which] yielded only small stakes in terms of extraction or governmentality'.[54] While large-scale, state-sponsored scientific expeditions have tended to be studied in detail, the inner-workings of smaller, independently organised collecting parties, such as those arranged by Krebs, provide a fresh perspective from which to examine the confluence of social, cultural and political factors at play in the early 19th-century Cape Colony. Although Krebs often presented himself as the sole collector of his specimens, his collecting expeditions involved teams of hunters travelling in advance of the rest of the party, assistants responsible for 'skinning, salting, and preserving' specimens, who would remain at camp for days or 'even weeks', as well as numerous further servants, usually Khoikhoi labourers, who were required to perform a range of tasks, including driving wagons, carrying equipment and preparing food. Indeed, as Krebs would concede after the experiences of his 1839 expedition, 'never again shall I journey with such a few companions'.[55]

Beinart has shown how both indigenous Khoikhoi knowledge and the 'amalgam of knowledge and techniques' born of contact between Khoikhoi and Europeans was invaluable for 18th- and early 19th-century farmers and travellers who employed Khoikhoi servants. This extended from their familiarity with 'routes, geography, water, plants and animals' to an ability to guide pack oxen, track lost oxen, hunt game, conserve meat, find water holes and start fire, all crucial skills for collectors such as Krebs on long expeditions by ox-wagon.[56] However, despite the key role which KhoikhoiKhoekhoe assistants would have played on his expeditions, Krebs was very dismissive of their efforts. Bemoaning the 1828 law, commonly known as Proclamation 50, which effectively placed so-called Hottentots on the same legal footing as Europeans,[57] Krebs complained that:

> under the new laws for the equal status of the Hottentots with the rights of the African citizen, a certain sense of freedom has come to these people, which makes them obstinate and unwilling to work. My best shots have almost all left me, in order to visit the praised land of the Kat River, where the Government has started a colony of Hottentots and Bastards, or to wander about, spending their lives free and thieving. . . . You will notice that the big bones of the rhinoceros in this collection are missing, through the carelessness of the Hottentot who was on night watch over the skeleton.[58]

Four years later, Krebs had still not accustomed himself to the new legal dispensation, bemoaning the 'unwilling, lazy, unfaithful and very stupid herdboys and other servants, whom so-called philanthropist missionaries humbug, particularly the Hottentots, about being as good as white people'. In the same paragraph, however, he was prepared to admit that 'I am glad to say that I always have a few Hottentots who, with a lot of indulgence on my part, do their work satisfactorily'.[59]

This begrudging concession notwithstanding, Khoikhoi servants, whom Krebs rarely named or individualised, would otherwise be mentioned only to bemoan their failures. These deficiencies, in turn, he always understood as the manifestation of a lesser intellect or flawed character and never as an act of indifference or defiance in what was still a tense, racially defined and hierarchical work environment and socio-cultural climate. Lamenting to Lichtenstein how 'very unlucky [I have been] with my employees', Krebs complained how they:

> almost never acted to my instructions. Even the young colonist, to whom I paid 300 thalers annually, did not come up to my expectations. . . . One cannot rely at all on Hottentots, so that with all the heavy expenses, I myself have collected most of what I have sent you.[60]

Krebs's letters were, as Essner similarly observes for the reports written by late 19th-century German travellers in East Africa, 'dominated by a quasi-capitalist point of view, in which . . . African companions . . . were perceived solely as part of a cost-benefit analysis'.[61] This attitude is most evident in Krebs's description of the death of one of his servants. Here, he reported in a letter to his brother Georg how in 'a sad scene. . . [my] host, by an unlucky shot, ended the life of my hottentot. However, the magistrate of Grahamstown was good enough to procure another one for me'.[62] For Krebs at least, the death of a Khoikhoi servant, although admittedly "sad", could be remedied simply by replacing him with another.[63]

Krebs, however, *was* reliant on his individual servants and other local peoples, not only for their physical labour, but, even more crucially, for the local information and knowledge which they could pass on to him, as suggested by the local names which he provided in his letters and collecting lists. For example, as Ffolliott and Liversidge note, Krebs listed numerous local river names – including the "Likwa River", which was already known by its European name as the "Vaal River" – as collecting localities.[64] Local knowledge also extended beyond geographical information. For example, Krebs wrote how:

> Tamboukies [abaTembu] described to me a snake of about 120 feet length which is supposed to live in the Kraal of the Tamboukie chief, Wasanna, beyond Kap Bay. I also saw a drawing of it on a rock wall which was of a terrible length compared with other animals. . . . They also told me of a unicorn [and] further described to me a scaly animal like a tortoise [an ant-eater], of which I confirmed the existence later.[65]

Krebs therefore did listen to and value local advice – indeed, even if it contained references to doubtful creatures such as unicorns – and was eager to follow up on what he was told. The difficulty is to recover the extent of such reliance on local knowledge, as Krebs did little more than hint at this in the above manner in his letters. Apart from his own feelings of intellectual superiority as a European and as a naturalist, this may also have arisen out of a concern to present himself to Lichtenstein as a knowledgeable expert and inimitable collector who, as he claimed, had alone 'collected most of what I have sent you'.[66]

Collecting practice

As Larsen Hollerbach argues, 19th-century natural history was 'a specimen-based science' in which sedentary naturalists sought out a grand design in nature by examining dead specimens – or, often more precisely, parts of dead organisms – not through 'observation in the field', but in distant and isolated studies or laboratories.[67] Similarly, Larsen notes the 'artificial nature of specimens', 'designed and constructed by naturalists to answer various scientific needs' and serve as 'convenient stand-ins for their dynamic, living cousins'.[68] For the prey of collectors, the implications thereof were ominous: in order to be transformed into examinable specimens, living plants and animals had to be converted into objects of natural history through a process which included their death and subsequent shipment to a metropolitan institution, where scientists could compare newly arriving objects with other, long-dead specimens. At the same time, as McCook has observed, colonial-era field sciences such as botany and zoology relied 'heavily on trust', as 'metropolitan scientists' built their reputations on the 'reports and collections from people who were on the social margins of the scientific world but who had privileged access to the field'.[69] Given the mutual respect in which they held each other, it is thus not surprising that Lichtenstein and Krebs would enjoy a prolific partnership, including the sending of 15 large collections from the

Cape to Berlin over the course of two decades. More important, however, was that both shared the specimen-based imperatives and aims of their complementary occupations: to claim natural history specimens as isolated objects removed from their natural environment in order to transform them into commodities to be gathered, packaged, analysed and compared, and ultimately traded for profit. For both Lichtenstein, whose museum relied on the money it made from trading duplicate specimens, and Krebs, who lived and collected off of part of that income, collecting was reduced to the goal of maximising transactions.

This is evident in a detailed letter which Lichtenstein sent in 1821 containing 'some hints [and] remarks' for the then still inexperienced collector Krebs.[70] Krebs, Lichtenstein advised, should plan to 'settle with your equipment in Uitenhage or Algoa Bay, and to explore [the southeastern coast of the colony] from there'. Whereas specimens from closer to Cape Town had 'very much gone down in price, because of the frequent consignments from there', those from 'beyond the Hottentot (Holland) Mountains', Lichtenstein suggested, 'will have a much higher value', while 'collectors to the Tzitsi-Kamma, which truly is *terra incognita*, should obtain extraordinary results'. 'I cannot give you specific instructions', he continued, 'but I can assure you that every piece which arrives here well preserved, carefully handled and packed will be paid for according to the value which it has in the trade at the time'. Because of a growing lay interest in ornithology in early 19th-century Germany, birds, Lichtenstein emphasised, 'will make your consignments most payable'. Conspicuous in these remarks is Lichtenstein's apparent lack of interest in scientific considerations: so conceived, collecting is about obtaining the highest quantity of specimens at the highest value.

Krebs, who had no formal training in the natural sciences, was also limited in his goals. In a letter to his brother Georg, he expanded on his understanding of the role of collector:

> It is not the main work of the collector to classify all the new objects offered by him. This would take too much time. One must therefore limit oneself in most cases to locality, habit, season and the local names (if any).[71]

However, Lichtenstein appears to have ignored the lists recording these details which Krebs provided with each consignment, instead assigning his own, often inaccurate, labels and descriptions to specimens.[72] Lichtenstein, it appears again, was interested in little more than each specimen's trade value: as he instructed Krebs regarding the collection of birds: 'as soon as you arrive in the wooded country east of Plettenberg Bay, you may shoot whatever carries feathers, whether beautiful or ugly, young or old, it makes no difference'.[73] This was not circumscribed, targeted killing for the benefit of science, but the unrestrained destruction of animal life for commercial purposes.

'I have declared war on the Cape plant and animal kingdom', Krebs's predecessor, Carl Bergius, had written enthusiastically to Lichtenstein from Cape Town in August 1817.[74] The 28-year-old Bergius, who died less than six months later, had insufficient time to fulfil this pledge to the full himself, but, as a call to arms for collectors, his statement evokes an attitude towards nature which was central to the collecting enterprise. Indeed, the very *raison d'etre* of collecting consisted in the fatal wrenching of life from its natural environment, the enactment of multiple daily episodes of death and displacement. However, Krebs himself was not unaware of the effects of overhunting on animal populations. For instance, he commented on the decrease in elephant and buffalo numbers near the colony's eastern frontier, a trend which he attributed to the residents of the mission stations of Enon, Bethelsdorp and Theopolis hunting the animals for their meat to such an extent that they were 'have almost disappeared from the Colony'.[75] Mirroring the disdain with which British "sport" hunters viewed their subsistence counterparts,[76] Krebs would display no self-reflection on his own role in this very process. His job, so he understood it, was to harvest as many specimens of natural history as he could – a stated aim in one letter, for example, was to obtain a 'good crop of insects'[77] – even, as we shall see in the next section, if this meant training his collector's eyes on the bodies of local human beings and seeking to collect human remains in ways which were deeply disrespectful to local peoples' customs and dignity.

Collecting human remains

Collections of human skulls, bones and skeletons – usually gathered in colonial contexts and still held in Western museums of science, anthropology or medicine – have been invested with increasing political meaning in recent decades as calls for the repatriation of human remains to their places of origin have grown ever louder. While a large scholarly literature has come to exist on the subject, its focus often tends to lie more on current politicised debates around the identification and repatriation of human remains than on the circumstances and narratives of their initial collection. Tracing the moment of collection is important, however, if we are to recover some of the humanity lost by human beings as their mortal remains were converted into objects of science and commerce. It is this dubious transformation which I aim to highlight in this section through an analysis of Ludwig Krebs's descriptions of his own attempts to claim human specimens.

The collection of human remains was relatively commonplace at the early 19th-century Cape, as Harries has shown in his article on 'Warfare, Commerce and Science: Racial Biology in South Africa'.[78] Bank has also written on how the Frontier Wars provided a steady supply of human skulls for local phrenologists and other racial scientists, who were, he argues, even 'as early as the 1820s' already concerned with theories of racial difference.[79] Both Harries and Bank show that a range of actors – from physicians to soldiers, but also naturalist-collectors – exploited the frequent outbreaks of violence and war on the colony's eastern and northern frontiers to increase their collections of human remains.[80] For collectors such as Krebs and Lichtenstein, there was little inconsistency in adding human specimens to their botanical and zoological collections; they did so through a range of actions, from the gathering of unburied corpses on battlefields to the digging up of graves and the seeking out of cadavers from prisons and other colonial institutions. For instance, as he described enthusiastically in his *Travels*, Lichtenstein claimed the skull of an executed San prisoner from Tulbagh prison, which he later donated to the influential Göttingen comparative anatomist and nascent racial theorist, Johann Friedrich Blumenbach.[81]

From the beginning of his collecting career, Ludwig Krebs was determined to claim human skulls of his own, and in 1821 he gathered the skull of an umXhosa man, which may or may not have been that of Makhanda, the umXhosa prophet and warrior-hero from the Fifth Frontier War (1818–19), who had drowned while trying to escape from Robben Island with seven fellow prisoners.[82] To the noticeable shock of his servants and slave, Krebs 'thought immediately to obtain a scientific advantage out of this unhappy incident' by digging up the body, which had washed ashore near Table Bay:

> Only two bodies of a kaffir and a hottentot were washed ashore . . . the kaffir's body was mostly eaten by wild animals and . . . the head of the hottentot was smashed on the rocks by the waves. [A fellow collector, Stadler] thought, however, that the head of the [other] kaffir was well preserved. He would dig this up for me. . . . After a good breakfast, which the kind daughter of Stadler had prepared for us, we went to work. Three slaves now began to dig, and I noticed with pleasure that the head was well-preserved, although the neck was already half-eaten. I decided immediately to separate it from the body. Those present, in particular the slaves, looked at me in horror! But I placed my conquest into a container and tied it up with a cloth.[83]

The Sixth Frontier War (1834–5), during which time Krebs lived close to the frontline on his farm "Lichtenstein", provided a further opportunity to collect human specimens amid a high amaXhosa casualty rate – as he wrote to Lichtenstein, 'in no previous war have so many Kaffirs lost their lives'[84] – and a toxic and dehumanising climate of hate and fear.[85] Recalling in a letter to Lichtenstein how a colonial commando had killed 22 amaXhosa warriors, Krebs's reaction had been to 'immediately [think] of fetching some of the bodies', a foray which resulted in his party finding 'all the skeletons destroyed by the hyenas except two heads; these I will send you'.[86] Krebs, who himself probably remained at home due to ill health, noted how his brother Georg, who made the collections, had been struck by the 'beautiful athletic Kaffir bodies lying about the rock caves'.[87] In his correspondence, Krebs made no secret of his views on the amaXhosa. The 1834–5 war, he argued in a long letter to Lichtenstein describing his experiences living very close to the frontline, was a 'radical lesson as to the danger of giving guns by barter to wild and semi-wild nations' and a direct result of missionaries' failed efforts at the 'civilisation' of the amaXhosa.[88] In the above description, however, the umXhosa body, hitherto so feared, is stripped in death of its danger, reduced to the status of a rare and beautiful object waiting to be gathered, packaged, catalogued and traded.

Krebs did not only display an obsessive enthusiasm for the collection of the remains of enemy amaXhosa warriors; the misfortunes of his own Khoikhoi and

San servants also provided welcome opportunities to expand his collection of human specimens. For example, he recounted how one night during the Sixth Frontier War when 'the Kaffirs were particularly impudent, they frightened the servants so much that a Bushman wife lost a five-month baby prematurely. My brother placed the foetus in alcohol and it will follow in the next shipment'.[89] Here, Krebs exhibits the collector's cold logic: the matter-of-fact statement confirming that the specimen has been put in storage for sending immediately follows the report of the foetus's death. For Krebs, death thus provided the happy opportunity for "conquest" and eternal storage. While Rau is correct to place colonial collectors' quests for human remains in the context of their time – an era in Europe in which public dissections of human cadavers were common and the bodies of criminals and members of other supposedly deviant groups were regularly collected for study[90] – the dehumanising gaze of the vigilant colonial collector always on the lookout for indigenous human specimens to place among his collections of exotic mammals, birds, reptiles, amphibians, fish, insects, plants and minerals is marked. In the auction of Krebs's fifteenth collection, for example, human specimens were presented as one rarity for sale among many others from all spheres of nature;[91] included in a list of mammal skins and skulls are three 'Schädel von Hottentotten aus einer Höhle am Umpukanie, ohne Unterkiefe, sonst gut erhalten. Jeder wird geschätzt auf 5 Thlr'.[92] These cost 1 Thlr. less than the skull of a Burchell's zebra (*Equus Burchellii*) and were far below the price of the most expensive object, an intact male lion (*Felis Leo*) skin at 80 Thlr.

The absorption of human specimens into auction lists of natural history specimens represented the final step in their conversion from *human bodies* to *objects* of science *and* commerce, fully stripped of their individuality and personal history, numbered and priced, stored, displayed and traded alongside animals, plants and minerals. Only in recent decades, with increased political attention placed on the repatriation of human remains to their countries of origin, is some effort being made to reinfuse human specimens with their humanity, although, as Katz and Kirby convincingly argue, this 'hardly erases the significance of the initial rupture'.[93]

Conclusion

European collectors of southern African plant and animal specimens have been generously commemorated, their names suffixing the scientific names identifying numerous species of flora and fauna. Krebs himself has a number of plant and animal species named after him, including the flowering plant *Gazania krebsiana*[94] and the Krebs's fat mouse (*Steatomys krebsii*).[95] These appellations may create the impression that collectors such as Krebs were pioneering figures forging their way alone through unchartered territories for the sole sake of science. However, as this chapter has sought to bring to attention, a number of considerations suggest otherwise. For one, Krebs did not collect alone but relied on collecting parties of usually unnamed servants as well as often unacknowledged external informants and helpers. Furthermore, although his collections undoubtedly played their part in the explosion of scientific knowledge being produced at contemporary European scientific institutions, his career was directed by other concerns, including covering the often significant costs of collecting and proving himself professionally as a willing, daring and able collector. Perhaps most important to note, however, is that collecting as practised in the Lichtenstein-Krebs partnership was of its essence a *destructive* activity which was not circumscribed by scientific imperatives, but driven by Lichtenstein's aim to sell as many specimens as possible and the corresponding arrangement which saw Krebs being paid per specimen collected. In its most obviously dubious light, this was evident in the practice of collecting human remains, but also in the daily pursuit, collection and trade of dead animals and plants – small and large alike – collecting was an inescapably macabre practice. A brief study of collecting practice such as this can thus cast significant – and often unsavoury – light on the broad ambit of activities which encompassed the profession as well as on the real-world social, institutional, cultural, political and commercial conditions of which natural scientific knowledge was born, often merely as a by-product.

Notes

1 Bregman, 'For Sale', 249.
2 Jahn, 'Zoologische Gärten – Zoologische Museen', 16; see also Farber, *Finding Order in Nature*, 1–45.
3 Farber, *Finding Order in Nature*, 25.
4 Bregman, 'For Sale', 247–63.
5 See Huigen, *Knowledge and Colonialism*, 3–8.
6 Ibid., 9.
7 Ibid., 9; Beinart, *Rise of Conservation in South Africa*, 65.
8 Bregman ('For Sale', 250) argues that collectors at the Cape from 1750 to 1850 can be divided into three separate yet occasionally overlapping categories: 'gentlemanly collectors', who collected and exchanged specimens, usually in their spare time, 'without direct payment'; 'salaried collectors', who collected for 'metropolitan institutions'; and 'commercial collectors', who 'openly gathered, bought, and sold specimens for money'.
9 See e.g. Lichtenstein, *Travels in Southern Africa*, 32. This and other extracts in this chapter from Lichtenstein's southern African travel account are from a translation of his original two-volume German-language edition, which was published in Berlin in 1811–12 as *Reisen im südlichen Africa in den Jahren 1803, 1804, 1805 und 1806.* While the English-language translation by Anne Plumtre has been criticised by Hoge (*Personalia of the Germans at the Cape*, 240–1) as 'extremely inaccurate and totally unreadable', I have found no grave inaccuracies or omissions in the passages which I cite in this chapter.
10 For a detailed account of the development of Lichtenstein's post-Cape career and the importance of his fieldwork in southern Africa for his subsequent rapid rise as a scientist in Berlin, see Näf-Gloor, 'A Naturalist's Career', in this volume.
11 See Rieke-Müller, 'Angewandte Zoologie', 475–8; Jahn, 'Zur Vertretung der Zoologie', 260–80.
12 Harries, 'Warfare, Commerce and Science', 171.
13 Stresemann, 'Hinrich Lichtenstein', 73–96.
14 Ibid., 76–7.
15 Spohr, 'Introduction', 11.

16 Stresemann, 'Hinrich Lichtenstein', 76–7.
17 Ibid.
18 Spohr, 'Introduction', 11.
19 Stresemann, 'Hinrich Lichtenstein', 81–3.
20 See Ffolliott and Liversidge, 'Carl Heinrich Bergius', 134–41, 183–92. Bergius's original correspondence is archived at 'Signatur S I, Bergius', Museum für Naturkunde, Berlin.
21 Bregman, 'For Sale', 258. Other German collectors who worked at Pallas & Polemann were Carl Drège and Christian Friedrich Ecklon. As apothecaries, they all followed in the vein of a rich collecting tradition. Chemists, as Jahn notes, had played a 'Pionierrolle [pioneering role]' in the development of natural history collections from the 16th to the 18th century. As itinerant traders, chemists enjoyed regular opportunities to collect new specimens and were well-suited in terms of their 'Kenntnisse und Materialen [knowledge and materials]' to identify and conserve objects of natural history. (See Jahn, 'Zoologische Gärten – Zoologische Museen', 12)
22 See the diary of Drège, 'MSC. 61: The Papers of the Drège Family', File 524, National Library of South Africa, Cape Town. For more on Drège, see Kirby, 'Early Professional Museum Collectors', 393–400 and Rau, 'Der Mainzer Quagga-Fötus', 245–57.
23 Judging by the experiences of both Bergius, his predecessor, and Drège, his replacement, Krebs would have grown frustrated by the high time demands which his position at Pallas & Polemann placed upon him. As Bergius complained in 1817, 'three of us are fully occupied and have our hands full from morning to night', including on Sundays, such that he had not been able to make 'one single excursion'. (Letter, Bergius to Lichtenstein, 1 April 1816, translated in Ffolliott and Liversidge, 'Carl Heinrich Bergius', 139–40). Drège recorded similar complaints in 1824. Like Krebs, Drège immediately left the store at the end of his five-year contract in 1826, also to establish a collecting business.
24 Lichtenstein had also sent the Prussian collectors Leopold Mund (1791–1831) and Louis Maire (biographical dates unknown) to the Cape in 1816 to take over collecting responsibilities from the struggling Bergius, but, after sending back only a few small collections, they seem to have absconded from their duties. Eventually recalled to Berlin to account for their dereliction, they ignored this order, spending the rest of their lives at the Cape. See Glen and Germishuizen, *Botanical Exploration of Southern Africa*, 284, 309–10.
25 Letter, Lichtenstein to Altenstein, 18 January 1821, translated in Ffolliott and Liversidge, *Ludwig Krebs*, 21.
26 Brüsch, 'Provisions Made for Prosperity and Affluence', 19–21.
27 Ffolliott and Liversidge, *Ludwig Krebs*, 100.
28 See Ibid., 138, 124–49.
29 Graham's *Town Journal*, 16 May 1844, quoted in Ffolliott and Liversidge, *Ludwig Krebs*, 138.
30 Ffolliott and Liversidge, *Ludwig Krebs*, ix.
31 In reality, however, their book represents little more than an annotated and translated collection of Krebs's correspondence and other related documents archived under 'Signatur S I, Krebs', Museum für Naturkunde, Berlin. For the purposes of this chapter, direct quotations from these documents are given in their translated form, as published in Ffolliott and Liversidge's book.
32 Ffolliott and Liversidge, *Ludwig Krebs*, 1.
33 Ibid., 1.
34 Huigen, *Knowledge and Colonialism*, 26.
35 Ibid., 27.
36 For example, see Pratt, *Imperial Eyes*.
37 Huigen, *Knowledge and Colonialism*, 28.
38 Dubow, *A Commonwealth of Knowledge*, 10.

39 Ibid., 10.
40 Beinart, *Rise of Conservation in South Africa*, 30.
41 Jacobs, 'The Intimate Politics of Ornithology', 564–603.
42 Huigen, *Knowledge and Colonialism*, 28.
43 Lichtenstein to Altenstein, 3 February 1823, translated in Ffolliott and Liversidge, Ludwig Krebs, 47–8.
44 Krebs claimed to have recorded a loss of 'at least £400' on his fifteenth collection, 'especially through the high wages which I had to pay on this journey to the servants' (Letter, Krebs to Lichtenstein, 26 December 1841, translated in Ffolliott and Liversidge, Ludwig Krebs, 129.).
45 Krebs to Lichtenstein, 26 October 1841, translated in Ffolliott and Liversidge, Ludwig Krebs.
46 Krebs to Lichtenstein, 21 June 1824, translated in ibid., 52.
47 Krebs to Lichtenstein, 20 May 1821, translated in ibid., 31.
48 Krebs to Lichtenstein, 29 July 1821, translated in ibid., 32–3.
49 Krebs to Lichtenstein, 12 October 1822, translated in ibid., 45.
50 Krebs to Lichtenstein, 29 July 1821, translated in ibid., 32. As a result of his precarious financial situation, Krebs was compelled to maintain other business interests, such as co-ownership of a pharmacy in Grahamstown. See ibid., 103.
51 Krebs to Lichtenstein, 29 July 1821, translated in ibid., 32. In this chapter, terms such as "Kaffir", '"Hottentot" or "Bushman", which are now regarded as pejorative or racist, are used only in direct quotations, with their accepted modern-day equivalents employed in the analysis.
52 Krebs to Lichtenstein, 1 August 1, 1839, translated in ibid., 108.
53 Krebs to Lichtenstein, 12 October 1822, translated in ibid., 45.
54 Jacobs, 'The Intimate Politics of Ornithology', 564–5.
55 Krebs to Lichtenstein, 1.8.1839, translated in Ffolliott and Liversidge, Ludwig Krebs, 109.
56 Beinart, *Rise of Conservation in South Africa*, 30.
57 Proclamation 50 abolished the 1809 pass laws, the so-called Caledon Code, which had effectively turned the Khoekhoe into indentured labourers. See Elphick and Malherbe, 'The Khoisan to 1828', 40.
58 Krebs to Lichtenstein, 8 July 1830, translated in Ffolliott and Liversidge, Ludwig Krebs, 71–2.
59 Krebs to Lichtenstein, 15 December 1834, translated in ibid., 90.
60 Krebs to Lichtenstein, 21 June 1824, translated in ibid., 52.
61 Essner, 'Some Aspects of German Travellers' Accounts', 200.
62 Krebs to Georg Krebs, 4 April 1823, translated in Ffolliott and Liversidge, Ludwig Krebs, 49. Georg, who acted as Krebs's agent in Berlin, would join him at the Cape in 1834. He ultimately settled in Graaf-Reinet as a doctor.
63 Krebs claimed to prefer white guides and assistants, whom he held to be more reliable than the 'negligent Hottentots'. One white guide, he reported, shot 'well over 100 birds', another example of how collecting for the collector Krebs was in reality a delegated activity (Krebs to Lichtenstein, 12 October 1822, translated in ibid., 43).
64 Ibid., 117.
65 Krebs to Lichtenstein, 1 July 1826, translated in ibid., 52.
66 Krebs to Lichtenstein, 21 June 1824, translated in ibid., 52.
67 Larsen Hollerbach, 'Of Sangfroid and Sphinx Moths', 201–3.
68 Larsen, 'Equipment for the Field', 358.
69 McCook, 'It May Be Truth', 196.
70 Lichtenstein to Krebs, 21 March 1821, translated in Ffolliott and Liversidge, Ludwig Krebs, 26–9.
71 Krebs to Georg Krebs, 2–3 October 1820, translated in ibid., 42.
72 Ibid., 149.

73 Lichtenstein to Krebs, 21 March 1821, translated in ibid., 26–9.
74 The original reads: "Einen allgemeinen Krieg habe ich die Capischen Their und Pflanzenwelt angekundigt" [my translation], (Bergius to Lichtenstein, 20 August 1817, 'Signatur S I, Bergius', Museum für Naturkunde, Berlin).
75 Krebs to Lichtenstein, 8 July 1830, translated in Ffolliott and Liversidge, Ludwig Krebs, 72.
76 See MacKenzie, *The Empire of Nature*, 25–53.
77 Krebs to Lichtenstein, 22 April 1822, translated in Ffolliott and Liversidge, Ludwig Krebs, 41.
78 See Harries, 'Warfare, Commerce and Science'.
79 Bank, 'Of Native Skulls and Noble Caucasians', 388.
80 See Harries, 'Warfare, Commerce and Science' and Bank, 'Of Native Skulls and Noble Caucasians', 387–403. See also Harrison, 'Skulls and Scientific Collecting', 289–90.
81 Lichtenstein, *Travels in Southern Africa*, 588.
82 See Harries, 'Warfare, Commerce and Science', 174. See also Wells, *The Return of Makhanda*, 229–30. In her otherwise detailed account of the escape and subsequent deaths of Makhanda and his fellow prisoners, Wells appears unaware of the skull-hunt among collectors which these events sparked.
83 Krebs to Georg Krebs, 4 April 1823, translated in Ffolliott and Liversidge, *Ludwig Krebs*, 22.
84 Ibid., 111.
85 In 1822, in a letter to his brother, Krebs had already endorsed provocative views about the amaXhosa: 'The farmers who are plagued by these barbarians are very glad when they manage to kill a kaffir' he reported, 'for what a kaffir cannot steal, he will kill' (Krebs to Georg Krebs, 1 February 1822, translated in ibid., 39).
86 Krebs to Lichtenstein, ca. 1835, translated in ibid., 95–6.
87 Krebs to Lichtenstein, 12 June 1835, translated in ibid., 96.
88 Krebs to Lichtenstein, 21 June 1824, translated in ibid., 92.
89 Krebs to Lichtenstein, ca. 1835, translated in ibid., 96.
90 Rau, 'Der Mainzer Quagga-Fötus', 248.
91 Lichtenstein, *Verzeichniss einer Sammlung*.
92 Three 'Hottentot skulls from a cave on the Umpukanie, without lower jaws, otherwise well-preserved. Each is valued at 5 Thlr'. [my translation] Ibid., 10.
93 Katz and Kirby, 'In the Nature of Things', 262.
94 Glen and Germishuizen, *Botanical Exploration of Southern Africa*, 249.
95 Moffett, *Biographical Dictionary*, 157.

Bibliography

Published sources

Bank, Andrew, 'Of "Native Skulls" and "Noble Caucasians": Phrenology in Colonial South Africa', *Journal of Southern African Studies* 22 (1996): 387–403.

Beinart, William, *The Rise of Conservation in South Africa: Settlers, Livestock, and the Environment 1770–1950* (Oxford: Oxford University Press, 2008).

Bregman, Leigh Davin, '"For Sale": Botanists and the Commercial Trade in Botanical Specimens at the Cape of Good Hope, 1750–1850', *Archives Internationales d'Histoire des Sciences* 56 (2006): 247–63.

Brüsch, Björn, 'Provisions Made for Prosperity and Affluence: Karl Sigmund Franz Freiherr von Stein zum Altenstein and the Establishment of the *Gärtnerlehranstalt* in Prussia', *Centaurus* 49 (2007): 15–55.

Dubow, Saul, *A Commonwealth of Knowledge: Science, Sensibility, and White South Africa, 1820–2000* (Oxford: Oxford University Press, 2006).

Elphick, Richard and Vertrees Canby Malherbe, 'The Khoisan to 1828', in *The Shaping of South African Society, 1652–1840*, ed. Richard Elphick and Hermann Giliomee (Middletown, CT: Wesleyan University Press, 1989), 3–65.

Essner, Cornelia, 'Some Aspects of German Travellers' Accounts from the Second Half of the 19th Century', *Paideuma* 33 (1987): 197–205.

Farber, Paul Lawrence, *Finding Order in Nature: The Naturalist Tradition from Linnaeus to E.O. Wilson* (Baltimore: JHU Press, 2000).

Ffolliott, Pamela and Richard Liversidge, 'Carl Heinrich Bergius, Cape Apothecary and Collector, 1816–1818', *Quarterly Bulletin of the South African Library* 18 (1964): 134–42, 183–92.

Pamela Ffolliott and Richard Liversidge, *Ludwig Krebs: Cape Naturalist to the King of Prussia 1792–1844* (Cape Town: A.A. Balkema, 1971).

Hugh F. Glen and Gerrit Germishuizen, *Botanical Exploration of Southern Africa* (Cape Town: South Africa National Biodiversity Institute, 2010).

Patrick Harries, 'Warfare, Commerce and Science: Racial Biology in South Africa', in *The Invention of Race: Scientific and Popular Representations*, ed. Nicholas Bancel, Thomas David, and Dominic Thomas (London: Routledge, 2014), 170–84.

Simon J. Harrison, 'Skulls and Scientific Collecting in the Victorian Military: Keeping the Enemy Dead in British Frontier Warfare', *Comparative Studies in Society and History* 50 (2008): 285–303.

John Hoge, *Personalia of the Germans at the Cape, 1652–1806* (Cape Town: Government Printer, 1946).

Siegfried Huigen, *Knowledge and Colonialism: Eighteenth-Century Travellers in South Africa* (Leiden: Brill, 2009).

Nancy J. Jacobs, 'The Intimate Politics of Ornithology in Colonial Africa', *Comparative Studies in Society and History* 48 (2006): 564–603.

Ilse Jahn, 'Zur Vertretung der Zoologie und zur Entwicklung ihrer Institutionellen Grundlagen an der Berliner Universität von ihrer Gründung bis 1920', *Wiss. Zeitschrift der Humboldt-Universität zu Berlin, Math.-Nat. Reihe* 34 (1985): 260–80.

———, 'Zoologische Gärten – Zoologische Museen: Parallelen ihrer Entstehung', *Bongo: Zoo-Zeitschrift des Zoologischen Garten Berlin* 24 (1994): 7–30.

Cindi Katz and Andrew Kirby, 'In the Nature of Things: The Environment and Everyday Life', *Transactions of the Institute of British Geographers* 16 (1991): 259–71.

Percival Robson Kirby, 'Early Professional Museum Collectors in South Africa', *South African Museums Association Bulletin* 16 (1942): 393–400.

Anne Larsen Hollerbach, 'Of Sangfroid and Sphinx Moths: Cruelty, Public Relations, and the Growth of Entomology in England, 1800–1840', *Osiris* 11 (1996): 201–20.

Anne Larsen, 'Equipment for the Field', in *Cultures of Natural History*, ed. Nicholas Jardine, James A. Secord, and Emma. C. Spary (Cambridge: Cambridge University Press, 1996), 358–77.

Hinrich Lichtenstein, *Reisen im Südlichen Africa in den Jahren 1803, 1804, 1805 und 1806*, vol. 1 and 2 (Berlin: Salfeld, 1811/1812).

———, *Travels in Southern Africa in the Years 1803, 1804, 1805 and 1806* (London: Printed for Henry Colburn, 1812).

———, *Verzeichniss einer Sammlung von Säugethieren und Vögeln aus dem Kaffernlande, nebst einer Käfer-Sammlung, welche am 14ten März 1842 durch den Königl. Gerichtlichen Auctions-Commissarius Rauch öffentlich meistbietend verkauft werden sollen* (Berlin: Königliche Akademie der Wissenschaften, 1842).

John M. MacKenzie, *The Empire of Nature: Hunting, Conservation and British Imperialism* (Manchester: University of Manchester Press, 1988).
Stuart McCook, '"It May Be Truth, But It Is Not Evidence": Paul du Chaillu and the Legitimation of Evidence in the Field Sciences', *Osiris* 11 (1996): 177–97.
Rodney Moffett, *A Biographical Dictionary of Contributors to the Natural History of the Free State and Lesotho* (Bloemfontein: SUN MeDIA Bloemfontein, 2014).
Sandra Näf-Gloor, 'A Naturalist's Career: Hinrich Lichtenstein (1780–1857)', in *Science, Africa and Europe: Processing Information and Creating Knowledge*, ed. Martin Lengwiler, Nigel Penn, and Patrick Harries (this volume).
Mary Louise Pratt, *Imperial Eyes: Travel Writing and Transculturation* (London: Routledge, 1992).
Reinhold E. Rau, 'Der Mainzer *Quagga*-Fötus', *Mainzer Naturwissenschaftliche Archiv* 42 (2004): 245–57.
Annelore Rieke-Müller, 'Angewandte Zoologie und die Wahrnehmung Exotischer Natur in der Zweiten Hälfte des 18. und im 19. Jahrhundert', *History and Philosophy of the Life Sciences* 17 (1995): 461–84.
Otto H. Spohr, 'Introduction', in *Foundation of the Cape: and About the Bechuanas* (Cape Town: A.A. Balkema, 1973), 4–13.
Erwin Stresemann, 'Hinrich Lichtenstein: Lebensbild des Ersten Zoologen der Berliner Universität', in *Forschen und Wirken: Festschrift zur 150-Jahr-Feier der Humboldt-Universität zu Berlin 1810–1960*, vol. 1, ed. Willi Göber and Friedrich Herneck (Berlin: Verlag der Wissenschaft, 1960), 73–96.
Julia C. Wells, *The Return of Makhanda: Exploring the Legend* (Pietermaritzburg: University of KwaZulu-Natal Press, 2012).

Unpublished sources

'MSC. 61: The Papers of the Drège Family', National Library of South Africa, Cape Town.
'Signatur S I, Bergius', Museum für Naturkunde der Humboldt-Universität zu Berlin, Historische Bild- und Schriftgutsammlung, Bestand Zoologisches Museum.
'Signatur S I, Krebs', Museum für Naturkunde der Humboldt-Universität zu Berlin, Historische Bild- und Schriftgutsammlung, Bestand Zoologisches Museum.

5 The African travels of Hans Schinz

Biological transfer and the academisation and popularisation of (African) Botany in Zurich*

Dag Henrichsen

The 'Introductory Volume to the Flora of Southern Africa', published in 1981 by the South African Botanical Research Institute, classifies Zurich as 'one of the leading centres for the study of African plants' and attributes this to the labours of Hans Schinz and the large herbarium he build up in that Swiss city between the late 19th century and the 1920s.[1] As early as in 1890, the journal *Fernschau* of the Mittelschweizerische Geographisch-Commercielle Gesellschaft in Aarau, one of a handful of recently founded geographical societies in the German-speaking part of Switzerland, pointed out that the then still private African herbarium of Hans Schinz, originating from his travels in southwestern Africa during the mid-1880s, had to be regarded as the 'most valuable one on the continent', a claim of perhaps purposeful exaggeration.[2] His numerous *Beiträge zur Kenntnis der Afrikanischen Flora* (Contributions to the Knowledge of African Flora)[3] as Professor of Systematic Botany over a span of 36 years in journals from Geneva and Zurich have been considered 'a basis for any study of African botany'.[4] Needless to say, many species bear his name in epithets such as Schinziana, Schinzianu, Schinzianus or Schinzii.

All this is remarkable when bearing in mind that, when Schinz became, firstly, Director of the Botanical Garden in Zurich in 1893, and, shortly afterwards, professor of the newly founded Institute of Systematic Botany at the University of Zurich, there were no institutionalised foundations for his research interests. This was not only true with regard to African Botany as such, but, more generally speaking, also held true with regard to comparative research at the existing Department of Botany. Importantly herbaria, and thus those collections of dried and systematically classified plant specimens necessary for taxonomic and comparative research, were hardly to be found and those available were old and not well preserved. A few decades later, however, Zurich would be regarded as the leading centre in Switzerland for "Vegetationskunde" (vegetation science) in general and systematic and special botany in particular.[5]

When looking at Hans Schinz, his African travels and his research in African Botany, one is effectively dealing with the establishment of systematic botany as a discipline in Zurich, and the modernisation and professionalisation of botanical and, to a lesser degree, biological systematics at a leading Swiss university at the turn of the 20th century. My argument in this chapter is that these developments were brought about, in part, by the specific career of Schinz, a career modelled on African travels on the one hand, and comparative taxonomic research in the context of Western imperial science on the other hand. Intense efforts aimed at managing and processing the transfer of biological data from Africa to Europe underpinned this development. For Schinz this meant gathering raw empirical material in the field, living flora specimens and the appropriation of local bio-medical and other knowledge. These efforts, importantly, were complemented by the popularisation of botanical knowledge through the botanical garden in Zurich which Schinz headed for more than three decades.

Hans Schinz and his African collections

Hans Schinz (1858–1941), who hailed from a Zurich family of scientists, artists and traders, received his academic education in Zurich and Berlin.[6] After having obtained his first botanical degree at the University of Zurich, Schinz moved to Berlin for his doctorate at the Berlin University (today the Humboldt University), by then a leading European university. With the advent of German colonialism in the 1880s, the Humboldt developed new fields of what would become "colonial sciences", notably in Geography, Cartography, Botany, Zoology, Anthropology and Linguistics. Already in the 1870s Berlin was regarded as the central 'Sprungbrett für Forschungsreisen' (springboard for scientific travels).[7] In the early 1880s, Schinz became a member of a group of academics in Berlin, whose careers were modelled on scientific travels. As Cornelia Essner pointed out with regard to the social profiles of German travellers to Africa in the 19th century, opting for an academic career in the social and natural sciences implied travelling to distant countries, engaging in collecting activities on a wide-ranging scale and building (new) collections and scientific specialisation.[8] The "new botany" to which Schinz was effectively introduced in Germany importantly moved beyond herbarium research, emphasising instead theoretical approaches and field research – for example research into 'the adaptation of plants to specific local environments'.[9] As such, these travellers, being already trained scientists, furthered their careers as professionals and deepened boundaries, though not very distinct ones, between themselves as academic experts and adventure explorers and amateur researchers.[10] Amongst young natural scientists in Berlin, travels were thus regarded as particularly career-enhancing.

Schinz's mentors in Berlin were the botanist, ethnographer and linguist Professor Paul Ascherson (1834–1913), who introduced him to systematic botany, and the very well-known "African botanist", ethnographer and geographer Professor Georg Schweinfurth (1836–1915). Through the latter, Schinz, after having travelled in 1882 to Greece and parts of the Middle East, received the opportunity to join the so-called 2nd Lüderitz-Expedition to the newly acquired colony of German South West Africa (modern Namibia), together with two other natural scientists, Hermann Pohle and Adolph Schenk.[11] Schinz was required to study the southern Namibian flora for classificatory purposes, but also in order to investigate their "technical and medical purposes", and their possible commercial value.[12] This particular research agenda pointed to the crucial concern of what for the "new botanists" was fast becoming an important field of research at centres in Berlin or Kew, namely economic botany.[13] Arriving in Angra Pequena (Lüderitzbucht) in late 1884 Schinz soon realised that the expedition was, for various reasons, including rivalry between the three scientists, doomed to failure. He cancelled his contract and from mid-1885 onwards funded his own travels.[14] As his letters, written during his travels to his mother and brother in Zurich, indicate, he did not want to return to Europe without having "completed the African tour".[15] As this phrase makes clear, the "African tour" was seen as a professional necessity, and southwestern Africa was to become the accidental field of his activities.[16]

His route took him from southern Namibia through central Namibia to northern Namibia where he spent several months in 1885–6, with a brief excursion into southern Angola. In mid-1886 he crossed the Kalahari and reached the Okavango swamps in modern Botswana before returning to central Namibia and, in late 1886, Cape Town. He reached Zurich again in early 1887.

For Schinz, who during his initial stay in Cape Town in 1884, roamed the local libraries and read widely on southern Africa, his travels to southwestern Africa implied travelling to "scientifically unexplored areas".[17] This statement is somewhat surprising given a certain amount of published output, notably on the ethnography and history on southern and central Namibia, since a couple of decades, whilst it does hold true with regard to botanical research. By that time, these areas had been the subject of numerous exploration expeditions and a certain scholarly output, notably by protestant German and Finnish missionaries, who had produced first ethnographies, historical overviews, a few dictionaries on regional languages and some detailed maps. A few scientifically interested traders had by then also accumulated first ethnographic or ornithological collections which had found their way to either Cape Town, Germany or Sweden.[18] Schinz readily acknowledged the vast knowledge and writings of the many missionaries whose libraries and personal papers he consulted whenever arriving on the many mission stations which dotted his travel route.[19] Yet, the missionaries, as he made clear throughout his travels, were not scientists. He regarded himself not only as a botanist but also as having been scientifically trained in general and thus equipped, again in contrast to missionaries or, for that matter, some other colleagues amongst the initial Lüderitz Expedition whom he disdainfully labelled "Autodidakten",[20] to collect anything, in a proper scientific way, and write about it. To prove the point, he taught some local missionaries how to collect plant species, read thermometers and record observations properly.[21]

During the travels that Schinz undertook with the support of only a few local men and by making use of a light ox-wagon in order to travel quicker, he collected some 1,000 different plant specimens, seeds and bulbs, including four 'whole' *Welwitschia mirabilis*. He had a good eye for 'unknown' species and a remarkable percentage of his botanical collection did eventually yield 'new' species and first plant descriptions.[22] Importantly, he likewise collected with feverish determination ethnographic artefacts and claimed already during his travels that his ethnographic collections for 'Hereroland' and 'Amboland' in central and northern Namibia respectively, were 'complete'.[23] Simultaneously he compiled, with the help of missionaries and, it seems, only a few local experts, grammars and lists of vocabularies. He furthermore collected anthropological specimens such as hair samples from 'Bushmen', as well as insects, mammals and minerals. Regularly he took what he called 'photographic-physiognomical' images and attempted to contextualise his photographic objects through information on his or her name, 'tribe', gender, age, skin colour and hair descriptions.[24] On a daily basis he took weather temperatures and mapped out geographical bearings. He regretted not having brought along instruments to measure human beings, but nevertheless did so with the instruments at hand.[25] When he found, on a battlefield in northern Namibia, a 'totally sceletonised Ambo', the sceleton of a young man, he was not only, as he wrote to his mother, 'very happy' but added: 'Man muss eben Alles sammeln' (You actually need to collect Everything).[26] All these artefacts, including human remains, were eventually shipped to Zurich.[27]

Working during his travels on his notes on local religious beliefs, Schinz remarked that he placed the information obtained 'into a system'.[28] Again, this seems to be a clear remark against the writings of the missionaries which he would later criticise in his book. Their well-stocked libraries provided an important if not the most crucial source for his own notes, and new titles which he found in their libraries were duely recorded.[29] All these collecting activities he regarded as being "tiresome": every piece of information needed to be "compared"; he had to "ask back", "compare again" and then "combine".[30] Importantly, it seems to have been mostly the missionaries themselves whom he had to ask "back".

His papers[31] make only a few references to consultations with African experts. His conversations with the Herero chief Tjiharine on the Rhenish mission station of Omburo in mid-1885 on local customs and the utility of ethnographic objects which Schinz bought here, and the recording of a brief grammar and vocabulary of the #Kx'ao//'ae language from the northern Khoisan language group whilst travelling in the Kalahari in mid-1886 are rare exceptions in his attempts to obtain local information and knowledge from local people.[32] Most of the botanical names which he recorded he seems to have obtained from the missionaries and the same seems to hold true with regard to local usages of plants.[33] He regarded Africans as "wenig naturkundig" (displaying little knowledge on nature) though he had to concede at times that local botanical plant names did, for example, emphasise specific plant characteristics.[34] He was eager to distinguish his work from that of earlier travellers whose observations he found to be subjective and thus also made efforts to record geographical names as provided by his guides and in his writings strove to provide what he regarded as an orthographically correct version, adding, if possible, some etymological explanations, again in contrast to earlier writings. And he could express, as in the case of a medical treatment which he observed amongst San between Etosha and Grootfontein, a 'desire to understand the Khoisan phenomena [of healing] with an objectivity and relativity not apparent in the rhetoric of deliberate deception . . . typical of earlier ethnographic reports'.[35]

Yet, Schinz's interests and activities during his travels reflect severe limitations in terms of contacts and knowledge accumulation. Frequently, his papers record his often very negative attitude towards Africans in general and reflect an inexorable behaviour in his collecting strategies and social skills. Notable are various incidents close to violent situations. Whether forcing his transport labourers under threats[36] to travel as quickly as possible, whether digging up a skeleton or chopping away pieces from sacred stones close to a chiefly burial site in the Ndonga kingdom or taking photographs from such a site[37] – he often triggered mistrust and hostility, which at times flared up into open resistance and several times resulted in desertion of his small labour force. In the end, he had to flee Ovamboland for his ruthless behaviour in pursuit of scientific research.[38] His often very negative comments about those few individuals he did ask for linguistic information[39] further indicate that he had no high regard of those people whose lands and culture he was studying, whilst his "friends" in the field were, perhaps not surprisingly, the missionaries![40] Often, his comments in his letters and notebooks evoke images of an "imperial" habitus which he resumed for himself, and reflecting determination and force.

A frequently used term in his letters written during his travels and with regard to his activities is "verwerthen" (to utilise).[41] Everything he collected was for the purpose of 'utilisation' for 'the next years to come'. He thought of putting 'together all the material in order to publish it as a whole'.[42] Whilst travelling, he already started to report on his travels for periodicals of the various Swiss geographical societies.[43] In 1891, four years after his travels, he published his '*Deutsch-Südwest-Afrika. Forschungsreisen*' which became the first comprehensive and illustrated "Länderkunde" (country monograph) on this newly acquired German colony. In it, his chronological travel account is frequently interrupted by detailed and structured ethnographies. It ends with brief overviews of recent German colonial politics in the region, the history of the activities of mission societies and then a list of 'animals collected during the travels', followed by lists of 'idioms of the natives'. The latter included, rather importantly, a very first outline of the #Kx'ao//'ae language which continues until today to remain poorly documented. Having read some of the then already available academic literature on Khoisan languages, Schinz was very much aware of the importance of this outline.[44]

"Utility", however, meant much more. As he put it in one letter: 'Gold and jewels I do not bring [back], . . . for ME my collections and notes are more valuable than these'.[45] Indeed, they were invaluable. They formed the basis of his lifelong career as a specialised botanist. In short: Whilst roaming newly acquired colonial territory and feverishly accumulating objects and knowledge, he in fact was mapping out his future scientific career in Europe.

Academisation of botanical research

Having returned from southern Africa in early 1887, Schinz displayed his '50 boxes' of collections, amongst them 400 parcels with African plants, at his parents' home to the general public of Zurich.[46] This private exhibition generated some wider public attention, was extensively reported on in the *Neue Zürcher Zeitung* and thus also brought across to a wider readership and perhaps not unintentionally the academic ambitions of Schinz which were likewise commented on in the newspaper.[47]

After the exhibition of his collections, his next step of "utilisation" was to sell off parts of them, notably the ethnographic and zoological ones. His ethnographic collection became 'the first important' purchase of the newly founded Ethnological Society in Zurich. It was acquired for no less than 3,500 Swiss francs.[48] Today, this particular collection is regarded as 'the first comprehensive ethnographical collection from the Southwest African region'.[49] Interestingly, the board of the Society, which brokered the deal, included two academics who by then had already close ties with Schinz and who both were natural scientists: Conrad Keller (1848–1930), the president of the Society, who had travelled to eastern Africa and Madagascar in the mid-1880s and who would become Professor of Zoology in Zurich in 1889, and Professor Carl Schroeter (1855–1939), Schinz's graduate teacher. The sale of his collection thus interwove financial aspects with personal networks and professional interests.[50]

Schinz then set about to classify his botanical collections but was soon to be severely disappointed. The Botanical Garden and the Chair of Botany at the local university were, according to him, inadequately equipped for any comparative work, not only with regard to the library and herbarium, but also with regard to the most basic technical instruments.[51] He thus moved again to Berlin, and also to Kew, and for the next two years seemingly did much basic research, before starting to publish on African Flora throughout the 1890s. Perhaps most notably, he published during this time together with the Belgian botanist Théophile Durand (1855–1912), the (incomplete) 'Conspectus Florae Africae ou énumération des plantes d'Afrique' (1895–8). He did, though, submit his Habilitation in Zurich, possibly for strategic reasons. It has to be born in mind, that, as Essner emphasises, competition was high amongst the young career-orientated natural scientists. Even whilst travelling in Africa, Schinz kept in close contact with academic networks in Zurich, including the above mentioned zoologist Conrad Keller, as well as his mentors in Berlin.

Shortly afterwards, in 1893, Schinz obtained the post as Director of the Botanical Garden in Zurich, interestingly amidst heavy criticism aimed at Professor Karl Cramer (1931–1901), then the Director. Cramer, who, as Schinz continued to point out even 40 years later, 'was not a Systematiker',[52] had resigned due to various accusations, one being that he had totally neglected the Botanical Garden, the herbarium and the library.[53] Schinz's private herbaria were well known by then, as mentioned in the introduction to this essay and he donated his large African plant and library collection to the Garden, as the basis for a Botanical Museum soon to be founded by him.[54] At the same time the interests of the Federal Polytechnic (ETH) also began to shape the new direction of botanical research in Zurich. It was agreed between the Polytechnic and the University that the Botanical Garden was to collect systematically in the fields of economic botany ("Nutzpflanzen",

"Industrie- und Wirtschaftspflanzen") and build up a herbaria.[55] In connection with this, the university decided in 1895 that Schinz, being already the Director of the Garden, should also build a newly created chair for Systematic Botany, a field which until then had been taught by Professor Arnold Dodel-Port (1843–1908) as only one of various botanical fields. As such, systematic botany became a special field of research at the University of Zurich and an institute in itself.[56] In order to accomplish this, Schinz founded, in 1895, the Botanical Museum. This institution became his major achievement and laid the foundations of Zurich becoming a leading house in Europe for African botany and, in general, in systematic botany.

In all this, Schinz followed earlier trends in Kew, Paris or Berlin to build 'huge herbaria' as the 'essential tool . . . for the circumscription of species' (see Figure 5.1).[57] As pointed out by Christophe Bonneuil, these scientific metropoles with their ever increasing herbaria from their colonial outposts became 'metrological institutions that imposed the broad species concept, by controlling the production of new names and setting standards in the practices of classification'.[58] Schinz, due to his African travels and feverish international collection practices, could soon add Zurich to the list of these few botanical research metropoles and engage in the powerful establishment of a "Eurocentric taxonomic order" (Bonneuil) and control of systematic taxonomic research, concepts, nomenclature and communication.[59]

Before I lastly look at the Botany Museum, let me point out here that the sources available to me clearly indicate that Schinz operated in a web of competition between personalities of different academic generations and, hence, different interests with regard to the directions to be taken in academic, university-based botanical research and, for a general public, modernising the Botanical Garden in Zurich. Schinz's intense labours of international collecting activities, of editing the *Beiträge zur Kenntnis der afrikanischen Flora* as well as his very successful efforts to produce a comprehensive compendium of Swiss Flora also have to be seen in the light of these struggles.

The Botanical Museum and garden in Zurich

As pointed out, Zurich lacked an adequate botanical library, adequate research laboratories, and comparative plant materials in the late 19th century. It seems that in order to redress the situation and especially rapidly to build up plant collections in order to have adequate comparative – and competitive – research material, Hans Schinz founded the Botanical Museum. It was initially based on his private library and herbaria, both of which he donated to the Museum. The Museum became the centre of his and the Institute's activities and Schinz build not only large collections on a global scale, but, importantly, many new and special collections with regard also to Swiss flora or, in general, seeds, fungi and many other dry or "Spiritussammlungen".[60] In the words of his successor after 1928, Professor Albrecht Däniker (1894–1957): 'In clear recognition that more extended systematic research is possible only through larger collections. . . , he embarked on building a herbarium of international scale. . . . Of particular and repeated

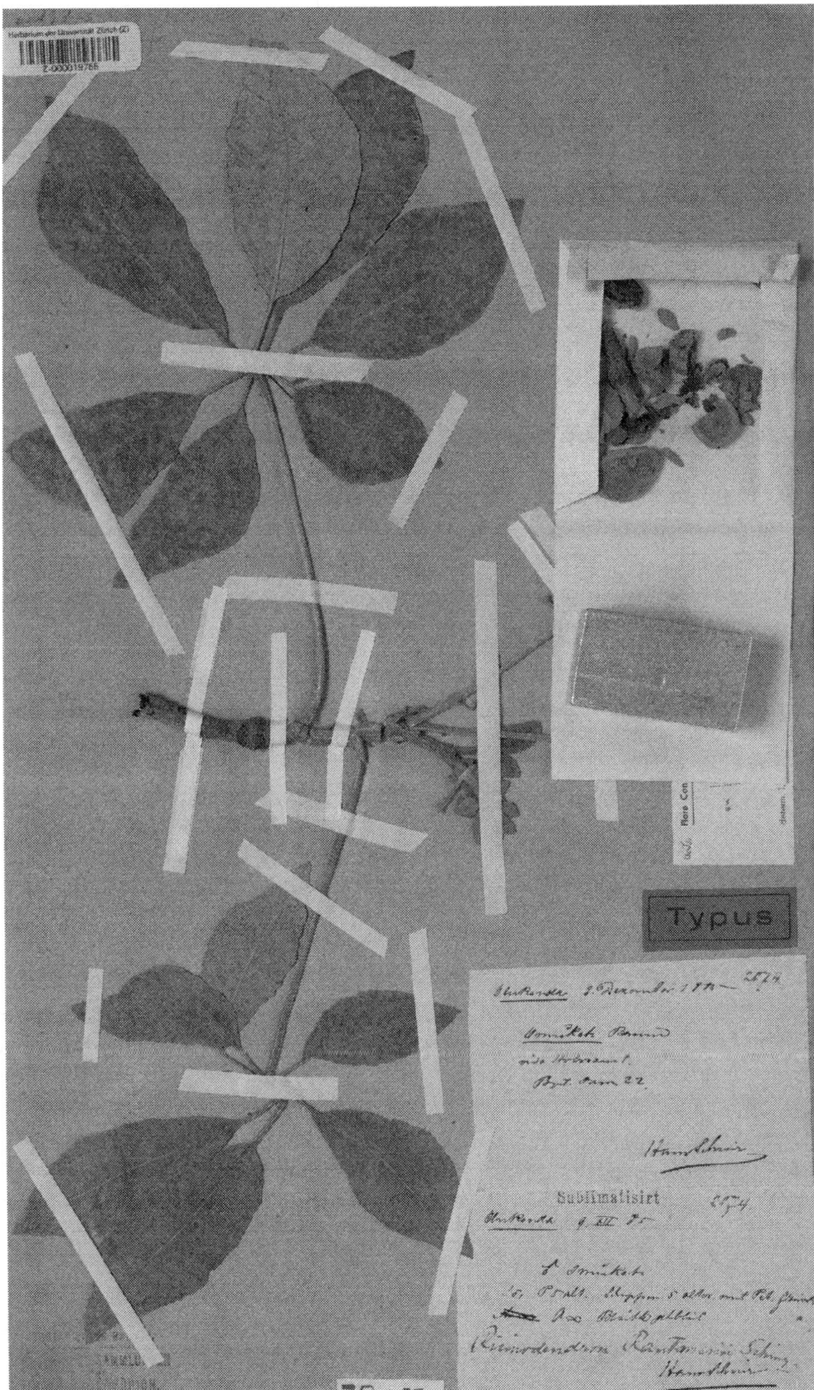

Figure 5.1 Excerpt with a specimen of the herbarium of Hans Schinz

Source: Library of the Botanical Institute of the University of Zurich

relevance were his own collections and those African collections from other collectors'. Däniker placed the Botanical Museum of Schinz squarely between 'Berlin, Kew, Paris, and South Africa'.[61]

Schinz claims to have used mainly his private means to build the Museum's collections.[62] His endeavours bore fruit when the canton of Zurich recognised the Museum after 1907 as a fully fledged museum, which in turn and rather importantly secured basic and long-term institutional funding.[63] Similar to ethnographic museums at the time, 'civic self-promotion' (Penny) played a role here, but also, of course, the necessity to be able to generate vast resources in order to build collections, house, manage and process them for research purposes and the production of knowledge.[64]

Schinz engaged in acquisition transactions of herbaria on an international scale, one important acquisition being, for example, the herbaria of the Polish botanist Professor Anton Rehmann (1840–1917), whose collections were built during his travels to South Africa in 1875–7 and 1879–80.[65] Missionaries in southern Africa were another important source for plant material, the Finnish missionary Martti Rautanen (1845–1926) from northern Namibia, whom Schinz had met in 1885, being only one of them; he sent him plant material but also weather reports over four decades.[66] In his many annual reports and in various published contributions Schinz for decades urged Swiss consulates, trading houses and travellers in distant countries to collect and as such contribute to the herbarium in Zurich. He kept up a large correspondence and exchange network in order to secure new and comparative research material and research results.

Schinz's garden and museum thus engaged in what Penny, with regard to municipal German ethnographic museums, has termed 'a cosmopolitan culture of collecting'. This combined research and 'civic self-promotion [was a means of] gaining prestige and often international recognition'.[67] Similar to ethnographic museums of the time, the Zurich institutions of Hans Schinz resonated with the belief 'that it was actually possible to collect, house and order such a vast number of artefacts' in order 'to function as a laboratory for the comparative analysis'.[68]

With the Museum, the Zurich University acquired a modern, and, as it turned out, competitive department. Competitive university departments, research institutes, botanical gardens and museums were, of course, part and parcel of the developing imperial scientific culture, all of which 'came to carry international prestige in the generation before the First World War'.[69] Clearly, the struggle over 'scientific hegemony' or 'superiority in science' is difficult to contextualise. Schinz expressed aspects of this in his 1904 publication concerning 'Swiss African travellers and the share of Switzerland with regard to the exploration and research of Africa in general'.[70] As the title indicates, he aimed to put Swiss exploratory and scientific contributions with regard to Africa on the imperial scientific map. In his introductory text, he bemoaned the fact that, 'the research and cultural missions of our countrymen' are usually not recognised as Swiss contributions by those two countries for whom these endeavours were often being undertaken and paid for, namely Germany and France. However, according to Schinz, one should reflect, 'not without pride' on 'the pioneers of our blood and our habits, who, despite the smallness of Switzerland, have contributed to the exploration of the dark continent'.[71] Such thoughts constructed an image of imperial culture and claimed a role for Switzerland in its construction. The vexed question for Schinz was the fact that Swiss contributions were not easily recognisable because of the usage of and publications in German or French. Obviously, then, Schinz was confronted with a classificatory problem and found the solution of how to classify Swiss contributions in a nationalist discourse and by means of the deployment of imperial images of scientific competition and excellency.

The public space of African botany in Zurich

With Schinz, Systematic Botany, which today forms part of the field of Historical Biology, established itself in Zurich and helped transform the university into a national and international centre. At the 1939 "Landi", the Swiss national exhibition, the Institute was well represented by the successor of Schinz, Professor Däniker, who, it seems, reaped some of the fruits of Schinz's labours. As Niklaus Stettler has observed in his study on the history of biodiversity science in Switzerland after the 2nd World War, Zurich remained the leading centre of "Vegetationskunde", and the herbarium of the Museum and the Institute of Systematic Botany played a pivotal part in this.[72]

It would be wrong to assume that if Schinz would not have travelled to Africa and if Schinz would not have obtained a post in Zurich, the modernisation of botany at Zurich University would not have taken place. Yet, without these travels, there would not have been an expert in African botany, and thus not the possibility of so profoundly establishing comparative and systematic botany in Zurich. What role, then, did African botany in particular play in Zurich and how was it contextualised?

Whilst African botany provided a specialisation, it was used by Schinz to conceptualise wider theoretical explanations for global plant evolution. In order to achieve this, African botany had to be developed. Without it, no comparative research, no generalisations about the global evolution of plant systems and their relationships, not even plant morphology, were possible.

For example, succulents, one important plant group from the Namib desert, were displayed in Schinz's Botanical Garden in order to illustrate varied biological forms, adaptations to different environments and, more precisely, to illustrate the morphological variations of transpiration as visibly reflected by the shapes of leafs or stems. They were compared, in the garden, with the local horsetail (Schachtelhalm) or gorse (Ginster). They were not necessarily displayed for the purpose of the exotic, the beautiful and decorative, or because they originated from Africa. Their role was to illustrate botanical and biological variety on a global scale.[73] In a guide for teachers and schools which Schinz wrote and regularly updated from 1908 onwards, he explained that the visitor could immerse himself or herself in a global panorama of plant varieties in all their forms for the purpose of both "Belehrung und Ergötzung" (instruction and enjoyment) and as such be taught the systematic ordering of plant variations, hierarchies and plant family relationships in a symbolic universe of global plant life.[74] Teaching and displaying global botanical variations and interconnections, differences, similarities and hierarchies to a general public was a major function of the Garden, whilst the Museum was the academic space of all this. Not surprisingly, the Garden thus also regularly organised "Schaustellungen" (literally: display installations).[75] Schinz's various teaching guides to public collections and museums in Zurich attest to his very personal interest in reaching out to a general public, apart from his intense scientific labours. This, in turn, made the Botanical Garden as important as other museums and collections in the city with regard to the production and dissemination of knowledge in the age of imperial scientific culture.[76]

Surprisingly, the contemporary visitor to Schinz's garden in Zurich would not, it seems, be able to have obtained much additional information on the exhibited plants such as any African names of plants or African knowledge supporting their medical utilisation. Schinz, in contrast to, for example, the German botanist Kurt Dinter (1864–1945) who started to work extensively on Namibian botany from the late 1890s onwards and was another of his important exchange partners,[77] rarely included African knowledge in his botanical research. African botany at this stage could already mean quite different things, and the scientific utilisation of plants from Africa in European centres of research, whilst being part and parcel of an imperial culture, built up not uniformly but in various strands of knowledge. In Zurich, perhaps because of the lack of Swiss colonies, herbaria provided the entry point for internationally competitive botanical research.[78] Hans Schinz remained, it seems, primarily an herbarium researcher and, particularly, a patron of science. Yet, when it came to African botany, his travels gave him hands-on experience and provided him with the very foundations which made it possible for the University of Zurich to establish itself as a leading institution in international botany. Simultaneously, Schinz saw to it that the garden served to display to the local population major themes of and trends in Western scientific culture. This in turn provided botanical research with broad civic support and gave Zurich, the economic capital of a country without colonies, a prime position in imperial research and the popularisation of knowledge.

Notes

* This paper originated from a larger project on the travels of Hans Schinz in southwestern Africa much encouraged by Carl Schlettwein and Patrick Harries, the latter inviting me to present a first paper on Hans Schinz at his seminal conference 'Imperial culture without colonies: Africa and Switzerland', University of Basel, October 23–5, 2003. Many thanks to Gesine Krüger for valuable comments on a draft version of this paper. All translations from German into English are my own.

1 Gunn and Codd, *Introductory Volume*, 313.

2 Introduction to Fernschau vol. 4 (1890), lvi.

3 Peyer, 'Hans Schinz', 5. Since 1893 Schinz edited the *Beiträge zur Kenntnis der Afrikanischen Flora* which initially formed part of the Bulletin de l'Herbier Boissier in Geneva and since 1897 appeared in the Mitteilungen aus dem Botanischen Museum der Universität Zürich. In the *Beiträge zur Kenntnis der Afrikanischen Flora*, Schinz mainly published descriptions of new species and often invited contributions from collegues from, amongst other centres, Kew or Berlin, in German, Latin or English.

4 A.J.G., 'Hans Schinz', 630.
5 Stettler, *Natur Erforschen*, 44, 55.
6 For an overview on Schinz's career, his African travels and his publications see the bio-bibliography compiled by Henrichsen, 'Hans Schinz, von Zürich', 134–8.
7 Honold and Simons, *Kolonialismus als Kultur*, 11. On the academisation of the natural sciences at the Humboldt during the 19th century and another academic career with reference to southern African collections see the contributions in this volume by Sandra Näf-Gloor and Patrick Grogan on the Berlin-based scientist Hinrich Lichtensein.
8 Essner, *Deutsche Afrikareisende*, 93–5, 121.
9 See Drayton, *Nature's Government*, 246–7.
10 See also MacKenzie, *Imperialism*, 4.
11 Pohle and Schenk, similarly to Schinz, followed academic careers in Germany after their return. Pohle became curator at the Zoological Museum in Berlin, Schenk professor for colonial geography in Halle.
12 Schinz, *Deutsch-Südwest-Afrika*, 1.
13 On the importance of economic botany in the late 19th century and its interconnections with imperial agricultural politics, see, for example, Drayton, *Nature's Government*, 248–51.
14 Schinz could rely on family funds though being very conscious about his expenses and seeing to it that he could sell off parts of his collections upon his return to Zurich in order to recuperate some money.
15 Zentralbibliothek Zürich (ZB), Handschriftenabteilung (manuscripts), Ms. Z IX 319, Hans Schinz (1858–1941) Botaniker und Ethnologe. Dokumente betreffend seine Forschungsreisen in Südwestafrika, 1884–1886. Including the letter Ms. Z IX 319.1, Hans Schinz to his mother Julie Schinz-Vögeli, letter dated 24 October 1885. See also Ms. Z IX 319.2, Hans Schinz to his brother Emil Schinz, letter dated 9 February 1886. Whilst this paper cites Schinz's travel letters with the original archives reference, they have also been published and annotated, together with his travel photographs, in Henrichsen, Hans Schinz. Bruchstücke.
16 ZB, Ms. Z IX 319.1, Hans Schinz to his mother Julie Schinz-Vögeli, letter dated Olukonda, 24 October 1885.
17 ZB, Ms. Z IX 319.2, Hans Schinz to his uncle, letter dated Omaruru, 16 July 1885.
18 Foremost amongst these early collectors were the Swedish traders and travellers Charles John Andersson and Axel Eriksson. See, for example, Henrichsen, 'Die Hegemonie der Herero', 44–59, esp. 58. Also Johansson, *The Trader King*.
19 Indeed, as his diary (ZB, Ms. Z IX 656, Reisetagebuch) indicates, it was in the missionary's libraries that he not only came across very different published works whose titles he would jot down, but where he at times started to study and accumulate notes for ethnographies, grammars and other topics on which he expanded in his monograph *Deutsch-Südwest-Afrika*.
20 ZB, Ms. Z IX 656, Reisetagebuch (travel diary) Hans Schinz, 27 October 1884.
21 See, for example, ZB, Ms.Z IX 319.1, Hans Schinz to his mother, letter dated July [January] 1886.
22 C.S., 'Die südafrikanische Sammlung'. On his botanical collecting strategies see Nyffeler, 'Einblicke'.
23 ZB, Ms.Z IX 319.1, Hans Schinz to his mother, letter dated July [January] 1886. On the crucial issue of what Schinz understood as being a 'complete' collection see Henrichsen and Krüger, 'Kreuz- und Querzüge'.
24 ZB, Ms. Z IX 656, Reisetagebuch, 15.
25 Ibid., 23 December 1884, 66.
26 ZB, Ms. Z IX 319.1, Hans Schinz to his mother, letter dated Olukonda 24 October 1885. Underlined in the original. See also ZB, Ms. Z IX 656, Reisetagebuch, 156. In February 1886 he was accused by the Ndonga king of profoundly violating sacred spaces and performing 'witch craft'; the skeleton and his photography were mentioned in particular. Schinz, confronted with increasing hostility, decided to bury the skeleton and 'only to keep the scull'. Realising that the father of the king mobilised the army, Schinz fled the kingdom. See ibid., 187.

27 On this and also the human remains which Schinz shipped to Zurich see Henrichsen and Krüger, 'Kreuz- und Querzüge'; Henrichsen, 'Die Skelettaffaire'.
28 ZB, Ms. Z IX 319.2, Hans Schinz to his brother, letter dated 9 February 1886.
29 As reflected in his book *Deutsch-Südwest-Afrika* where, for example in the ethnographical sections, he discusses conflicting views by different authors. See also ibid., viii, or ZB, Ms. Z IX 319, Hans Schinz to his mother, letter dated 16 to 27 December 1884.
30 See esp. ZB, Ms. Z IX 319.1, Hans Schinz to his mother, letter dated 13 March 1886.
31 These include, apart from what has been mentioned before, his botanical field diary Hans Schinz. Reise nach Great Namaqualand 1884. Damaraland, Ovamboland, Kunene, Ngamisee, Kalahari 1884–1886, Institut für Systematische Botanik, Library, Zurich, G 1347.
32 On the conversations with chief Tjiharine see ZB, Ms. Z IX 319.2, Hans Schinz to his uncle, letter dated July 1885. With regard to his recording of the #Kx'ao//'ae language see ZB, Ms. Z IX 319.1, Hans Schinz to his mother, letter dated 1 June to 27 August 1886.
33 See, for example, ZB, Ms. Z IX 319.1, Hans Schinz to his mother, letter dated 16 to 27 December 1884.
34 See, for example, Staatsbibliothek Berlin, Sammlung Darmstaedter, Afrika 1884 (8), Hans Schinz to Prof. Ascherson, letter dated 5 April 1885.
35 See Low, 'Khoisan Healing', 103. On 16 March 1886 Schinz described in detail a treatment on one of his wagon labourers by 'an old Bushman' and concerning 'foreign objects' in the labourer's body and involving various healing techniques. See ZB, Ms. Z IX 656, Reisetagebuch, 16 March 1886, 194.
36 For example, his diary entries in ZB, Ms. Z IX 656, Reisetagebuch, 14 May 1886, 204; 30 May 1886, 211.
37 Ibid., 16 February 1886, 184–5; 18 February 1886, 186.
38 See Staatsbibliothek Berlin, Sammlung Darmstaedter, Afrika 1884 (8), Hans Schinz to Prof. Ascherson, letter dated 18 December 1886. Also C.S., 'Die südafrikanische Sammlung'. For a more extensive analysis see Henrichsen, 'Die Skelettaffaire'.
39 ZB, Ms. Z IX 656, Reisetagebuch, 4 May 1886, 201. Also ZB, Ms.Z IX 319.1., Hans Schinz to his mother, letter dated 1 June 1886.
40 Schinz, *Deutsch-Südwest-Afrika*, vii.
41 See, for example, ZB, Ms. Z IX 319.1, Hans Schinz to his mother, letter dated 16 to 27 December 1884.
42 Ibid., Hans Schinz to his mother, letter dated 13 March 1886.
43 See Henrichsen, 'Hans Schinz, von Zürich'.
44 Schinz, *Deutsch-Südwest-Afrika*, 540.
45 ZB, Ms. Z IX 319.1, Hans Schinz to his mother, letter dated 13 March 1886. Emphasis in the original.
46 C.S., 'Die südafrikanische Sammlung'. Also Anon., 'Lokales'.
47 On this see Henrichsen and Krüger, 'Kreuz- und Querzüge'.
48 Seemingly by means of private contributions, see Völkerkundemuseum Zürich, Archives, Ethnographische Gesellschaft, *Vorstands-Protokoll, Protokollbuch, I.-V. Sitzung*.
49 Müller, Gerber, and Jud, *100 Jahre Völkerkundemuseum*, 31.
50 On the sale of his zoological collection see Zoologisches Museum der Universität Zürich, Archives, Anschaffungen für Rechnung der vereinigten Sammlungen, Säugethiere, 1886.
51 For example Hans Schinz, *Der Botanische Garten*, 1937a, 40, 46–7. On the situation at the Department of Botany in Zurich in the 1880s and early 1890s see also the very negative comments by the then Professor of Zoology, Keller, Lebenserinnerungen, 37–55, especially 38, 50–1. See also the critical remarks in C.S., 'Die südafrikanische Sammlung'.
52 Schinz, *Der Botanische Garten*, 1937a.

53 Ibid., 46–7. In one of his letters during his travels, Schinz revealed in very explicit terms that he felt having been treated unfairly by Cramer (and Schröter) during his MA examination. ZB, Ms IX 319.1, Hans Schinz to his mother, letter dated 13 March 1886.
54 Schinz, *Der Botanische Garten*, 1931.
55 Schinz, *Der Botanische Garten*, 1937a, 41.
56 Ibid., 41, 46, 48. Compare this with Schinz, *Mein Lebenslauf*, 36.
57 Bonneuil, 'The Manufacture of Species', 189. See also the contribution of Sandra Näf-Gloor in this volume.
58 Bonneuil, 'The Manufacture of Species', 190.
59 For more details on the role of the few leading European herbaria in this respect see ibid., 206–9.
60 See the detailed explanations in Schinz, *Der Botanische Garten*, 1937a, 46–51.
61 A.U. Däniker, 'Professor Hans Schinz', 62–3. On the importance of herbaria for botanical research see also Pyenson, *Cultural Imperialism*, 10–11.
62 Schinz, *Mein Lebenslauf*, 48.
63 An important aspect with regard to the recognition of the Museum by the city seems to have been its function, officially destined to the Museum by the canton, of the municipal mushroom control. Also enhancing the character of the museum was an agreement, in 1909, between the canton of Zurich and the Museum that the latter would act as the legal deposit for pre-historic botanical finds. See Schinz, *Der Botanische Garten*, 1937a, 49.
64 For ethnographic museums at the time see the succinct essay by Penny, 'Municipal Displays'.
65 Gunn and Codd, Introductory Volume, 292, 313. For an overview of the (African) holdings of the Botanical Museum see Schinz, Führer, 1934a, 19–20.
66 See, for example, the letter of missionary Rautanen to Schinz, 13.5.1926, in Institut für Systematische Botanik, Zurich, Autographensammlung. See also Ruatiala and Junikka, 'Martti Rautanen – kasviharrastaja Ambomaalta'. Apart from Rautanen, Schinz in 1934 also emphasised the collections of the Swiss missionary Henri-Alexandre Junod for southeastern Africa, in Schinz, Führer, 1934a. On the collaboration of Schinz and Junod see Harries, 'Field Sciences', 21.
67 Penny, 'Municipal Displays', 158.
68 Ibid., 159.
69 Pyenson, *Cultural Imperialism*, 312–15.
70 Schinz, 'Schweizerische Afrika-Reisende'.
71 Ibid., 3.
72 Stettler, *Natur Erforschen*, 44, 55, African botany, it seems, did not continue, after Schinz, to remain a focus in Zurich.
73 Schinz, *Der Botanische Garten*, 1937b, 14–15. A more geographical ordering of plants (geobotany) which gained popularity in European botanical gardens in the late 19th century could, for the lack of space in Zurich, not be accomplished. However, in the Museum itself, Schinz had arranged large vitrine's which displayed vegetation groups from German South West Africa. See Schinz, *Der Botanische Garten*, 1913, 19.
74 Schinz, *Wegleitung*, 1–2.
75 Schinz, *Führer*, 1934b, 17. See also Schinz, *Der Botanische Garten*, 1937a, 43–6.
76 See, for example, Schinz, *Wegleitung*; Schinz, *Führer*, 1934a.
77 Schinz and Dinter at times worked closely together and shared collections. See, for example, Gunn and Codd, Introductory Volume, entry on Dinter. Dinter, who became the official government botanist in German South West Africa from 1900 onwards, worked closely with African assistants and incorporated not only African plant names (often in two local languages) in his publications more often than Schinz. He also paid particular attention to the utilisation of local veld fruit and vegetables. See Dinter, *Deutsch-Südwest-Afrika*; Dinter, *Die vegetabilische Veldkost*.
78 On the debate regarding (post-)coloniality and Switzerland as a country without colonies see esp. Purtschert, Lüthi and Falk, *Postkoloniale Schweiz*.

Bibliography

Anon., 'Lokales: Hans Schinz', *Neue Zürcher Zeitung* (*NZZ*), 24, April 1887.

A.J.G., 'Hans Schinz', *Dictionary of South African Biography* (Cape Town: Tafelberg, 1972), II.

C.S., 'Die südafrikanische Sammlung von Dr. Hans Schinz', *Neue Zürcher Zeitung* (*NZZ*), 4 May 1887.

Bonneuil, Christophe, 'The Manufacture of Species. Kew Gardens, the Empire, and the Standardisation of Taxonomic Practices in late 19th Botany', in *Instruments, Travel and Science: Itineraries of Precision from the 17th to the 20th Century*, ed. Marie-Nielle Bourguet, Christian Licoppe, and H. Otto Sibum (London: Routledge, 2002), 189–215.

Däniker, A.U., 'Professor Hans Schinz', in *Jahresbericht 1941/42* (Universität Zürich: Art. Institut Orell Füssli A.-G., 1942), 61–4.

Dinter, Kurt, *Deutsch-Südwest-Afrika. Flora, forst- und landwirtschaftliche Fragmente* (Leipzig: Theodor Oswald Weigel, 1909).

———, *Die vegetabilische Veldkost Deutsch-Südwest-Afrikas* (Okahandja: Selbstverlag, 1912).

Drayton, Richard, *Nature's Government. Science, Imperial Britain and the 'Improvement' of the World* (New Haven: Yale University Press, 2000).

Essner, Cornelia, *Deutsche Afrikareisende im neunzehnten Jahrhundert: Zur Sozialgeschichte des Reisens* (Stuttgart: Steiner, 1985).

Gunn, Mary and L.E. Codd, *Introductory Volume to the Flora of Southern Africa: Botanical Exploration of Southern Africa* (Cape Town: Balkema, 1981).

Harries, Patrick, 'Field Sciences in Scientific Fields: Entomology, Botany and the Early Ethnographic Monograph in the Work of H.-A. Junod', in *Science and Society in Southern Africa*, ed. Saul Dubow (Manchester: Manchester University Press, 2000).

Henrichsen, Dag, 'Hans Schinz, von Zürich, 1858–1941, Botaniker', *Nachrichten/Newsletter der Basler Afrika Bibliographien (BAB)* 13, no. 4 (1989): 134–8.

———, 'Die Hegemonie der Herero in Zentralnamibia zu Beginn der deutschen Kolonialzeit', in *Namibia-Deutschland: Eine geteilte Geschichte. Widerstand, Gewalt, Erinnerung*, ed. Larissa Förster, Dag Henrichsen, and Michael Bollig (Wolfratshausen: Edition Minerva, 2004), 44–59.

———, *Hans Schinz. Bruchstücke. Forschungsreisen in Deutsch-Südwestafrika* (Basel: Basler Afrika Bibliographien, 2012).

——— and Gesine Krüger, '"Kreuz- und Querzüge in Afrika": Darf man alles sammeln?', in *Man muss eben Alles sammeln. Der Zürcher Botaniker und Forschungsreisende Hans Schinz und seine ethnographische Sammlung Südwestafrika*, ed. Gitte Beckmann, (Zurich: NZZ Libro, 2012), 127–35.

———, 'Die "Skelettaffaire"und andere 'Geheimnisse' – Sammlungsstrategien, Grenzüberschreitungen und Wissenskonzeptionen des Zürcher Botanikers Hans Schinz', in *Sammeln, Erforschen, Zurückgeben? Menschliche Gebeine aus der Kolonialzeit in akademischen und musealen Sammlungen*, ed. Holger Stoecker, Thomas Schnalke, and Andreas Winkelmann (Berlin: Ch. Links, 2013), 121–9.

Honold, Alexander and Oliver Simons, eds., *Kolonialismus als Kultur. Literatur, Medien, Wissenschaft in der deutschen Gründerzeit des Fremden* (Tübingen: Francke, 2002).

Johansson, Peter, *The Trader King of Damaraland Axel Eriksson. A Swedish Pioneer in Southern Africa* (Windhoek: Gamsberg Macmillan, 2007).

Keller, Conrad, *Lebenserinnerungen eines schweizerischen Naturforschers* (Zurich: Orell Füssli, 1928).

Low, Chris, 'Khoisan Healing: Understandings, Ideas and Practices' (DPhil thesis, University of Oxford, 2004).

MacKenzie, John, *Imperialism and the Natural World* (Manchester: Manchester University Press, 1990).

Müller, Verena, Peter R. Gerber, and Peter Jud, *100 Jahre Völkerkundemuseum der Universität Zürich 1889–1989* (Zurich: Völkerkundemuseum der Universität Zürich, 1989).

Nyffeler, Reto, 'Einblicke in die botanische Sammeltätigkeit von Hans Schinz während seiner Reise durch Deutsch-Südwest-Afrika', in *Man muss eben Alles sammeln. Der Zürcher Botaniker und Forschungsreisende Hans Schinz und seine ethnographische Sammlung Südwestafrika*, ed. Gitte Beckmann (Zurich: NZZ Libro, 2012), 119–26.

Penny, Glenn, 'Municipal Displays. Civic Self-Promotion and the Development of German Ethnographic Museums, 1870–1914', *Social Anthropology* 6, no. 2 (1998): 157–68.

Peyer, Bernhard, 'Hans Schinz, 1858–1941', *Verhandlungen der Schweizerischen Naturforschenden Gesellschaft* 121 (1941): S407–21.

Purtschert, Patricia, Barbara Lüthi, and Francesca Falk, eds., *Postkoloniale Schweiz. Formen und Folgen eines Kolonialismus ohne Kolonien* (Bielefeld: Transcipt, 2012).

Pyenson, Lewis, *Cultural Imperialism and Exact Sciences. German Expansion Overseas 1900–1930* (Berne: P. Lang, 1985).

Ruatiala, Marjatta and Leo Junikka, 'Martti Rautanen – kasviharrastaja Ambomaalta', *Luonnon Tutkija* 99, no. 5 (1995): 152–6.

Schinz, Hans, *Deutsch-Südwest-Afrika. Forschungsreisen durch die deutschen Schutzgebiete Gross-Nama und Hereroland, nach dem Kunene, dem Ngami-See und der Kalahari 1884–1887* (Oldenburg: Schulze, 1891).

———, 'Schweizerische Afrika-Reisende und der Anteil der Schweiz an der Erschliessung und Erforschung Afrikas überhaupt', *Neujahrsblatt der Naturforschenden Gesellschaft in Zürich auf das Jahr* 106 (1904).

———, *Wegleitung für Lehrer und Schulen zum Besuch des botanischen Gartens und botanischen Museums in Zürich* (Zurich: Verlag der Erziehungsdirektion, 1908).

———, *Der Botanische Garten und das Botanische Museum der Universität im Jahre 1912* (Zurich: Leemann, 1913).

———, *Der Botanische Garten und das Botanische Museum der Universität Zürich in den Jahren 1929 und 1930* (Zurich: Leemann, 1931).

———, *Führer durch die naturwissenschaftlichen und medizinischen Anstalten, Institute, Kliniken, Sammlungen und Bibliotheken Zürichs* (Zurich, 1934a).

———, *Führer durch die naturwissenschaftlichen und medizinischen Anstalten, Institute, Kliniken, Sammlungen und Bibliotheken Zürichs, die Graphische Sammlung der E.T.H. und die Archäologische Sammlung der Universität, sowie durch einige naturwissenschaftlicht interessante Werke und Einrichtungen der Stadt Zürich*, 3rd edition (Zurich, 1934b).

———, 'Der Botanische Garten und das Botanische Museum der Universität Zürich', *Beiblatt zur Vierteljahrsschrift der Naturforschenden Gesellschaft in Zürich* 82, no. 29 (1937a).

———, 'Der Botanische Garten und das Botanische Museum der Universität Zürich', *Zürcher Monats-Chronik* 6 (1937b).

———, *Mein Lebenslauf* (Zurich: Fretz, 1940).

Stettler, Niklaus, *Natur Erforschen. Perspektiven einer Kulturgeschichte der Biowissenschaft an Schweizer Universitäten, 1945–1975* (Zurich: Chronos, 2002).

Part II

Knowledge practices between colonial and local actors

6 *Mission from Cape Coast Castle to Ashantee*

One work's significance for European knowledge production about the Asante Empire

Sonia Abun-Nasr

Introduction

Past and present debates about "colonial knowledge"[1] often follow predictable patterns. This is evident nowadays in academic stances taken with respect to the history of European research about the African continent. Following Valentin Yves Mudimbe, one often points out that "Africa" is a construct,[2] or, citing Edward Said's "orientalism" thesis, that the treatment of an unfamiliar Africa is a discourse about the "Other".[3] If one follows the literary scholar Marie Louise Pratt's analysis of early European travel accounts, then Europeans through their texts have for centuries created a non-European "rest of the world", and employed a scientific systematisation which proved an influential instrument of domination.[4]

Given the uniformity of postmodern opinions, it almost seems naïve and anachronistic to examine a 19th-century European travel account from a different perspective. Nonetheless, the following analysis is of just such an account. It is assumed that on the one hand, it has considerable value as a source, and that, on the other, it was by no means written from a colonial perspective.

There is no question that Thomas Edward Bowdich's *Mission from Cape Coast Castle to Ashantee*, published in London in 1819 by John Murray, is a classic work. Specialists describe it as a rich and vivid source on the history of the Asante Empire. 'It is', Nigel Barley writes, 'a lively and colourful first-hand account, full of accurate detail and observation, and has proved an invaluable source for both anthropologists and historians'. In a 1977 British Museum yearbook, Malcom McLeod notes that '*Mission from Cape Coast Castle to Ashantee* . . . in many ways remains the best account we have of the Ashanti state at the height of its power'. And Tom C. McCaskie, the author of a more recent standard work on the Asante Empire, described Bowdich's account as 'insightful ethnohistorical reportage'.[5]

The following discussion is based on the conviction that travellers, regardless of era or destination, do not remain unmoved by the experiences they have in places strange to them.[6] That is why it is unrealistic in their travel accounts to only see European mentalities, stereotypes and prejudices reflected. It is far more necessary to recognise the empirical basis for such texts, and to differentiate observations from their accompanying, and subjective, interpretations.

Thus, the following will show how Bowdich reflected the methods of his era in compiling his findings, as well as how his book helped disseminate those findings in Europe. Some of the details of social and political relations in the Asante Empire which he conveyed through text and illustration are analysed to show the interpretation of this society he wished to convey to his readers. The argument here roughly follows the stages of Bowdich's book: (1) its genesis; (2 – 3) the contents as an interplay between text and illustration; and finally (4) an overview of the publication history and reception of the work. *Mission from Cape Coast Castle to Ashantee* is treated as a weighty work of research, in which Bowdich, due to specific patterns of perception, formulated – for his time – an unusually positive evaluation of the Asante Empire.

On the genesis of the book

In *Mission from Cape Coast Castle to Ashantee*, Bowdich describes a journey he undertook into the hinterlands of the Gold Coast (today's Ghana) in 1817. This journey should be seen in a historical context determined by supraregional as well as regional (meaning those specifically related to the Gold Coast) factors.

After centuries of trade and cultural contact confined largely to the coastal regions, the early 19th century was an era when European travellers pushed into the interior of the African continent. With the assistance of African middlemen, commercial trade had been carried out at European forts and settlements spread around the Gulf of Guinea since the late 15th century. But by the early 19th century, British, Danish and Dutch traders on the Gold Coast were competing for access to the hinterlands and its valued goods: slaves, gold and ivory. To obtain them, they traded European products and weapons. The British forts at the time were not run by the government but rather by the African Company of Merchants.[7]

In the hinterlands of the Gold Coast, the Asante Empire acted as a powerful adversary to the European trading companies. Since the early 18th century, it had evolved out of a federation of local rulers to become a major power, doing so by subjugating neighbouring peoples and expanding in all directions from the capital city of Kumase. The last expansionary surges, when the Asante in the south advanced as far as the ocean and when the Empire was at its largest, took place between 1806 and 1816. As a result, the Europeans settled on the coast had to make the effort to establish direct trade with Kumase itself.[8]

For this reason, the British African Company sent an expedition to Kumase in 1817, its mandate to conclude a trade and peace agreement with the Asante Empire's ruler. The British slave trade had been prohibited in 1807, and the Dutch were generating stiff competition, so there was a desire to open up new markets.[9] Other than its leader, Sir Frederick James, and numerous Africans, three additional Englishmen took part in this expedition, one of them the 25-year-old, Bristol-born, Thomas Bowdich, a writer for the African Company who was to evaluate what they found from a scientific point of view. His employment on the Gold Coast was due to family connections. His uncle, John Hope Smith, had worked for the African Company as a writer as well, and was, at age 30, appointed Governor-in-Chief of the Gold Coast; he evidently sponsored his nephew.[10]

The expedition set out in April 1817 from Cape Coast Castle and took four weeks to reach Kumase. Negotiations with the ruler of the Asante Empire, the *Asantehene* Osei Tutu Kwame,[11] also called Osei Bonsu, dragged on, and the more Frederick James proved himself an indecisive leader, the more determined Bowdich became. Bowdich soon took control of the negotiations with the *Asantehene*, displacing James and even managing to have James be recalled to Cape Coast Castle and have himself formally named leader of the expedition.[12] On 7 September 1817, it was Bowdich who signed the agreement in Kumase with Osei Bonsu and with the ruler of the Dwabin kingdom.

The agreement emphasised the mutual desire for peace 'between the British subjects in this country, and the subjects of the Kings of Ashantee and Dwabin', and stressed that both kings wished to encourage their subjects to trade with the British then operating on the Gold Coast.[13] However, the agreement was vague about Kumase's right to rule over African coastal cities, and the version Bowdich presented to the African Company differed from the version Osei Bonsu was given. As a result, relations between Great Britain and the Asante Empire remained tenuous, and Joseph Dupuis, the envoy of the British government sent to Kumase in 1820, was unable to make it more solid.[14] By the second half of the 19th century, diplomatic efforts were succeeded by military force, and led to a series of wars between the British and the Asante, together with their various African allies. In 1874, a British victory led to the destruction of Kumase, and in 1902 to placing the entire region under British rule, a colonisation which would only end in 1957.[15]

Bowdich wrote his report about the expedition during his return to England from West Africa, and had it published in 1819 in London.[16] Among other things, this publication can be understood as an effort at justification, as Bowdich was criticised back in England for his having replaced Frederick James as leader of the expedition (in addition, the African Company paid him less for his efforts than he had expected, about which he complained with some bitterness in a polemic published in 1819).[17] He found himself excluded from scientific circles as a result, and as it seemed unlikely he would receive support in England for future projects, he and his wife Sarah moved to Paris.

There, their luck took a turn for the better, at least for a while. They cultivated their relationship with George Cuvier, an influential naturalist who supported their writing and research, as well as with Alexander von Humboldt and other scholars of the time. During their sojourn in Paris, the couple engaged in studies of mathematics, physics and natural history. Bowdich himself translated and wrote natural history, ethnography and geography works, and the proceeds realised thereby enabled the two of them to launch a second expedition in 1822. The goal was Sierra Leone, but before reaching it, Bowdich took ill while surveying at the mouth of the Gambia River, and died at a hospital at Bathurst (today's Banjul) on 10 January 1824 – prematurely ending what might have been a stellar research career.[18]

The book: travelogue and scientific documentation

Bowdich's travel account was commissioned by the African Company with clear expectations. The Company wanted documentation and information about the Asante Empire that they could use either for expanding British trade contacts or in exercising direct political influence over the Gold Coast. In his account, Bowdich himself names the specified objects of his investigation, which include geography, ethnography, politics and social relations.[19] In its utilitarian orientation, this travel account reflected contemporary trends: 'The 1817 inquiries were thus part of the general surge of interest in Africa which was occurring at this time'.[20] Beyond this, however, the text reflects Bowdich's genuine interests. His precise documentation and methods he employed place him in the tradition of Enlightenment-era researchers who were interested in everything, and whose expeditions were intended to expand humankind's knowledge of the world. In that sense, Bowdich's account should not be understood solely in terms of expanding "colonial knowledge" to serve political or economic interests.[21]

Mission from Cape Coast Castle to Ashantee is 512 pages long, and is in three parts, the first two text, the third an appendix. In the first part, Bowdich describes the course of the expedition to Kumase, from the departure from Cape Coast Castle in April 1817 until it returned in September of that year. The six chapters (or 160 pages) devoted to this account are extremely heterogeneous both in content and in style.[22]

Bowdich begins with a discussion of the literature, in which he surveys existing research about the Asante region, before turning to the mandate for the expedition given by the African Committee (the letters from this committee and from the Governor of Cape Coast Castle are reproduced)[23] and then continues with a description of the journey.[24] Content, subject and style alternate in rapid succession, though one can distinguish between two types of text. One is a straightforward and sober geographic inventory in which Bowdich gives the names of settlements, the distances travelled and the direction of the paths the expedition took. These alternate with colourful and extravagant observations about African landscapes and villages.[25] The account has a vibrant character which results from the unexpected combination of scientific dispassion and poetic effusiveness.

With the arrival in Kumase, the character of the text changes. After describing the reception ceremony at court,[26] the negotiations with the *Asantehene* Osei Bonsu – and the conflict between Bowdich and his travel companions (including Frederick James) – take centre stage, and are documented by various sources.[27] During their stay in Kumase, which lasted several months, James, his English travel companions and the *Asantehene* all sent letters to the British Governor at the coast. These letters are reproduced, together with the answers and instructions the Governor sent back, as are excerpts from the diaries kept by the travellers; the trade and peace agreements reached between Great Britain and the Asante Empire are also included. It is not wrong to see all of this, in fact, as Bowdich's effort to give an account to those who had employed him – as well as to an interested British public – which might legitimise his behaviour during the journey.

Overall, the first part of his work is characterised by a kind of source-based transparency. It permits a detailed understanding of the conditions under which quite varied knowledge was acquired, whether this was geographic, landscape observations or information about the course of the expedition.

There are 14 chapters (or 300 pages) in the second part of the book; it provides the reader with an exhaustive description of the living condition in the Asante Empire. It is here one finds the proper research and evaluation work of the expedition, one separated from the travel account as such. Bowdich addresses a very wide variety of topics under the headings of Geography, History, Constitution and Laws, Superstitions, Rituals and Celebrations, Architecture, Art and Handicrafts, Trade, Language and Music. Owing to this broad scope, one cannot readily describe this part of the text, though one can summarise some of the typical or formal characteristics of the contents.

He begins with a chapter on geography[28] which serves as the basis for subsequent topics, and as it is the longest chapter, it also carries particular weight. However, Bowdich's research opportunities were very restricted as the *Asantehene* Osei Bonsu did not actually allow him to leave Kumase during the negotiations.[29] His description of the geographic conditions in the Asante Empire were thus based on what information he could gather on the trips to and from Kumase, together with what he could glean from conversations while in the town itself.

Kumase, in fact, lay at the intersection of a number of larger paths (or roads) which radiated out in all directions and connected the Asante capital with its subject territories and important trading centres. Bowdich asked knowledgeable travellers about the towns and territories to which the paths or roads led, along with how long it took to reach the respective settlements. He attached considerable importance to comparing the information he gleaned in separate conversations. One can almost speak of his use of a scientific technique in his interviews:

> I shall pass over a mass of memoranda recorded on individual report, and only select such, wherein Moors and natives, unknown to each other, have agreed; describing their travels in their own way, without my questions anticipating or directing them. These routes and observations were further confirmed by the evidence of children, recently arrived as slaves from the various countries, whose artless replies decided my credence. . . . [T]heir evidence, therefore, was a genuine and acceptable check on the Moorish and Negro adults.[30]

Bowdich also converted the travel times he was told, following a particular formula, so as to have a result in miles, and in this manner obtained a quite accurate idea of Kumase's location in the middle of the 'nine great paths'.[31] He documented this information in his work,[32] and where possible, compared it with what he could find in the contemporary literature.[33] Given his precision and transparency, it seems appropriate to call his an empirical approach – which is not to say that his figures were error-free. Yet in 20th-century research, his work did achieve the status of being the first source-based information about the Asante Empire. Ivor Wilks, for example, in his *Asante in the Nineteenth Century* begins – in this comparable to Bowdich – with a chapter on the *Spatial Aspects of Government* in which he includes an extensive description of the 19th-century network of paths around Kumase, adding a sketch based both on Bowdich's account and on the accounts of later European visitors.[34]

In the chapter on the history of the Asante Empire,[35] Bowdich also uses a research method one could call "oral history". However, here one finds the limits of what possibilities for gaining knowledge were open to him.

Bowdich's history begins with the genesis of the original community around 1700 and continues up to the early 19th-century present. It includes assertions about the close historical and linguistic affinity of the peoples now known as the Akan. The historical account is organised around the succession of Kumase rulers since the Asante kingdom was established, focusing not just on the dates when power was assumed but even more importantly on wars associated with the expansion of the territory under the control of the Asante since the early 18th century. A text structured in this manner can readily be seen as a reflection of an official presentation of history of the kind passed down – in the interests of the Empire – at the court of the *Asantehene*.[36] It is no coincidence that Bowdich cites members of the ruling families[37] as interlocutors and his source of information, including even the *Asantehene* himself.[38] Yet one searches in vain in Bowdich for historical accounts of subjugated peoples, with their tales of injustice and suppression.[39] Bowdich conducted his research at the centre of the Empire and thus received correspondingly tinged information.

The subsequent chapters reflect a great earnestness, for Bowdich treats the objects of his research with respect. This is most clearly visible in the European-oriented terms he uses, such as "architecture" or "music" to describe the Asante culture, or when he entitles a chapter "Constitution and Laws". Of course one can criticise such usage as the expression of Eurocentric perspective,[40] but given that he was the first European to spend a longer time in Kumase recording his impressions, his approach was quite remarkable for his time. He simply assumed European and African cultures were comparable. Even more strikingly, he assumed they were of an equal value that permitted them to be thought of, and described, using the same terms. The only exception is when he discusses Asante religion; here he entitled the chapter using the pejorative term "Superstitions".[41] In his understanding, this was the veneration of the many African deities, which – following European usage at the time – he refers to as "fetishes", though after all he calls them "subordinate deities".[42]

Bowdich's efforts to give a serious character to his work can be seen the book's appendix, which contains supplementary documentation including a list of the settlements to be found along the Asante trade routes, a list of the temperatures he measured, and a vocabulary list.[43]

If one ventures an interim assessment, based on Marie Louise Pratt's thesis about European travellers' accounts and their "imperial meaning-making" classifications of regions (peripheries) far from Europe,[44] then one can say of Bowdich's book – or at least of what has thus far been presented – that one can draw a clear conclusion. His work uses a perspective oriented to systematising the knowledge he acquired, but it was done without a specifically hegemonic intent to demean the other culture.

There are certainly passages in Bowdich where he contrasts a self-evidently superior European civilisation with conditions in the Asante Empire.[45] Moreover, he assumes the Gold Coast will become increasingly subjected to British influence (certainly more so than in 1817) and accordingly gives advice on how to expand British trade.[46] Yet his description of the Asante Empire is not significantly coloured by this orientation. Rather, his account reflects the political situation in the Gold Coast at the time, with a weak and almost non-existent British presence and influence and a completely entrenched and secure power in Kumase. During his stay, Bowdich was on alien territory and dependent on the goodwill of *Asantehene*, who he experienced as an awe-inspiring ruler of an imposing empire. Ensconced at the heart of the Asante Empire, my thesis argues, he allowed himself to be coopted by the imperial perspective, one he was not in a position to question and one which was reflected in his respectful description of the Asante Empire and its history. This is exemplified by the illustrations which accompany Bowdich's text.

Bowdich's illustration of the Asante *Odwira* festival

Mission from Cape Coast Castle to Ashantee contains 16 illustrations, which include illustrations in colour, two geographic maps, a city map of Kumase and inserted scores of African music. Although all these illustrations accompanying the text are remarkable, the coloured illustrations, also as decorative elements, are particularly striking. These were hand-colored aquatints[47] on separate pages and based on drawings from Bowdich's own hand.[48]

Four of these panels, containing seven images,[49] show interior and exterior views of buildings in Kumase. The palace of the *Asantehene*, as well as the houses of high officials at court, is depicted, and in their beauty and clarity of line, the illustrations reflect Bowdich's admiration for the 'ornamental architecture of Coomassie'.[50]

Bowdich's book is particularly well-known for its largest illustration, a transverse insert sheet folded four times[51] showing a festival in Kumase called *Odwira* – at which Bowdich and his companions were present. This annual festival took (and still takes) place on the occasion of harvesting the new yams, and which Bowdich and other Europeans in the 19th century called the "yams custom".

While this description suggests it is a harvest festival, and hence carrying religious overtones, the *Odwira* festival was actually one of considerable political significance, as it emphasised the unity of the Asante people and served to strengthen the Empire's foundations. The chiefs of the various subjugated peoples came to Kumase on this occasion, paid homage to the *Asantehene*, affirming their loyalty through oaths of allegiance, and by this also affirmed their subordination to the Asante Empire. The opulent festivities, lasting several days, in which thousands participated, also included legal proceedings and executions. In this way, the Asante Empire celebrated its power and demonstrated its juridical authority over its subjects.[52]

Bowdich's *The First Day of the Yam Custom* illustration shows a procession of these subject chiefs appearing in Kumase before the *Asantehene* Osei Bonsu and the high officials at court. In this print, the chiefs process from left to right, together with their respective entourages, past the *Asantehene.* He sits between two trees under an umbrella symbolising the unity of the Asante people, though he almost disappears among all the figures shown.

Bowdich included a plethora of details while also showing individuals or groups representative of the Kumase court and court society, and accompanying the print is Bowdich's text describing the individuals and groups. In the lower left corner, a group of military leaders (called "captains") dance and fire off guns. To their right, a young chief is carried by a slave, and farther right, another chief, on foot, rests his hand – one laden with gold – on the head of a slave boy. A high official of the court, smoking, is carried in, lying on what looks like a stretcher or chaise. Bowdich also shows the *Asantehene*'s criers, who, dressed in fur hats, sit as a group on the ground, and to their right, a seated group of 'children of the nobility' with their eunuch minder. At the right-hand edge of the print, one finds Muslims living in Kumase, identifiable by their turbans.[53]

Bowdich's detailed description of the procession – he notes clothing, adornments, weapons and hand gestures – shows an ethnographic interest, one of importance for my argument. Bowdich's presence at this ceremony, and his description, reflects a perspective which ethnographers call that of a participatory observer. On the one hand, Bowdich is a stranger, an outside observer; on the other hand, he is participating in the events as a guest, and the significance of the event is changed by his presence.[54] This is even expressed in Bowdich's sketch, as he integrates himself and his two remaining companions into the picture: three Englishmen sit to the right of the *Asantehene.*[55] Behind them stand two uniformed white soldiers who had accompanied them, and to the left, on either side of the *Asantehene*, large Dutch, Danish and British flags wave (a smaller British flag also waves immediately above the white men). These were the flags of those European nations which maintained trade relations on the Gold Coast with the Asante Empire. One can conclude from this scene that the mighty *Asantehene* Osei Bonsu used the presence of the British at court to demonstrate his supremacy – not just over his subjects but over Europeans.

Bowdich clearly was deeply impressed by the ritual he observed, as well as its underlying intent to express the hegemonic position of the *Asantehene*: it is directly reflected in the drawing Bowdich made that was the basis for the print.[56] In his text, Bowdich emphasises his sensory impressions that seem to have nearly overwhelmed him. He describes the noise created by the music and cries as 'awful and distressing', and that the hangmen who appeared, with skulls, in the middle of the picture, had orchestrated their appearance in various ways, 'some with the most frightful gestures'. He seems worn out by 'the splendor' on display, by 'the tumult', the shooting of guns and the music.[57] The verbal description of the process serves as a counterpart to the illustration in which the variety of all the objects and instruments on display, as well as the movements of the dancing and shooting people, evoke an impression of the dramatic character of the scene.

Whether Bowdich intended it or not, the interplay of image and text send a clear message, one his book also carried back to Europe: the awe-inspiring power and size of the Asante Empire.

Publication history and reception

Bowdich's documentation of the Asante Empire ended once the illustrations and text were completed. The dissemination of his findings, however, would only come through the publication of his work, and according to Felix Driver, the reputation of explorers and travellers in the 18th and 19th centuries only partly came about as a result of their own reporting.[58] No differently than today, it was the printing of a work and its ensuing reception by various groups in society which determined the success of a book and the fame of its author. Differing motivations lay behind the various subsequent publications of Bowdich's account, and the reception after 1819 was inconsistent, as will be discussed somewhat cursorily below. Particularly important is that the unusual feature of the work already noted, namely that the implicit perspective was that of the Kumase court, was one which was reflected in its reception.

Mission from Cape Coast Castle to Ashantee was published in London in 1819 by John Murray. This meant it was superbly positioned in the book market at the time, and not just because London had, since the late 18th century, developed into a centre of the publishing and book trade.[59] It was also because John Murray, a London company established by a Scotsman in 1768, and 'London's most distinguished publisher'[60] by the early 19th century, supported new developments in publishing, in particular inexpensive editions in relatively large quantities. These were meant to satisfy the thirst for knowledge of an expanding reading public.[61] Particularly relevant here was that John Murray published the accounts of many important explorers of Africa in the course of the 19th century, starting with Mungo Park, through John and Richard Lander, and including David Livingstone.

So at John Murray's, now run by the second generation, Bowdich's work was in good hands. A total of 750[62] copies were published in a sumptuous edition only affordable by well-to-do readers, and no costs were spared in producing the book's illustrations. Bowdich's drawing of *The First Day of the Yam Custom* was transformed by professional copperplate engravers into a print, as one can see from the legend 'Engraved by R. Havell & Son' under the plate. Robert Havell and his son ran a printing shop in London and were famous for their craftsmanship in creating copper prints and aquatints during the first half of the 19th century.[63] Bowdich's drawing was transformed using an aquatint process – a form of intaglio printmaking technique that creates pictures with a gradation in the grey tones,[64] thus clearly separating them – and it once printed was then coloured. This technique was widely used in England for book illustrations, particularly in the late 18th and early 19th centuries, and the spectrum of motifs depicted was quite broad. A survey of illustrated books of this era explicitly notes that "foreign lands"[65] books were frequently illustrated. One can conclude from this that the visual impressions of West Africa conveyed by Bowdich in a volume issued by John Murray were reproduced in a manner and style typical for his times.

Bowdich's book was reviewed in England soon after its publication, and evaluated in an incisive manner. Extensive reviews[66] in *The British Critic* and in *The British Review*, and *London Critical Journal* appeared already in 1819, and show how the book was received at the time by readers versed in natural history and literature. These two anonymous reviews express both praise and critique, and in the case of the latter, both negatively evaluate the form the travel account takes, apparently not just for its collage-style structure but also for the density of facts presented.[67] Nevertheless, both were also impressed by Bowdich's work, calling it 'by far the most important addition to our knowledge of Africa which we have received since the time of Bruce'.[68]

Both reviews support the thesis argued here that Bowdich had unexpectedly placed the grandeur of the Asante Empire at the heart of his account, and had conveyed this view to Europe. The reviewer in *The British Critic* reproduces Bowdich's sympathetic description of the *Asantehene* Osei Bonsu:

> The present king is represented as an amiable and also an able man. He has increased his dominion, his prerogative, and his revenue. He has shewn himself inclined to humanity, by limiting the human sacrifices at his mother's funeral . . . he is courteous, inquisitive, ambitious, and just. Surely these are all characteristics of a great king.[69]

However, his colleague at *The British Review* does refer to the barbaric customs at the Kumase court, meaning the human sacrifices Bowdich notes,[70] arguing that in the religion of the Asante, 'if religion it can be called', and the African slave trade, one finds the cause 'of that inhumanity which has ever been the opprobrium and the calamity of Africa'.[71] Nevertheless, with reference to Asante architecture, which Bowdich portrays as vastly superior to that of other African peoples, the reviewer also accepts Bowdich's implicit argument about Asante culture,[72] writing that 'the representations given [in Bowdich's book] convey an idea of a higher degree of civilization than we had imagined to have existed in any part of Africa'.[73]

In contrast to these largely positive reviews, a scathing review appeared in 1820 in *The Quarterly Review*[74] ridiculing Bowdich's book. Its reviewer questions the veracity of his account in a sarcastic and incredulous tone, and demeans it as a collection of 'very silly stories'.[75] Critique of Bowdich's behaviour during the Asante expedition, in particular in removing and replacing its experienced leader Frederick James[76] is mixed together with this negative evaluation. Yet even this dismissive review, today ascribed to John Barrow, Second Secretary to the Admiralty and an influential sponsor of British expeditions,[77] fits into the argument made in this paper. Barrow, who had travelled in southern Africa and had written a highly praised account of his journey,[78] makes fun of Bowdich in his review, remarking that he had painted the Asante Empire 'in such glittering colours' as to make it appear to be the 'true El Dorado'.[79]

So it is ironic that this reviewers' refusal to give the account credence is rooted precisely in Bowdich's depiction of the splendour and lustre of the Asante court.

Not entirely without reason, Barrow criticises Bowdich for having been so fascinated by the person of the *Asantehene* Osei Bonsu that he almost forgot 'to give his readers any information concerning the "lower orders" [of the Asante Empire]'.[80] Though Barrow is not named, his authorship of the 1820 review in *The Quarterly Review* was probably known; he was among the most renowned authors of this periodical – one, interestingly enough, also published by John Murray. Barrow's reputation as a specialist on the subject of exploration and expeditions was based precisely on his reviews for *The Quarterly Review*, through which he had considerable influence over the public perception of such travels.[81] In a reply to Barrow's remarks, Bowdich defended himself against the charges raised, which he saw as pure viciousness on the part of the reviewer.[82]

In fact, the effect of this review must have been considerable. *The Quarterly Review* was regarded as the mouthpiece of the Tory establishment, its authors recruited from influential parliamentary, church and university circles.[83] This periodical also had large print runs, which meant that it had multiplier effects on public opinion: more than 12,000 copies of the issue with Barrow's review were sold, and it was certainly read by many more people than that.[84] One can therefore assume that Bowdich's reputation was considerably tarnished in scholarly circles – and well beyond them – by Barrow's negative assessment.[85]

This episode also makes clear that on the long path of disseminating research results – from their genesis through publication and to their reception by the broad public – men like Barrow play an important gatekeeper function. They help determine whether or not a scientific work should be regarded as credible, and whether a young researcher like Bowdich should be accepted into the ranks of the scientific establishment.[86]

It is difficult to assess what the long-term effects of Barrow's review were. But it is noteworthy that Bowdich's book was only republished in 1873, at a time when the interest of the British public had turned to West Africa in the context of the Third Anglo-Asante war. Bowdich's daughter, Tedlie Hale, put out the book in a small, inexpensive edition which lacked the appendix and, with one exception, had no illustrations.[87] The plate showing *The First Day of the Yam Custom* is missing in this version, along with its accompanying verbal description, though the gap is not noticeable in the text.

Bowdich's book, however, proved to be of great interest in other European countries during the 19th century, as can be seen from its many translations and editions. It was published already in 1819, the year the original appeared,[88] in French in Paris, and in German in Jena. In the following year, two more German editions appeared, printed in Brünn and Weimar, and 1820 saw Danish and Dutch translations being published in Copenhagen and Amsterdam, respectively. Further German translations appeared in 1825 and 1826, from two different Viennese publishing houses.[89]

Given the available bibliographic information, an interpretation of this publication history is only possible in part, though it is not without significance. The first striking detail is that these later editions were either abridged versions or they were assembled from excerpts taken from the original. In the French translation, for example, the appendix is missing, as are the illustrations (with the exception of the general map of West Africa), and both the sequence and number of chapters were changed. The 1819 (Jena) and 1820 (Brünn) German, as well as the Danish, translations, carry the note 'edacted and excerpted' in library catalogs. The Dutch translation is only 165 pages long, from which one can conclude that it too must have been a highly abbreviated version of the original text. The 1825 Viennese version, in turn, appeared in three small volumes without illustrations or appendix, and with a division into chapters which diverges from the original.

Such details suggest that unlike in the case of the English original, these translations were intended for a broader public which demanded reading material at lower prices. Nevertheless, the readership certainly included various social strata. In a copy of the French translation available at the University of Basel library, one finds a note that it was given as a gift by Councilman Peter Merian to the library of the Natural History Museum in 1841.[90] In this context, one can assume the European readership for Bowdich's book was educated and interested in natural history. The translations which appeared in Copenhagen and Weimar in 1820, as well as the small 1825 Vienna volumes, appeared in series with titles such as 'Library of the Latest Expeditions'. Such series evidently appealed to a broad public who were interested in reading about alien and exotic lands and the peoples who lived there.[91]

Bowdich's findings about the Asante Empire spread not merely through new editions of his own work but also through the work of other authors who referred to him. Their works appeared in different genres, and reading them gives insight into the multi-faceted reception Bowdich's book enjoyed. It also allows one to obtain an idea of what the image of the Asante Empire was in the 19th century and how it was inscribed into European consciousness as part of a classic 'repertoire of interpretation'.[92]

References to Bowdich's work can be found in a work published in 1841 in Leipzig and bearing the title *Malerische Reise in Asien und Afrika*. This was an early popular science book which contained summaries of the accounts of well-known travellers, ordered in various geographic categories, and presented in both text and images. The two authors refer extensively to Bowdich's book in the section entitled 'Guinea', particularly the reception ceremony at the Kumase court (which included dancing military chiefs) but also the illustration of the bedroom of the *Asantehene*.[93] In the section containing the illustrations, one finds the corresponding images from Bowdich's work, miniaturised and placed between images from the accounts of other African travellers.[94]

The connection to Bowdich is more distorted in Heinrich Heine's autobiographical *Geständnisse* (Confessions), which appeared in 1854, namely as the basis for Heine's ironic and jocular reflections about human nature. Heine tells of a 'King of the Ashantis, of whom he had recently read in a travel description much that was delightful', who had, in this travel description, wanted to be drawn as a white man. Heine used this episode, which he invented, to come to the conclusion that it was not only 'the Negro king' but everyone who 'wishes to appear to the public in a different color than that which exists'.[95] Heine's laconic and matter-of-fact reference to Bowdich would seem to indicate his work was read among the educated classes in German-speaking Europe.

The reception of *Mission from Cape Coast Castle to Ashantee* even extended beyond Europe. Essential facts about the Asante Empire made their way into Jorge Issacs's *Maria*, a Columbian romantic novel published in 1867 and counted among the best-known works in Latin American literature.[96] The biography of an African maid is integrated into the story of the two main characters in the novel. This maid comes from the Asante Empire, and is the daughter of an important military leader; her character is introduced in language which hearkens back to Bowdich. Here one has to assume Isaacs had his information from the Italian historian Cesare Cantú, whose world history extensively cites Bowdich.[97] Isaacs writes:

> Magmahú was from the time of his youth one of the most respected commanders of the army of Achanti, one of the mightiest states in West Africa. He proved his daring and his experience often enough in the wars he fought for King Tuto Kuamina against the warlike Achimis. . . . He was the father of the victory over all the tribes of the coastal area.[98]

The history of the reception of Bowdich's work in the 19th century, as incomplete as it is presented here, permits key insights into European knowledge production about the Asante Empire. It is clear that the reception of the work, whether directly in the book reviews or indirectly by the references made to it by other authors, always reflects Bowdich's own focus on the royal court at Kumase as well as the hegemonic position of the Asante Empire in the Gold Coast. As many-layered as Bowdich's book was with respect to its contents, his core message was clearly understood in the 19th century: his portrait was of an unusual African polity, which was an impressively powerful state with an equally impressive ruler. As the book's illustrations were barely mentioned in the early reviews, and also were not included in the numerous translations and new editions, the reception took place largely without the help of "visuals", so that one needs to speak of a dissemination of imagined images, or of influential and effective mental pictures.

This respectful approach, following Bowdich's work, to the Asante Empire differs markedly from the political propaganda unleashed, with the help of the press, in England as part of the colonialist "Anglo-Asante Wars" of the 1860s and 1870s. In this, the Asante Empire, not coincidentally from the perspective of the missionary societies active on the Gold Coast, is portrayed as the embodiment of an African state built on brutality and contempt for human life. This was a discourse which diverged sharply from Bowdich's insights and literary sensibility; it was an alternate discourse about the Asante Empire waged in the press and meant to influence the politics of the day.[99]

In the 20th century, Bowdich's book has been used primarily as a source of information about living conditions in the pre-colonial Gold Coast. This has also been the aspect most emphasised in this work since African Studies emerged as an independent discipline in the 1960s. A reprint of the original was issued by the historian W.E.F. Ward in 1966, though with a reduced number of illustrations. This reprint has been repeatedly digitally reproduced and is thus available as a source document for an interested readership.[100]

It is not by accident that current standard works on the Asante Empire by Ivor Wilks (1975) and Tom C. McCaskie (1995) rely extensively on Bowdich's book.[101] In his *Asante in the Nineteenth Century*, Wilks depicts Asante rule in a manner which emphasises the political, economic and social structures of the Empire, an interpretation typical of the – at the time still new – discipline of African History. This discipline was driven by an emancipatory impulse to "give back" to the African continent a history it had been long denied. In the course of so doing, research interest focused on the formation of states and empires, and here historians thought they recognised comparable developments to those they knew from European history.[102] One needs to see Wilk's remarks about the Asante Empire in this context, as he draws on analytic concepts such as "bureaucracy", "mercantilism", and "modernization"[103] used in European historiography and sociology. Remarkably, this has certain parallels with Bowdich's own conceptualisations and language.

T.C. McCaskie, writing 20 years after Wilks, claims instead to want to go beyond the description of socio-political conditions so as to identify the ideological structure of the Asante Empire, or put differently, to identify the underlying ideological and religious factors which held it together.[104] Bowdich's depiction of the *Odwira* festival is thus central for him. In McCaskie's interpretation of *Odwira*, to which he devotes an entire chapter, he refers not just to Bowdich's text but also to the illustration of *The First Day of the Yam Custom*. One of the appendixes of McCaskie's book contains a reproduction of this plate, along with Bowdich's explanation of the festival and his description of the illustration.[105] Both Wilks and McCaskie cannot avoid using Bowdich's as a central source document, but McCaskie's book would be unthinkable without it. The question for today's reader is to what extent both works can be regarded as late echoes of the admiring view of the splendour and might of the Asante Empire which Bowdich provided more than 150 years ago. It implies nothing less than that a unscientific, implicit discourse has been integrated into the contemporary academic analysis.

McCaskie's work also shows that the ethnographic value as well as significance of *The First Day of the Yam Custom* illustration has been recognised in the 20th century. In his *Religion and Art in Ashanti* (1923, reissued 1927), the ethnologist Robert Rattray included Bowdich's entire text about the *Odwira* festival, and his description of the illustration, but did not reproduce the plate itself.[106] In the 1990s, by contrast, McCaskie recognised the value of the illustration as a source of information in and of itself.

Prior to this time, the image had been reproduced either entirely or in part in numerous publications. Philip Curtin's 1964 *The Image of Africa*[107] reproduced *The First Day of the Yam Custom* on the dust jacket, for example, pointing out the relevance of Bowdich's work in coining imaginary pictures of Africa. One can find a similar use on the cover of *Painting Africa White*, published in 1971, in this case showing only the British spectators at the *Odwira* festival (namely Bowdich and his companions), and thereby illustrating the main theme: the presence of white Europeans in Africa.[108] Even the relatively recent anthology *Themes in West Africa's History*,[109] edited by Emmanuel Akyeampong, is graced by Bowdich's illustration, elevating it and treating Bowdich's work implicitly as a classic in West African history. Finally, as a search for *The First Day of the Yam Custom* in the Web indicates, the public reception of this particular print continues: it serves as a standard illustration for topics related to the history of the Asante Empire, and more generally, to topics in African history.[110]

Conclusion

It is, in every respect, paradoxical – and not just with respect to the Asante Empire – to speak of "knowledge production". From the perspective of information science, knowledge is linked to the cognitive structures of human understanding, and thus does not exist in the material form of a book, a text or an image. In a book, data is presented as individual facts, or information is presented as data placed in relevant contexts. Additionally, the knowledge of an individual or a group is enlarged through the acquisition of information.[111] Still, the fuzziness in the term "knowledge production" may be mitigated somewhat by the type of analysis used in the humanities, since it blurs the boundary between data (which is generated from the most objective documentation possible) and interpretation. However, it is only by drawing this distinction that one can explain that one and the same scientific or artistic work is subject to different readings depending on era, location and the perspective used.

If one applies these insights to Bowdich's book, then it should have become clear that he by no means presented findings solely from a Eurocentric viewpoint. Rather, Bowdich strove – very much in the spirit of the mandate he had been given – to accurately document the living conditions he observed in the Asante Empire. He also conveyed an interpretation of the local political structures whose scope and significance even he perhaps did not recognise. He brought an image of the Asante Empire to Europe which, though it corresponded to the imperial self-image of the Asante elite, collided with the expectations of British scholars (and of those in influential political positions like Barrow) at the time, and which can only be called "colonial knowledge" in a highly paradoxical manner.

The immediate reactions to Bowdich's work, both positive and negative, were also marked by astonishment at what he reported. During the 19th century, the research establishment in England refused him recognition, but in other European countries, the reception crossed literary genres and thus reflected the vibrant character of the original and the many ways it could be read and interpreted. It is also thinkable that the nationalist discourse during the 19th century, particularly in German-speaking Europe, may have encouraged interest in the book. The end of European colonial rule over Africa may have also provided a frame of reference which allowed for more emancipatory historical research.

The history of the reception of Bowdich's work has not ended. Beyond using his work as a historical source, the symbolic significance of *The First Day of the Yam*

Custom illustration has been uncovered during the 20th century. As its use in various contexts shows, it can be simply understood to express the realisation that the African continent has produced great cultures. But like any interpretation, this one also depends on the perspective of the reader and the content of the work itself.

Notes

1 van den Bersselaar, 'Acknowledging Knowledge', 389–94. The article has been translated by John Bendix.
2 See, for example, Reichart-Burikukiye, 'Erinnerungsräume und Wissenstransfer', 11.
3 'Orientalism', in Ashcroft, Griffiths and Tiffin, *Key Concepts*, 167–9.
4 Pratt, Imperial Eyes, 5.
5 Barley, 'Africa', 272; McLeod, 'T.E. Bowdich', 82–3; McCaskie, 'Asante Origins', 129.
6 Fabian, *Out of Our Minds*, 8–9.
7 Gocking, *Ghana*, 25–7, 30.
8 Ibid., 22–3, 30.
9 Ibid., 30; Barley, 'Africa', 272.
10 Ward, 'Introduction', 11–12; Barley, 'Africa', 272; McCaskie, 'Asante Origins', 125.
11 In Bowdich, *Sai Tootoo Quamina.*
12 Ward, 'Introduction', 12; McLeod, 'T.E. Bowdich', 87.
13 Bowdich, Mission, 126–7.
14 Ward, 'Introduction', 44–6; McLeod, 'T.E. Bowdich', 88.
15 Gocking, Ghana, 33–4, 46.
16 Strickrodt, *Wild Scenes*, 26, 32. See Bowdich, *Reply to the Quarterly Review*, 90.
17 Ward, 'Introduction', 12–13, 16; McLeod, 'T.E. Bowdich', 87; Bowdich, *African Committee.*
18 Westby-Gibson, 'Bowdich'; Strickrodt, *Wild Scenes*, 21–6; McCaskie, 'Asante Origins', 127–8.
19 See Bowdich, *Mission*, 4–6.
20 McLeod, 'T.E. Bowdich', 88.
21 Sloan, 'Aimed at Universality and Belonging to the Nation', 13; Curtin, *Image of Africa*, 209–11.
22 Bowdich, *Mission*, part I, chap. I – VI, 3–158.
23 Ibid., part I, chap. I.
24 Ibid., part I, chap. II, 14–31.
25 For an example, see the description of the village of Acroofroom: Ibid., 15.
26 Ibid., 31–41.
27 Ibid., chap. III–VI, 42–158.
28 Ibid., 161–227.
29 McLeod, 'T.E. Bowdich', 89.
30 Bowdich, *Mission*, 161–2.
31 Ibid., 162.
32 Ibid., 162–227.
33 See in particular ibid., 165–7.
34 Wilks, *Asante in the Nineteenth Century*, 1–18.
35 Bowdich, *Mission*, 228–51.
36 McCaskie, 'Local Knowledge', 184–5.
37 Bowdich, *Mission*, 232.
38 Ibid., 250.
39 See, for example, *Reindorf, History of the Gold Coast and Asante*, 162.
40 See Joseph K. Adjaye, 'Ghanaian Historiography', 2–3.
41 Bowdich, *Mission*, part II, chap. IV.

42 Ibid., 262–72.
43 Ibid., 463–512.
44 Pratt, *Imperial Eyes*, chap. 2.
45 Bowdich, Mission, 275: 'I never felt so grateful for being born in a civilized country', and 338–9.
46 Ibid., 338–43.
47 Göhring (Atelier druckwerk, Basel), in discussion with the author.
48 They bear the note 'Drawn by T.E. Bowdich Esqr'.
49 Bowdich, *Mission*, part III, 3–8, 10.
50 Ibid., 304.
51 Ibid., 274.
52 For a detailed interpretation of Odwira, see McCaskie, *State and Society*, 144–242.
53 Bowdich, *Mission*, 275–8.
54 Schuster, 'Probleme', 624, 632.
55 McCaskie, *State and Society*, 269.
56 Ibid., 204–5.
57 Bowdich, *Mission*, 275.
58 Driver, *Geography Militant*, 9.
59 Carpenter, *Seven Lives of John Murray*, 12.
60 Bennett, 'John Murray's Family Library', 308.
61 Ibid., 307–9.
62 Written response received from the John Murray Archive, National Library of Scotland, Edinburgh, 22 June 2007.
63 Deane, 'Copper-Plates', 406.
64 Weber, 'Alte Drucke', 17.
65 Ray, *The Illustrator*, 29.
66 *British Critic*, 372–88; *British Review*, 340–71.
67 *British Critic*, 373; *British Review*, 371.
68 *British Critic*, 373.
69 Ibid., 379–80.
70 *British Review*, 351.
71 Ibid., 353.
72 Ibid., 362.
73 Ibid., 345.
74 *Quarterly Review*, 273–302.
75 Ibid., 273.
76 Ibid., 282–3.
77 'Quarterly Review', 4; Driver, *Geography Militant*, 31–2. For John Barrow's biography, see Fleming, *Barrow's Boys*, 1–12.
78 Barrow, *Account of Travels*; Fleming, *Barrow's Boys*, 5, 6–7.
79 Barrow, *Review*, 284.
80 Ibid., 289.
81 Fleming, *Barrow's Boys*, 7–8.
82 Bowdich, *Reply to the Quarterly Review*, 1.
83 'The Quarterly Review under Gifford'.
84 'Quarterly Review'.
85 See Fleming, *Barrow's Boys*, 7–8.
86 Driver, *Geography Militant*, 31–2.
87 Bowdich, *Mission.*
88 Voyage dans le pays d'Aschantie ou la relation de l'ambassade envoyée dans ce royaume par les Anglais (Paris: Librairie de Gide Fils, 1819) sowie Geschichte der Brittischen Gesandschaft an den König von Ashantee auf der Goldküste im Jahr 1817 nebst statistisch-geographischen Nachrichten über dieses Reich u. mehrere andere Theile d. innern Africa's (Jena: Bran, 1819).

89 Geschichte der Brittischen Gesandschaft an den König von Ashantee auf der Goldküste im Jahr 1817 nebst statistisch-geographischen Nachrichten über dieses Reich u. mehrere andere Theile d. innern Africa's (Brünn: Trassler, 1820); Geschiedenis van het britsche gezantschap, in het jaar 1817 aan den koning van Ashantee, in de binnenlanden van Afrika, in de nabijheid van de Goudkust gezongen: benevens eenige statistieke, aardrijkskundige en andere berigten, wegens dit rijk en andere naburige gedeelten des lands (Amsterdam: Van Kesteren, 1820); Beskrivelse over det Brittiske Gesandtskabs Reise til Kongen af Ashantee i Aaret 1817 samt statistisk-geographiske Efterretninger om dette Rige og flere andre Dele af det indre Africa, Magazin for de allernyeste og interessanteste Reisebeskrivelser (Kopenhagen: Goldin, 1820); Mission der Englisch-Afrikanischen Compagnie von Cape Coast Castle nach Ashantee, mit statistischen, geographischen und anderen Nachrichten über das Innere von Afrika, Neue Bibliothek der Erd- und Völkerkunde, vol. 21 (Weimar: Verlag des Gr. H.S. priv. Landes-Industrie-Comptoirs, 1820); Bibliothek der neuesten Entdeckungsreisen, vols. 18–20 (Wien: Anton Strauss, 1825; Wien: Kaulfuss und Krammer, 1826).

90 The book is catalogued here as Hx VIII 7.

91 See endnote 89.

92 Dejung, 'Oral History', 103.

93 Eyriès and Diezman, *Malerische Reise*, 62.

94 Ibid., Tafel 11.

95 Cited in Heine, *Memoiren und Geständnisse*, 94.

96 *The New Encyclopaedia Britannica*, 396.

97 Musselwhite, 'Colombia of María', 47.

98 Cited in Isaacs, *María*, 126–7.

99 Haenger, *Basler Mission*, 40–6.

100 See endnote 10.

101 Wilks, *Asante in the Nineteenth Century*; McCaskie, *State and Society*.

102 Adjaye, 'Ghanaian Historiography', 10.

103 See the review of *Asante in the Nineteenth Century*, 138; Haenger, *Basler Mission*, 7–9; McCaskie, *State and Society*, 13–16; Adjaye, 'Ghanaian Historiography', 10–11.

104 McCaskie, *State and Society*, chap. 1: Varieties of the Asante past.

105 Ibid., Appendix 1, 268–71.

106 Rattray, *Religion & Art*, 122–6.

107 Curtin, *Image of Africa*.

108 Lewis and Foy, *Painting Africa White*.

109 Akyeampong, *Themes in West Africa's History*.

110 See, for example, http://encyclopedia.farlex.com/Ashante; http://home.wlu.edu/~patchw/His_224/slideshows/1960_Decolonization%20of%20Africa_files/frame.htm#slide0105.htm; www.bbc.co.uk/worldclass/freedom_culture.shtml (accessed 29 May 2011).

111 Kuhlen, 'Information', 11–15.

Bibliography

Adjaye, Joseph K., 'Perspectives on Fifty Years of Ghanaian Historiography', *History in Africa: A Journal of Method* 35 (2008): 1–24.

Akyeampong, Emmanuel Kwaku, ed., *Themes in West Africa's History* (Athens: Ohio University Press, 2006).

Ashcroft, Bill, Gareth Griffiths, and Helen Tiffin, eds., *Key Concepts in Post-Colonial Studies* (London: Routledge, 1999).

Barley, Nigel, 'Africa: In the Shadow of the Enlightenment', in *Enlightenment. Discovering the World in the Eighteenth Century*, ed. Kim Sloan and Andrew Burnett (London: The British Museum, 2003), 270–5.

Barrow, John, *An Account of Travels Into the Interior of Southern Africa, in the Years 1797 and 1798*, 2 vols (London: T. Cadell, 1801).

———, 'Review of Thomas Edward Bowdich, Mission from Cape Coast Castle to Ashantee', in *Quarterly Review* 22, no. 44 (1820): 273–302.

Bennett, Scott, 'John Murray's Family Library and the Cheapening of Books in Early Nineteenth Century Britain, *Studies in Bibliography* 29 (1976): 139–66', in *The History of the Book in the West: 1880–1914*, vol. IV, ed. Stephen Colclough and Alexis Weedon (Farnham: Ashgate, 2010), 307–34.

Bowdich, Thomas Edward, *The African Committee: A Complaint of the Author's Treatment By That Body* (London: H. Long, 1819).

———, *A Reply to the Quarterly Review* (Paris, 1820).

———, *Mission from Cape Coast Castle to Ashantee, with a Descriptive Account of That Kingdom*, new ed. with Introductory Preface by Mrs. Tedlie Hutchison Hale (London: Griffith & Farran, 1873).

The British Critic, New Series 11 (1819): 372–88.

The British Review, and London Critical Journal 13, no. 26 (1819): 340–71.

Carpenter, Humphrey, *The Seven Lives of John Murray. The Story of a Publishing Dynasty 1768–2002* (London: John Murray, 2009).

Curtin, Philip D., *The Image of Africa: British Ideas and Action, 1780–1850*, vol. 1 (Madison: University of Wisconsin Press, 1973).

Deane, Ruthven, 'The Copper-Plates of the Folio Edition of Audubon's "Birds of America", with a Brief Sketch of the Engravers', *The Auk: A Quarterly Journal of Ornithology* 25, no. 4 (1908): 401–13.

Dejung, Christof, 'Oral History und kollektives Gedächtnis. Für eine sozialhistorische Erweiterung der Erinnerungsgeschichte', *Geschichte und Gesellschaft* 34, no. 1 (2008): 96–115.

Driver, Felix, *Geography Militant: Cultures of Exploration and Empire* (Oxford: Blackwell, 2001).

Eyriès, Jean Baptiste Benoît and August Diezman, *Malerische Reise in Asien und Afrika* (Leipzig: Baumgartner, 1841).

Fabian, Johannes, *Out of Our Minds. Reason and Madness in the Exploration of Central Africa* (Berkeley: University of California Press, 2000).

Fleming, Fergus, *Barrow's Boys* (New York: Grove Press, 1998).

Gocking, Roger S., *The History of Ghana* (Westport, Connecticut: Greenwood Press, 2005).

Heine, Heinrich, *Memoiren und Geständnisse* (Düsseldorf: Artemis und Winkler, 1997).

Haenger, Peter, *Die Basler Mission im Spannungsbereich afrikanischer Integrationsversuche und europäischer Kolonialpolitik* (Lizentiatsarbeit, Universität Basel, 1989).

Isaacs, Jorge, *María* (Norderstedt: Sonrrie, 2003).

Kuhlen, Rainer, 'Information – Informationswissenschaft', in *Grundlagen der praktischen Information und Dokumentation*, vol. 1: Handbuch zur Einführung in die Informationswissenschaft und -praxis, ed. R. Kuhlen, T. Seeger, and D. Strauch (München: Saur, 2004), 1–24.

Lewis, Roy und Yvonne Foy, *Painting Africa White. The Human Side of British Colonialism* (New York: Universe Books, 1971).

McCaskie, Tom C., *State and Society in Pre-colonial Asante* (Cambridge: Cambridge University Press, 1995).

———, 'Asante Origins, Egypt, and the Near East. An Idea and Its History', in *Recasting the Past: History Writing and Political Work in Modern Africa*, ed. Derek R. Peterson and Giacomo Macola (Athens: Ohio University Press, 2009), 125–48.

———, 'Local Knowledge: An Akuapem Twi History of Asante', *History in Africa. A Journal of Method* 38 (2011): 169–92.

McLeod, Malcolm, 'T.E. Bowdich: an early collector in West Africa', *Collectors & Collections: The British Museum Yearbook* 2 (1977): 79–104.

Musselwhite, David, 'The Colombia of María: "un país de cafres"', *Romance Studies* 24, no. 1 (2006): 41–54.

Pratt, Marie Louise, *Imperial Eyes. Travel Writing and Transculturation* (London: Routledge, 1997).

'Quarterly Review 22, no. 44 (January 1820)', in *Quarterly Review Archive* 22, no. 44, Index (1820), ed. Jonathan Cutmore, www.rc.umd.edu/reference/qr/index/44.html (accessed 10 May 2011).

'The Quarterly Review under Gifford: An Overview', in *Quarterly Review Archive*, ed. Jonathan Cutmore, www.rc.umd.edu/reference/qr/founding/intro.html (accessed 16 May 2011).

Rattray, Robert Sutherland, *Religion & Art in Ashanti* (Oxford: Clarendon Press, 1927 [1923]).

Ray, Gordon Norton, *The Illustrator and the Book in England from 1790 to 1914* (London: Oxford University Press, 1976).

R. C. C. L., 'Review of *Asante in the Nineteenth Century. The Structure and Evolution of a Political Order*, by Ivor Wilks', *The Journal of African History* 17, no. 1 (1976): 137–9.

Reichart-Burikukiye, Christiane, 'Erinnerungsräume und Wissenstransfer. Die Erfindung Afrikas', in *Erinnerungsräume und Wissenstransfer. Beiträge zur afrikanischen Geschichte*, ed. Winfried Speitkamp (Göttingen: V&R unipress, 2008), 11–34.

Reindorf, Carl Christian, *The History of the Gold Coast and Asante. Based on Traditions and Historical Facts Comprising a Period of more than Three Centuries from about 1500 to 1860*, 2nd ed. (Basel: Basel Mission Book Depot, 1951).

Safra, Jacob E., chairman of the board, *The New Encyclopaedia Britannica, Micropaedia*, vol. 6, 15th ed. (Chicago: Encyclopaedia Britannica, 2007).

Schuster, Meinhard, 'Probleme des Beschreibens fremder Kulturen', in *Beschreibungskunst – Kunstbeschreibung. Ekphrasis von der Antike bis zur Gegenwart*, ed. G. Boehm and H. Pfotenhauer (München: Wilhelm Fink Verlag, 1995), 616–32.

Sloan, Kim, '"Aimed at Universality and Belonging to the Nation": The Enlightenment and the British Museum', in *Enlightenment. Discovering the World in the Eighteenth Century*, ed. Kim Sloan and Andrew Burnett (London: The British Museum, 2003), 12–25.

Strickrodt, Silke, *Those Wild Scenes. Africa in the Travel Writings of Sarah Lee (1791–1856)* (Berlin: Galda + Wilch Verlag, 1998).

van den Bersselaar, Dmitri, 'Acknowledging Knowledge: Dissemination and Reception of Expertise in Colonial Africa', *History in Africa* 33 (2006): 389–93.

Ward, W.E.F., 'Introduction' to *Mission from Cape Coast Castle to Ashantee, with a Statistical Account of That Kingdom, and Geographical Notices of Other Parts of the Interior of Africa*, by Thomas Edward Bowdich (London: Routledge, 1966), 9–60.

Weber, Bruno, 'Alte Drucke unter der Lupe', *Sammeln* 9, no. 2 (1985): 12–21.

Westby-Gibson, John, 'Bowdich, Thomas Edward (1791? – 1824)', in *Dictionary of National Biography*, rev. by Felix Driver (Oxford: Oxford University Press, 2004), online edition, May 2011, www.oxforddnb.com/view/article/3027 (accessed 22 July 2011).

Wilks, Ivor, *Asante in the Nineteenth Century. The Structure and Evolution of a Political Order* (London: Cambridge University Press, 1975).

7 Tropical soldiers?

New definitions of military strength in the colonial context (1884–1914)[1]

Heinrich Hartmann

At the end of the 19th century, the demographic discourses of many countries in Europe began to differentiate, and complex practices and approaches emerged with respect to population policy. The colonies, and their populations, were only very indirectly a part of this European demographic self-definition. In fact, a mercantilist concept of population continued to exist in the colonies, a concept that saw a large population as the basis for, and expression of, economic strength and political and military power.

The policy of colonial expansion, however, resulted in the growing importance of a different problem for the Europeans: European settlers and, still more, European soldiers suffered under the changed living and deployment conditions. They were reflected in dramatic morbidity and mortality rates mainly in the European tropical colonies (as shown for Dutch Batavia, French West Africa or the British West Indies),[2] with an especially high rate of morbidity within the often poorly immunised military population. Many soldiers arrived for a short stay, had travelled in salubrious conditions for weeks and months, had little experience with local adaptation strategies and, for these reasons, often did not develop body's defences, in the same way settler and permanent populations did.[3] John McNeill has argued that maintaining colonial empires, therefore, faced very different 'costs in manpower' for the colonial powers, which, at times, helps to explain geopolitical reshapes of these empires.[4]

This essay will not add to the literature on military morbidity in certain colonial parts of the world, nor will it take up McNeill's arguments about the agency of mosquitos and germs in shaping the Empire itself. Instead, it will seek to analyse the military response to that threat, and replace the debate of military fitness as a core concept for the understanding of demographic discourses in European populations in the late 19th century. The military strength of the European troops was no longer defined exclusively by the size of the army corps sent out, but also by their ability to adjust to the varied geographic, climatic and sanitary environments of their deployment. The question of whether or not soldiers could "acclimatize" to a supposedly hostile environment quickly evolved as a major battlefield of ongoing anthropologic discussions about the origins of mankind. At the same time, these debates did not enter into a vacuum but linked to a wide and more general discussion about the existing military fitness criteria in the European armies, themselves dependent on a wide range of demographic discourses in Europe. Discussing a separate set of criteria for military service in the colonies indirectly called into question the anthropological European self against the backdrop of colonial experiences and transformed the colonial situation into an experience of an enduring socio-scientific crisis and imperial helplessness.[5] Colonial contexts led to a reconfiguration of the scientific criteria used to examine fitness and of the social practices associated with them.

In recent years, the colonies have been increasingly understood as laboratories for the European powers. Dirk van Laak, for example, has drawn attention to the significance of the colonial territories as a sphere of activity for European scientists and engineers, and thereby emphasised the importance of the colonial contexts for scientific communities and individual careers.[6] Yet the colonies were not merely meant to serve as a "feasibility space". To a greater degree, they were a kind of "resonance chamber" for many scientific controversies that had their origins in Europe. Scientific discussions were reflected there, and their societal importance was illuminated from a different perspective. The reverberations produced by the colonies in this respect were altered not only by various political contexts, professional interests and institutional dynamics, but also, and especially, through the confrontation with unexpected observations.

Colonial population discourse was imprecise and characterised by a general lack of knowledge and data, shifting between the two poles of a hostile and thus underpopulated tropical environment, and a growing Malthusian fear of the uncountable colonial livestock. Definitions of colonial suitability and fitness thus fed into this picture and replaced a more accurate statistical discourse about the demographic dynamics in the colonies. These shifts in demographic thinking are most readily grasped in the reconfiguration of the European population discourses themselves. However, the history of demographics as colonial history has thus far barely been explored.[7] Whereas we have seen recently more and more research on the colonial medicine, statistical and demographical knowledge as a "tool of empire" has been widely neglected. I will argue in the following section, that statistical knowledge – or the lack of this knowledge – related quite directly discourses on "colonial" fitness to social and demographical debates inside Europe. Despite the fact that arguments for colonial engagement by European nations were always guided by the recurrent theme of population growth, the specific significance of these debates for scientific practices in Europe has been little investigated to date.[8] Andrew Zimmerman has noted the degree to which, in Germany, medical and racialised anthropological research – a demographic research field of growing significance – asserted itself in the academic landscape as an alternate model to humanistic traditions.[9] In this respect, medical and anthropological knowledge about the colonies shaped a new "epistemic community" that was organised in far-reaching transnational networks.[10] But it also called into question some demographic shared assumptions, as European anthropologists and physicians also came in contact with the fundamental discrepancy between their own self-perceptions, often racially motivated, and the negative experience of what were often dramatically high mortality rates among European soldiers in the colonies.

Especially, military strength, which had become a key factor in demographic discourses in the 19th century, now seemed no longer universally valid and definable.[11] The present essay intends to present the various levels of debate that linked the issue of pathologising the tropics and constructing a criteria catalogue to determine fitness for service in the tropics on the one hand, with defining and scientifically discussing key demographic categories in the European countries on the other. By doing this, it also tries to historicise the pathological perception of "tropical environments" by replacing these discourses in the context of European degeneration fears.[12] It takes a comparative look on material of the European colonial powers, taking a closer look to transnational networks of tropical and colonial military medicine.

In a first step, the general discourse on fitness for military service in the tropics, in the wake of the massive colonisation of sub-Saharan Africa from the 1880s to World War I, is briefly sketched. Then the state of statistical and demographic research on recruits in Europe is presented, as it provided the backdrop for debates about soldiers in the colonies. The third section discusses the practices and political mobilisation of military service by soldiers in and from the tropics. Finally, the debates over categorising fitness for duty in the tropics are summarised, as it is a point where demographic and anthropological knowledge and colonial "experiential spaces" intersect.

Colonialist discourse at a time of crisis: the fitness of Europeans for military service in the tropics

Europeans, even in the early phases of colonial rule, were confronted with the fact that in many parts of the world, their equipment and medical knowledge often was inadequate. In the early colonial territories, whose purpose actually was a mercantilist spreading of natural resources and the settlement of national populations,[13] mortality among European settlers and soldiers rapidly reached an alarming level. In some areas, including Batavia (Java), colonized by the Dutch in the 17th century, mortality was so high that the colony became known as the "Dutch graveyard".[14] This permanently compromised the ability to establish military control, let alone of create functioning governing and administrative structures.[15] Adaptive strategies and opportunities to receive medical care developed only slowly among the European settlers.

The European researchers who crisscrossed large stretches of Africa early in the 19th century in search of the sources of the Nile, of the Congo and of colonial wealth through natural resources, had similar experiences. Johannes Fabian has shown the extent to which the experience of tropical Africa as unwholesome was part of the edifice of knowledge gained by European researchers.[16] These researchers often set out on their journeys with a simplistic understanding of disease, and that made Africa a place where fever was a constant companion in life and a common metaphor; dealing with fatal threats was part of daily experience.[17] The search for adequate instruments and "prostheses" thereby became an omnipresent matter of survival, leading travellers to fall back not only on the most recent medical findings but also on cultural techniques used by Africans themselves.[18] At the same time, it pointed the researchers towards the European cultural techniques of "hygienic routines", including those outside the corresponding social frames of reference of their native countries. Given the widespread ignorance of the environment and unreliable medical knowledge, standardised forms of hygiene, such as hand washing, good nutrition, regular sleeping were interpreted as the only reliable attitudes that helped to survive in a supposedly hostile surrounding.

This pathologising of the African continent became more acute with the colonisation of sub-Saharan Africa after the 1880s. The few European settlements established in this region before the 19th century had mortality rates, in some instances, of up to 90 percent in the first seven years of their existence.[19] It was just this dramatic situation that provided the main motivation for the efforts of European physicians to expedite the establishing of tropical medicine as a specific, productive field.[20] The lack of knowledge was omnipresent in these attempts: medical history has shown that contemporary diagnoses misinterpreted not only causes and effects of pathologies, but also struggled with defining disease patterns correctly.[21] In many cases, this went hand in hand with a general lack of medico-statistical figures, be it of their own or of comparable European colonial populations.

Demographic arguments played an important role at the beginning of this colonisation process in sub-Saharan Africa and even before in forging arguments for an engagement in adventurous colonial endeavours. Most conspicuous were the Malthusian arguments, which repeatedly regarded Africa as a kind of "overflow basin" for the societies of industrialised Europe and thus made colonialism an inevitable consequence of European economic development. In the German case, the masses had chosen to emigrate to the United States in the previous decades. Now, by populating what appeared to be unimproved land in the German colonies, they were to be given new space to live in without being lost forever to the German Reich. Though these arguments did not even begin to correspond to the demographic reality,[22] advocates of a new colonial empire of this kind found conclusive legitimation for their arguments in such discourses.

Thus, Hans Ziemann, medical adviser to the German colonial government in Cameroon, argued at the 14th International Congress for Hygiene and Demography in 1907 that

> The population mass continues to swell in Europe. The need to find new land for this teeming population is becoming increasingly urgent for us. More and more, the white race shows its eagerness to advance to the Equator. We want to rule over all of Africa, however, by virtue of the law of the higher race, in order that through trade, agriculture, and industry, and with the aid of a multitudinous native population, we may develop the land. . . .
>
> To the conquest of Africa by the white race appertains . . . that, to begin with, only the best human material must be sent out as trailblazers of European culture, healthy to the core in body and soul, and medically examined. Much nuisance is still being created by sending out unsuited private persons. Our effort must be to cultivate a stock of select pioneers whose colonizing characteristics may be inherited by their descendants.[23]

Yet the practical implementation of colonial policies and their associated settlement projects deflated optimistic expectations of this kind. The medical and hygienic problems associated with the tropics raised serious questions for physicians and anthropologists with respect to the ability of Europeans to adapt to the very different climatic, social and biological paradigms in the tropics.[24] At the same time, these problems drew attention to the lack of reliable statistical and demographic data needed to address such questions.[25] Lacking knowledge thus opened spaces for wide-reaching interpretations that quickly showed a negative bias. "The colonies", in demographic research, became a term associated with crisis, particularly in Germany with its slight colonial experience, and brought the blending of superseded categories and thought patterns into plain view.

In this context, the new direction medical and anthropological research began to take in the 1890s seems especially significant to me, as does the increased scientific interest demonstrated between 1900 and 1914, especially at the transnational level. In particular, it focused on categorising the fitness for settling the tropical colonies. This, combined with the general popularity of demographic discussion, contributed decisively to the efforts to reach a scientific determination of the European's limits.[26]

The first elements of such a discussion go back to the French naturalist Guillaume T. Raynal (1713–96), who in the late 18th century held the view that northern Europeans, in particular, had a lower chance of survival in tropical countries, owing to their biological and racial dispositions. The further south the person in question was born, however, the more likely he or she was to be fit for the tropics.[27] Further, Raynal was of the opinion that women were far less affected by the general dangers of the tropics. In so arguing he laid the groundwork, *ex negativo*, for a new scientific categorisation of the white colonist who was not a priori determined as the "male European". Raynal descriptions showed an astonishing sense for patterns of differentiation among Europeans and underpinned their capacity to react and to adapt.

About a hundred years later, the French physician Orgéas came to the conclusion that important demographic markers, notably mortality and fertility, were significantly worse among the colonists than they were in the French metropolis, and argued much more on the ground of closed categories of a European "race".[28] This first effort to link the medical particularities of the tropics to a new, racially charged image of the European was followed by numerous other studies. The researcher S. Friedmann argued that the status of the tropics as harmful to health was related to the relatively low oxygen content of the tropical air, due to its high humidity. This led, he believed, to a poor blood flow to the brain, with concomitant hyperactivity in the abdominal and genital areas, which the European could not cope with.[29] Though short-term acclimatisation was possible for Europeans, long-term settlement was unthinkable, as

> among descendants of the third and fourth generation of European immigrants [one notes] a certain atrophy of the intellect, a lack of joie de vivre and drive, and even a bodily weakness, such that they definitely no longer resemble their European ancestors.[30]

Only the Jews provided an exception, he said, for in racial terms, they had an ascribed capacity for social and biological adaptation. Even these few examples make clear the extent to which the medical and anthropological discussions about European colonists can be understood as a mirror and echo of a newly emerging racial differentiation among populations and ethnically defined groups in Europe itself. At the end of the 19th century, such discourse had a fixed place on the research agenda of European scientific communities.

In the context of growing colonial ambitions, however, and in particular those of the German Reich after 1884, new interpretations and reconceptualisations of the question of fitness for the tropics became increasingly urgent. Ambitious physicians, over-represented in the colonialist movements in Germany, did not want to accept that the new colonies in East and West Africa were fundamentally uninhabitable for Europeans.[31] Purely anthropological paradigms evolved, often supported by cultural arguments. A year after Germany began its policy of colonisation, geographer and journalist Hugo Zoller (1852–1933) argued that from a geographical point of view, the settling of the tropics thus far had proceeded in nearly the worst fashion possible. Instead of seeking out the most inhabitable areas, the colonists, for trading reasons, had established settlements in the unhealthy river deltas.[32]

In the eyes of ethnologist Reinhold Pallmann (1834–96), such adverse conditions rendered it impossible to make clear statements about the racial ability to acclimatise, because, in the migration of ethnically diverse groups, such conditions always resulted in negative outcomes.[33] Certain diseases routinely broke out after resettlement to a different climate zone, and this had been no different when Indian soldiers of the British Army were relocated to Egypt.[34] It was "a fable" that the tropics were uninhabitable. Rather, it was necessary, through myriad hygienic, cultural and social measures, to make these areas accessible for settlers of the European race.

Against the backdrop of such discussions, a sense of agency formed in medical and social anthropology circles, one seemingly correlated with the social hygiene movements in the colonising nations, but it was at the same time also understood as a cross-boundary, transnational scientific project.[35] The need to intervene provided the basis for scientific missions to the tropics, with which European physicians often tried to create new room for manoeuvre as well as to find legitimation. Here the best-known example is Robert Koch, who used his research on malaria and sleeping-sickness in the tropics as an opportunity to draw attention away from the disappointing results of his work on tuberculin in Germany.[36] Koch, whose expeditions included a military physician, used these experiences not just to make his mark in research but, as in his investigations of malaria, also to develop theories in Europe about the origin, course of the disease, and how to combat it. In doing so, he gave important impetus to tropical medical research in Berlin and far beyond Germany. Colonial medicine thus never was detached of scientific discourses in the homeland but often played a complementary role in the career of physicians.

Categorisations of fitness and references to discourses on the demographic crisis in the home countries came to the tropics in the years around 1900. In contrast to the racist topos that had gradually become the dominant paradigm in anthropological research in the European home countries, those concerned with tropical hygiene became increasingly fixated on cultural and social explanatory models.[37] This notion of fitness depended as well on the new questions raised about psychic and emotional resilience. All post-1900 guidebooks for assessing fitness for the tropics see these criteria as central for evaluating the ability of Europeans to survive in the tropics. The physiological interaction between emotional strain and, for example, increased danger of epidemic infection was one of the main issues in medical research at the turn of the century. It became part of the catalogue of criteria for determining fitness for the tropics.

With growing frequency, this catalogue was defined with respect to lifestyle, in which, for example, the use of tobacco and alcohol, or even the wrong diet, was increasingly construed as a disqualifier.[38] At the same time, however, such a catalogue of criteria served to set Europeans apart from a local population that was regarded as culturally inferior. Against a backdrop of what were at times catastrophic experiences of morbidity and mortality on the part of white settlers, however, the biological argument of racial superiority could hardly be sustained, was called into question and in the end had to take a circuitous route by appealing to cultural values and regulated patterns of behaviour.

The tropical physician Hey attempted, in this regard, to educate the colonial settlers to take responsibility for themselves:

> Now, the emigrant will see matters with other eyes than before, and if he does not have the best intentions in his newly chosen home to look to and preserve himself there as well, to be "salt and light" to his environment, then I can only enjoin him with the utmost seriousness: stay where you are, for out here, sooner or later, you will degenerate, decay, be brutalized and dehumanized.[39]

The question of degeneration, one of the most-discussed controversies among German and European anthropologists in the early 20th century,[40] experienced a change in meaning in the colonial context. The changed climate and the lack of social discipline led, for the majority of the scientists, to the allegedly observable phenomenon of degeneration, most tellingly with the pathology of tropical neurasthenia, the so-called "Tropenkoller".[41] The causes of this degeneration lay not only in basic racial and biological dispositions but also in a combination of highly diverse explanatory patterns based on social hygiene and social anthropology.[42] This heterodox grid of arguments might be partially explained by different conceptions of heredity, deeply divided by the contemporary debates between "monogenists" and "polygenists".[43] But in the complex circumstances of colonial situations, many physicians did deviate from the paths of theory. The tropics became not just a site where biological differences were evident but also a space where the traditional values operating in human interaction were, in part, suspended, and it was only through strict disciplining of the individual that a complete moral collapse could be avoided.[44]

Nevertheless, among German scientists, an astonishingly fixed catalogue of criteria developed out of these combinations of arguments. According to one such catalogue, compiled by the physician Kohlstock in 1905, the person best suited for the tropics was

> a man of middling stature, aged about 24, with a healthy complexion, strong body build, and well developed. He should have a strong, firm musculature tempered by military service as well as through training and profession, and normal deposits of fat, and not be predisposed to any illness. His body organs, outer and inner, should be healthy and not have been subjected to any intrusive illness. Above all, he should possess a healthy, strongly beating heart and healthy, fully breathing lungs. He should always have a good appetite, a regular and normal digestion, and be of cheerful and quiet temperament, coupled with an energetic character.[45]

The development of such a catalogue of criteria may not have played much of a role in everyday life in the colonies, as there was no official regulatory body that would have screened the colonists according to such criteria, to weed out the unfit. However, such criteria gained far more relevance in borderline cases, as in the problem of risk assessment, on whose basis financial means were allocated and insurance policies finalised.[46] Precisely these questions of assessing health and economic risks, and the economic interests associated with them, provided one of the chief motives for the quest for international comparability of data and the experiences of other countries in their demographic surveys of what were, in part, considerably larger colonial populations.[47]

Despite all attempts to craft a positive definition of fitness criteria, the colonies remained a space defined by disease, and its dangers deterred potential settlers. As a result, the "tropics" were in danger of being lost to German settlement projects. The sole and original motivation for the colonial projects was, as the German tropical medicine specialist Külz put it, the exploitation of their resources to relieve the overpopulated home country. If it proved impossible for the German settlers to open up this area permanently, no alternative remained, in his eyes, to the 'improvement of the natives . . . to work for us'.[48] As a result, tropical medicine specialists in Germany increasingly argued for improving the hygiene of African societies (*Negersanität*, "Negro hygiene") in order to 'increase the number and productive efficiency of the inhabitants' [of the African continent].[49] New rationales of social engineering blended together with concepts of a presumed scientific racism. Nevertheless, these "missions" on the part of German doctors remained only a starting point. In its early phase, tropical medicine as a tool of the modern world for the development of Africa for the Europeans continued to be a significant theme.

A discourse of demographic crisis: military statistics in Germany and France before World War I

One should note that demographic knowledge at the turn of the century cannot be understood as an academically canonised science. The demographic discourses that affected public opinion at that time in various European countries owed their existence not to a solid research programme but to an eclectic multitude of methods, internal institutional dynamics and political legitimation strategies. They also fed on the strong transnational connections and relations[50] that characterised this scientific field after the International Congresses for Hygiene and Demography began in 1878, if not even earlier. These congresses provided an opportunity for military statisticians and military physicians to exchange information about statistical and demographic research on military recruits and about military statistics, and increasingly also were a venue for anthropological researchers. It was thus in the context of these congresses, and later also the International Medical Congresses that began in 1894, that the Commission for International Military Health Statistics was founded, and in this framework the international comparability of military statistics was to be encouraged through the use of a common categorisation scheme.

Demographic knowledge depended on generating a stable database. Efforts were already made in the United States in the wake of the Civil War to obtain an idea of the living conditions of the population, as reflected in military statistics. Similar efforts were undertaken soon thereafter in Great Britain and in other European countries.[51] The "recruit" as an object of study increasingly became the focal point of demographic interest, owing to two factors: the significant number of soldiers and the particularly detailed information that was gathered on the individuals.

At the same time, new paradigms were employed as guidelines. Since the time of Napoleon's mass armies, a close correlation between the military and the demographic discourses had emerged.[52] The drop in the supply of "soldier material", as well as the increasing deterioration of individual combat fitness, assumed to be influenced by a degeneration of industrial society,[53] linked demographic knowledge to the perception of military crises. In light of the declining birthrates in many European nations, population, as a reservoir of military clout, gained significance in public debates.[54] In particular, the falling birthrates in France, together with the defeat of 1870–71, played a decisive role.[55] They also fed into debates on theories on demographic degeneration.[56] Studies from other European countries, too, caused these interpretations to extend far beyond their intended national frames of reference.[57]

These networks of relationships, undoubtedly a significant part of the epistemic foundations of demographic research around 1900, gave central significance to the venues where the demographic information was acquired and where, in addition, the criteria for suitability for the tropics were defined: these were the medical exams for military service performed by the recruiting boards. The history of the induction of recruits, thus far unwritten, with the exception of a few initial attempts,[58] comes to the fore in this new perspective based on the history of ideas. As a result, this institution takes on the dimensions not only of a *rite de passage* in the mind of the militarised citizen but also of a central space in the exchange between science and social practices. That is also reflected in the negotiations regarding the establishment of new categories of physical, as well as psychological, suitability.

Military and political practices of recruitment in and from the colonies

Many aspects of the significance of the "soldiers from the tropics" have been studied in recent years. In the foreground, however, were primarily issues of counter-constructions of enemy stereotypes, and discourses about over-foreignisation in contact with these soldiers[59] and with regard to the construction of manly images of soldiers.[60] However, the military connotations of discourses of demographic crisis in many European countries gave a new dimension to the discussions of the fitness of the troops from the colonies. France – the country where, to some degree and after 1870, the discourse about the crisis in the birthrate originated[61] – relied quite early on these colonial troops. After the 1870s, recruitment of Zouaves and the Tirailleurs Sénégalais was seen in France as by no means just a picturesque way of replenishing troop strength, as it was seen by foreign observers in particular, but as a tool for offsetting the falling birthrates in the metropolis. Starting around 1900, the French War Ministry intensified its efforts to forecast future armed-forces strength figures as precisely as possible.[62] The unanimous conclusion was that further reduction was inevitable because of a corresponding drop in the birthrates. In 1911, the army leadership became alarmed and asked the War Ministry for a suitable package of measures to respond to this danger. The reply was quick and explicit: 'increase the size of indigenous forces'.[63] The "indigenous element", that is, the inhabitants of the colonies, was to be called upon to compensate for the weakness of the French army relative to the German.[64]

As a result, in the political constellation of the French state, the colonies took on a quite different role than they had, for example, in the German Reich. From the perspective of military men and politicians, the systematic incorporation of the colonial population into the army was meant to guarantee troop strength on a permanent basis. The recruiting of colonial troops was based overwhelmingly on volunteers, men hoping for a lucrative income in the French army. Only in a few isolated cases, such as that of Réunion Island, did the French administration introduce compulsory military service. The *tirage au sort* system of conscription introduced there in 1895 still provided for random selection of potential recruits, but otherwise was largely set up parallel to the conscription system in France.[65] The report of Dr. Théron, the military physician in charge, about these recruitment measures indicates the extent to which French military demography discourses were echoed in the colonies. While military physicians and anthropologists in Europe toiled away at a biologistic, racist classification of the continent, based on data about conscripts,[66] the military physicians on Réunion wrestled with applying such racist categories. In Europe, classifying recruits by whether they came from the mountains or the lowlands furnished the basis for classifying various human races – and on Réunion, this geographic distinction was used, among other things, to subdivide the island's 170,000 inhabitants into five different races. Characteristic of such a discussion of the colonial population was the fact that it also permitted phenotypic isolation of certain attributes and of specific combat power. Thus the doctors were able to characterise the "whites from the heights" in terms of their 'endurance and resistance to the tropical climate that the European lacks' and in this way make them the 'basis for recruitment on Réunion'.[67] By contrast, the 'kefirs from the coast . . . are less robust and have adopted the vices of the towns'. Such anthropologising views, marked by typecasting and a lack of historical perspective, attributed unambiguous qualitative characteristics to geographic origin. This served the military physicians as an explanatory model, and thus helps to explain the erosion of anthropological self-description in the tropics:

> In the operations of the inspection commission, we have noted that the more the conscript resembled the white race in his color, the more reasons the white creole gives for why he should be exempted from doing military service. When the conscript shows traces of African blood, the more likely he is judged physically suited and declared fit for service. Signs of racial degeneracy (congenital debilities, hernias, edemas, swollen lymph conditions, various forms of tuberculosis, etc.) are evident, for the large part, among young men or whites who are more and more light-skinned.[68]

In the opinion of the military physicians, it was precisely in the corresponding manoeuvres and operations that reflected the colonial troops' greater hardiness. In addition, this robustness was expressed in the respective morbidity statistics, when compared to the European soldiers.[69] This points to a popular practice among French and other European military physicians, who often used their deployments for detailed studies of health statistics. In the campaigns in Indochina or Madagascar, detailed records of morbidity rates were kept.[70] Such military operations provided virtually the only opportunities for systematic statistical study of the differences in morbidity rates between Europeans and "indigènes" or "natives". Such statistical reports were also used to define, to some extent after the fact, the concepts of 'indigenous fitness for military service' and European 'fitness for service in the tropics'.[71]

By World War I, utilisation of colonial troops was quite significant in scale. Of the 485,000 soldiers from the colonies who were deployed in France during the war, 134,000 came from the tropical colonies of West Africa alone.[72] This deployment was at first derided in Germany as a kind of "zoo-like display" as people assumed these soldiers, in keeping with the contemporary stereotype of the superiority of the white race, did not reflect a genuine escalation in French military potential. Nevertheless, it quickly raised the old questions about the perception of European national military strength. Though German intellectuals initially relied on arguments that reflected culturally laden images of the pre-war era to denounce the "African savages" as incapable of civilised warfare, and to argue their use was therefore a violation of international law, the discussion soon reanimated and included biologistic and racist components.[73]

German practices with respect to the recruitment of "indigenous" soldiers differed fundamentally from those of the French in the pre-war period as well. German colonialism did not think of the colonies as a place to recruit new soldiers. In keeping with the adage of 'white minds and black hands',[74] the main concern was to open up cheap commodity, labour and sales markets, and to find an outlet for the demographic pressure mounting in the German Reich.[75] Even the concept of the 'colonial economy of human beings',[76] increasingly important in the medical discourse after 1907, by no means included giving the duty of citizens to perform military service to residents of the colonies. German discourses about the colonial populations made them primarily the objects of economic logic or treated them as a latent danger. They did not play a constructive role in calculations of political power.

In consequence, German enlisted soldiers of the *Schutztruppe*, the German Empire's African colonial force, thus who were to be deployed in the colonies as well, in a departure from the general practice of filling these slots primarily with members of the local population. This goal, however, was fully achieved only in German South West Africa. In tropical Africa (German East Africa, Cameroon, Togo), it was realised soon after the founding of the *Schutztruppe* in 1891 that German soldiers had an insufficient capacity to adapt to the tropics, so that here, too, Africans were recruited into the service.

Despite the above-mentioned differences in recruiting African soldiers, despite the much smaller space they occupied in the military discourse of German colonialism, and despite the far slighter colonial experience of the German Reich as compared with France and Great Britain, the specifics of African military strength remained an acrimonious issue in the German army. It was precisely the difficulty in recruiting men for the *Schutztruppe*, and the dangers native soldiers were thought to pose, that formed a suitable backdrop for new military and scientific questions. In a kind of "field trial", the German government recruited 89 Sudanese in 1894 and then sent them to Cameroon. The trial was intended to test the operational fitness of soldiers from East Africa for service in West Africa, an experiment for which there was, in the strict sense, no precedent in either the German or the English colonies. The plan was to test the soldiers' adaptability to the tropical climate; the hope was to ensure the recruitment of better fit and loyal soldiers presenting a reduced risk in the calculus of military officials. Both the army command and the government of the Reich were disappointed to learn that the ability of soldiers from East Africa to adapt to the damp climate of Cameroon was basically just as poor as that of the German troops, if not even worse.[77] The experiment failed, and with it the quest for a specific and comprehensive "African suitability". In this case too, however, the military leaders interpreted the reasons for failure as being not only race-specific, but also decisively culturally determined. Above all, the substantial consumption of brandy by Muslim soldiers far from home was blamed for a failure. An environing monotheistic culture seemed of crucial importance for an appropriate human behaviour. In this perspective, Muslim faith was a little less strange to the German militaries then the presumably hostile spiritual religious environment in Cameroon itself.

The construction of suitability for military service in the tropics – discourses and actors

In the colonial context, the military was the only realm that, in principle, allowed the collection of statistics and yet simultaneously was subject to the hygienic surveillance mechanisms of the state. Thus, much like the situation in Europe, the military represented an object of study, a kind of laboratory in which abstract demographic, hygienic and racial hypotheses concerning the tropical colonies were actually manifested. At the same time, many physicians hoped to make use of the military to gain a clearer understanding of the anthropological specifics of the colonies.

In some cases, the statistical investigations reached a highly differentiated understanding of the pathological features of the colonies. For a while, the statistical studies primarily reproduced and affirmed the physicians' racist preconceptions, who relied on "racial" patterns of disease in their reasoning. The 1894 report on disease statistics for the Dutch army in today's Indonesia was quite convinced that soldiers from the colonies and from Europe could be differentiated based on special patterns of infection and courses of disease, and on that basis spoke of two distinct races.[78] Dr. H. Gros, a French specialist in colonial and tropical medicine, felt compelled to extensively revise these results, which in his opinion did not go far enough in questioning the diagnosis of relevant diseases such as beriberi or malaria. Symptoms of disease were indeed expressed in accordance with population group, he conceded, but one still could not generalise this and conclude that the morbidity rates were fundamentally different.[79]

Two aspects of the controversy exemplified the efforts to produce a medical assessment of the tropics and their associated dangers:

1 The physicians and statisticians generally had to contend with a serious shortage of data. Because of this particular difficulty, they used data from other European colonial powers for their research. Implicitly, European populations in the tropics were transformed into one homogeneous "race" by this practice, and finding fixed criteria for their survival in these regions was characterised as a pan-European project, for 'only through joint use of theoretical research findings is it possible to do justice to the difficult task of keeping European troops healthy when fighting in the tropics'.[80]

Statistics, and with it an internationally valid scientific method, was interpreted here as a tool with which the presence of the Europeans in the tropical countries was made possible. In the process, international teamwork, especially in the eyes of the German doctors, seemed in order, for the reason that 'the statistics are still letting us down badly in Africa',[81] as Oberstabsarzt Dr. Ziemann, a German navy physician and the medical adviser in Cameroon, commented.

The 1907 Berlin International Congress for Hygiene and Demography, at which the more general topic of compiling statistics on recruits already occupied a prominent place, was ultimately the place where such transnational ambitions to unify the suitability criteria and categories of international statistics on recruits were formulated. In the section specifically established to assess European soldiers' suitability for the tropics, the German speaker, Oberstabsarzt Dr. Steudel,[82] the French military physician Reynaud,[83] and the British representative Davies[84] agreed that cooperation among the military physicians of the respective colonial powers must be arranged so as to ensure the comparability of the sparse data available. Finally, a suggestion was made to draw up an internationally binding scheme for ascertaining suitability criteria, to be worked out by a newly established "international colonial society", the *Internationale Kolonial-Gesellschaft*.[85] In this same section, however, voices criticising the implementation of such a project were already to be heard. There were assertions that the screening criteria in Germany already were too diverse, especially when undertaken in individual hospitals for tropical medicine. Against this backdrop, international comparability was alleged to be not feasible.

2 In their efforts to describe geographic variations in disease, scientists tried hard for a long time to relate pathologies to the concept of race. Charles Woodruff, a US military physician, during his deployment in the Philippines made a lengthy statistical study of the morbidity rate based on the soldiers' respective skin and hair colour.[86] His findings led him to conclude that long-term survival in the tropics was virtually impossible for European and North American peoples. A German military physician, Th. zur Verth, saw these findings somewhat differently, but also concluded that they indicated 'the superiority of brown-haired people in the tropics' and advocated the inclusion of appropriate criteria in recruitment programmes.[87]

In this context, military specialists in tropical medicine planned to take recourse in the most modern methods of research, in order to substantiate the categories of suitability for the tropics. They proposed a thorough examination of the recruits, including analysis of the blood count and blood groups, and also employing the newly introduced technique of serum diagnosis.[88] The physicians believed this technique could be used not only as a new, objective method for determining race but also as a way of making reactions of "European bodies" in tropical climate understandable.[89] Yet carrying out such a proposal in a comprehensive fashion remained impossible.

The concept of race underwent a substantial shift in meaning in these colonial contexts, as compared with how the term was used within European contexts before World War I. In the contexts of colonial and military medicine, this notion was given a strong geographic component. With the first large-scale military campaigns in the tropics, military physicians also attempted to pathologise the geographical area as such.[90] The device they employed was the open and largely undefined concept of race, which was used to group together populations and regions. It is precisely in the early writings before 1900 that such a notion of race as a spatially ordered concept of population can frequently be found.[91] But after the turn of the century, increasing doubt was cast on such a uniform understanding of race.[92] A purely biological, abstract, definition of fitness for military service in the tropics did not substantiate in colonial campaigns, which fed into ongoing doubts among European elites about a simplistic biological sociology arguing that colonial rule was just a consequence of racial supremacy.[93]

Instead, the notion of suitability for the tropics was reduced once again to the level of the individual. Individual behaviour and the acquisition of cultural techniques, along with family disposition and nervous disorders, subsequently were central in the catalogue of criteria. Steudel, in his list of criteria for the screening of suitable soldiers, wrote:

> During the examination, the nervous system requires the greatest consideration, because it most commonly suffers in the tropics. Individuals with a tendency to neuralgias, headaches, or hypochondria, or individuals who suffer from nervous strain even in the home country when the demands of military service escalate, and then require convalescent leave, are absolutely unsuited to serve in the tropics, in common with individuals who have a hereditary taint or are mentally inferior.[94]

This opinion was shared by zur Verth, who suggested, 'besides objective signs, consideration of the personal history and mental state of the nearest relatives' for the purpose of adequately judging the predisposition of a recruit.[95] The principal argument of the physicians continued to be that the special circumstances and duties of the soldiers required a stable nervous system, and not that the tropical countries themselves had produced specific types of diseases. In particular, the drug and alcohol consumption of many European soldiers triggered "psychic irritations", they contended. As early as 1894, at the International Congress for Hygiene and Demography in Budapest, British Surgeon-General Charles Richard Francis condemned the use of nicotine and alcohol in tropical countries.[96] In the following years, several of the most notable experts in community medicine and social hygiene concurred in this advice.[97] Psychic irritations, but most especially the dreaded "Tropenkoller" (tropical neurasthenia), were induced or at least accelerated by such use, it was said. As a consequence, a penchant for consuming alcohol became a disqualifying criterion for service in the tropics.[98]

In a more extended sense, this applied to the issue of "tropical hygiene" as well.[99] Medical experts now insisted on the application of hygienic principles to private life for tropical colonies as well. Both physicians and hygienists were to be sent to these countries to develop them.[100] 'The question of using European troops is primarily one of practicing military tropical hygiene on the greatest possible scale'.[101] Individual hygiene and the internalisation of a catalogue of appropriate measures thus were regarded as essential for survival,[102] and discipline took on an important role in the understanding of suitability for the tropics. Technical aids such as the pith helmet, or medical resources such as quinine, became indispensable tools that made survival in the tropics even appear possible. At the same time, they also defined the abilities individual soldiers had to possess to be deployed in the tropics. German reports repeatedly stress that fitness for the tropics was to be defined, among other things, by the ability of soldiers to swallow quinine tablets of the correct size – a criterion that resulted in eliminating a rather large number of applicants for service in the tropics.[103]

Conclusion – suitability as a condensation of the demographic discourses

The mortality rates in European armies dropped rapidly after the 1860s and 1870s.[104] The systematic use of quinine for treatment and prevention, well established since mid-century, substantially defused a situation among European troops that had appeared alarming. With the systematic colonial appropriation of territory in Africa, and thus with the genuine beginnings of a German colonial empire in the 1880s, the situation had already eased to a large degree.

The pathological threats for European soldiers in the tropical colonies remained tangible, but their reality was not based exclusively on the miasmic theories of germs circulating there. Meanings were ascribed to the diseases and to the risks of contagion, and in these meanings European fields of knowledge such as climatology, aetiology, biological genetics, and social hygiene were put to the test. They used their particular explanatory potential to construct an image of the "tropics" as a pathologically laden space,[105] and simultaneously renegotiated demographic heuristics in this resonance chamber.

The early demographic discourses and scientific practices in this era may be understood on the basis of just such eclectic methodological expertise. With regard to the colonies, where systematically collected data were practically unavailable, these bodies of knowledge played a decisive role in assessing the demographic perspectives on life and survival in the tropics. The concept of suitability served to merge the various scientific approaches together into a system that shaped recruitment practices, among other things.

In the medical investigation of climatic, cultural and racial interactions between the tropics and any European race – no matter which one – scientists saw a joint project of the colonial powers, in which the varying contexts of scientific and political discussions were reflected nonetheless in the respective home countries. At the end of the 19th century, the French had far greater colonial experience than the young German colonial power. But in France too, the hygienic reforms in Europe placed tropical medicine and tropical hygiene in a different light. The concept of acclimatisation complemented and replaced an inflexible concept of racist categories.[106] The image of the inhabitants of tropical regions, whether they hailed from Europe or from the colonies themselves, was more differentiated in the French contributions to the discourse. Some of the far too simplistic racist dichotomies in the medical discourse of the early colonial period cannot, or can no longer be, found in this form in France. One might speculate that the more general crisis of physical anthropology in French scientific thinking had its causes also in this colonial dimension.[107]

As for the specifically military aspects of this discussion, however, it is important to note that the French army very early on relied on incorporating soldiers from the colonies. From that, pressure arose to formulate viable standards for recruitment. But in the German Reich too, the discussion of soldiers' fitness for the tropics remained vigorous, although only a relatively small number of soldiers was at issue in this case. With the growing wealth of experience of German experts in tropical medicine, it became apparent that clear biological categories for assessing suitability for service in the tropics could not be found. To an increasing degree, the debate turned towards examining individual behaviour and cultural practices, and the medical and technical fostering of the appropriate dispositions.

Yet the dynamic change to which these discourses on suitability for the tropics were subjected did not mean the end of stereotypes and investigatory paradigms motivated by racism. Colonial dichotomising of the world and implicit racism remained the backdrop against which debates about the suitability for the tropics evolved, both in the German Reich and in other European states. Even so, the attempt to tie this specific colonial racism[108] to clear biological categories failed. But that this pattern of argument was a scientific failure by no means signified that the racist concept itself moved into the background; it merely meant that its outward form was altered.

Ultimately, these discussions had a lasting effect on the European colonial discourse. The question of survival in the tropical colonies continued to be posed long after these colonies ceased to exist. Wolfgang Eckart has worked out to precisely what extent the field of tropical medicine became a sphere of vigorous colonial revisionism even after the end of colonial rule.[109] Corresponding research institutes and scientific programmes ensured that such an interest was lastingly associated with the notion of the tropics.

Notes

1 This is a slightly reversed version of an article, originally published as 'Soldaten in den Tropen, Soldaten aus den Tropen. Neudefinitionen der Wehrkraft im kolonialen Kontext zwischen 1884 und 1914', in Chatriot and Gosewinkel, Koloniale Politik / Politiques et pratiques coloniales, 223–46. I want to thank John Bendix for the faithful translation.
2 Curtin, 'White Man's Grave'; Van der Brug, 'Unhealthy Batavia'; Geggus, 'Yellow Fever'; Blais et al., *Les sociétés coloniales*, 312–13.
3 As shown for the Carribean by the works of David Geggus; Geggus, *Slavery, War, and Revolution*; Geggus, 'Yellow Fever'.
4 McNeill, *Mosquito Empires*.
5 Mass, 'Welcome to the Jungle'.
6 van Laak, *Imperiale Infrastruktur*, 195ff.
7 For an important exception see Ittmann, Cordell and Maddox, *Demographics of Empire*.
8 Foucrier, 'Populations Coloniales', 5–11.
9 Zimmerman, *Anthropology and Antihumanism*, 111–45.
10 Neill, *Networks in Tropical Medicine*, 26–9.
11 Ittmann, 'Where Nature Dominates Men', 69–70; Blais et al., *Les sociétés coloniales*, 312–13; Mass, 'Welcome to the Jungle'; Zweiniger-Bargielowska, *Managing the Body*, 69–104; Streets, *Martial Races*.
12 Maddox, 'Disease and Environment', 203.
13 Süssmilch, *Die göttliche Ordnung*, 418.
14 Peter van der Brug suggests that after 1733, when more than 50 percent of the new arrivals were dying within six months of arrival, the cause was malaria. Van der Brug, 'Unhealthy Batavia'. I am grateful to John Bendix for this hint.
15 Etemad, 'Pour une approche démographique', 17.

16 'Knowledge of Africa began with the know-how needed to survive the climate'. Fabian, Im Tropenfieber, 88; Surun, 'Le terrain de l'exploration', 64–5.
17 Curtin, 'White Man's Grave'.
18 Fabian, *Im Tropenfieber*, 103–5; Pesek, *Koloniale Herrschaft*, 140.
19 Etemad, 'Pour une approche démographique', 17.
20 Eckart points out, for example, how important the involvement of German doctors was, from the first moment on, for the establishment of a colonial empire. Eckart, *Medizin und Kolonialimperialismus*, 25–40.
21 Geggus, *Slavery, War, and Revolution*, 347–72.
22 van Laak, *Über alles in der Welt*, 81–2.
23 Ziemann, *Wie erobert man Afrika*, 17, 26–7.
24 Virchow, 'Akklimatisation'.
25 Ittmann, 'Where Nature Dominates Men', 61. Foucrier, 'Populations coloniales'.
26 Lipphardt and Kretzschmar, 'Europäer in Übersee'.
27 Raynal, *Histoire philosophique et politique*, 231.
28 Orgéas, *Contribution à l'étude*.
29 Friedmann, *Niederländisch Ost- und Westindien*, 122.
30 Ibid., 169.
31 Eckart, *Medizin und Kolonialimperialismus*, 25–7; Virchow thought that the tropics were simply unsuited for colonization because they were uninhabitable for the white race. For that reason, Germany's colonial ambitions should limit themselves to subtropical and more temperate zones. Virchow, 'Akklimatisation', 202.
32 Zöller, *Die deutschen Besitzungen*.
33 Pallmann, *Bewohnbarkeit der Tropen*.
34 Ibid., 10–13.
35 Neill, *Networks in Tropical Medicine*, S. 16–17.
36 Koch, *Reiseberichte*, 125–33; Gradmann, *Krankheit im Labor*.
37 Wulffert, *Akklimatisation*. Nevertheless, the author concludes 'that the necessary preconditions for establishing larger Germanic farming colonies in the hot countries will also not be fulfilled, even over the course of the entire twentieth century' (176).
38 Ibid., 175. "Drinking habits" as the major impediment to "acclimatizing" to the tropics is an argument reflected in a wide array of publications on the topic of fitness for the tropics. They reflect the slow transition from a purely biological to a cultural definition of such fitness. Thus, the physician H. Sunder drew the conclusion that 'the ventilating fan . . . is just as important for the white race in the tropics as is the oven in colder climates'. Sunder, Die weise Rasse. In his guidebook for tropical physicians, Dr. Fr. Hey also provides the corresponding recommendations for a healthy diet: 'Spicy food, as well as eating and drinking together, are not good habits but the bad habits of a backwards human race'. Hey, *Der Tropenarzt*, 6.
39 Hey, *Der Tropenarzt*, 12.
40 Soloway, *Demography and Degeneration*.
41 Bischoff, 'Tropenkoller'; Livingstone, 'Tropical Climate'.
42 In the context of "training for work" in German colonialism, Conrad also emphasises this particular mix of racist and social anthropological arguments. Conrad, *Globalisierung und Nation*, 79–84.
43 Paligot, *République Rraciale*, 33–88; Blais et al., *Les sociétés coloniales*; Blais, *Societies coloniales*, 313.
44 Livingstone speaks of a 'moral economy of the climate'. Livingstone, 'Tropical Climate', 114.
45 Kohlstock, *Ratgeber für die Tropen*.
46 Braun, *Behandlung*.
47 Ibid., 34–5.
48 Cited in Eckart, *Medizin und Kolonialimperialismus*, 71.
49 Külz, *Volkshygiene für Eingeborene*, cited in Eckart, *Medizin und Kolonialimperialismus*, 59; on the concept of the "colonial economy of human beings" behind it, see Külz, *Grundzüge*.

50 Herren, 'Erweiterung des Wissens'. On the politicization of International Congresses: Herren and Zala, Netzwerk Aussenpolitik.
51 Gould, *Investigations in the Military*; Bircher, *Rekrutirung und Ausmusterung*.
52 Krumeich, 'Zur Entwicklung'; Becker, 'Bewaffnetes Volk'; Becker, 'Synthetischer Militarismus'.
53 Alsberg, *Militäruntauglichkeit*.
54 Ferdinand, 'Debatte Agrar- vs. Industriestaat'.
55 Ronsin, *Grève des ventres*; Dienel, *Kinderzahl und Staatsräson*.
56 Ferdinand, 'Debatte Agrar- vs. Industriestaat'; Hartmann, *Volkskörper bei der Musterung*. These debates showed to be very comparable between Germany and Great Britain; Bischoff, 'Tropenkoller'; Zweiniger-Bargielowska, *Managing the Body*.
57 For Austria: Weisbach, Körpermessungen verschiedener Menschenrassen; Weisbach, Herzegowiner, verglichen mit Tschechen; for France: Collignon, Anthropologie de la France; Vacher de Lapouge, L'Aryen; first beginnings for Germany: Ammon, Zur Anthropologie der Badener; for Italy: Livi, Antropometria Militare; for Sweden and Norway: Fürst and Retzius, Anthropometria suecica; Arbo, Sveriges anthropologi.
58 Roynette, *Bons pour le service*; Benecke, *Militär, Reform und Gesellschaft*, 32–7.
59 Jeismann, *Vaterland der Feinde*, 280–3; Koller, *Von Wilden aller Rassen*.
60 Bock, 'Geschichte, Frauengeschichte, Geschlechtergeschichte'.
61 Ronsin, *Grève des ventres*; Overath, 'Zwischen Krisendeutung'.
62 'Calcul de l'effectif probable de l'Armée en 1912', 1902, Archiv des Service Historique de l'Armée de Terre, Vincennes 7 N 100.
63 'Note du ministère de la guerre à la demande de la direction du contrôle', 12.8.1911, SHAT Vincennes 7 N 108.
64 Michel, *Les Africains*, 15ff.
65 Théron, 'Le récrutement à la Réunion'.
66 Hartmann, *Volkskörper bei der Musterung*, 40–8.
67 Théron, 'Le récrutement à la Réunion', 13.
68 Ibid., 15.
69 Ibid., 17.
70 In the case of Madagascar, these studies were made only in hindsight. Inadequate preparation for using the French army in colonial conquest in 1895 led to the death, by various infectious diseases, of one-third of the approximately 13,000 French soldiers. Rapport médical d'inspection générale; Rapport médical sur la colonne dirigée contre Bossi, 62–3; Notes recueuillies à l'hopital de Nossi-Bé; Cathoire, 'Relation de deux observations'; see also Curtin, *Disease and Empire*, 175–201.
71 zur Verth, *Zur Hygiene europäischer Truppen*.
72 Koller, 'Farbige Truppen'.
73 Waldeyer, *Die im Weltkrieg stehenden Völker*.
74 A slogan of Las Casas, cited in Wulffert, *Akklimatisation*, 165.
75 Van Laak, *Über Alles in der Welt*, 33–7, 56–60, 81–6.
76 Külz, *Grundzüge*, 14ff.
77 Immediatbericht des Reichskanzlers Caprivi an den Kaiser über den misslungenen Einsatz von Sudanesen in Kamerun vom 24.08.1894; Bundesarchiv – Militärarchiv Freiburg, MSg 101/147.
78 Geneeskundi Tijdschrift voor Nederlandsch Indie.
79 Gros, 'Les enseignements'.
80 Ibid., 7.
81 Ziemann, *Wie erobert man Afrika*, 6.
82 Steudel, 'Die Beurteilung der Tropendiensttauglichkeit'.
83 Reynaud Paligot, 'Jugement quant à l'aptitude', 136.
84 Davies, 'How to Judge'.
85 This paralleled Ziemann's call for an 'international society for tropical medicine and hygiene', which was to be made up of "separate national curia"; Ziemann, *Wie erobert man Afrika*, 29.

86 Woodruff, *The Effects of Tropical Light.*
87 zur Verth, *Zur Hygiene europäischer Truppen*, 10.
88 Reynaud Paligot, 'Jugement quant à l'aptitude'.
89 Bruck, 'Die biologische Differenzierung'; Spörri, 'Reines Blut'.
90 Here one should point, for example, to the grand project of the French hygienist Layet, who described the health conditions in the tropical countries in each specific case as dependent on certain magnetic fields, or on the electrical and hygrometric state of the atmosphere. Layet, *La santé des Européens.*
91 Friedmann, *Niederländisch Ost- und Westindien*, 122; Raynal, *Histoire philosophique et politique*, 231; Pallmann, *Bewohnbarkeit der Tropen.*
92 'The questions of medical geography have much lost the importance attributed to them, at a time when one believed every region of the earth, in accordance with their latitude and longitude, had a characteristic ethnography, zoology, or phytogeography, and so forth. Such conceptions were facilitated by an idea that was too large, and as such poorly delimited, of climatic zones'. Moreira and Peixoto, *Les maladies mentales*, 175.
93 Tilley, *Africa as a Living Laboratory*, 223–4.
94 Steudel, 'Die Beurteilung der Tropendiensttauglichkeit', 74.
95 zur Verth, *Zur Hygiene europäischer Truppen*, 11.
96 In contrast to the consuming of hashish and opium, which he recommended, though in moderation; Francis, 'On Opium'.
97 Külz, *Zur Hygiene des Trinkens*; Deherme, 'L'alcoolisme dans les colonies'.
98 Steudel, *Die Beurteilung der Tropendiensttauglichkeit*, 75. 'In tropical campaigns, alcohol drinks in the hands of the soldiers are equivalent to self-destruction', zur Verth, *Zur Hygiene europäischer Truppen*, 42.
99 These tropical hygiene measures could result in reshaping entire 'artificial lifeworlds' for the members of the colonial powers, as Eric Jennings has recently shown in his examination of the French hydrotherapy spas operated in the colonies. Jennings, *Curing the Colonizers.*
100 On this, see the speech by Schilling, a specialist in tropical medicine, at the plenary session of the Colonial Congress in 1910, entitled: 'Of What Significance Are the Recent Developments in Tropical Hygiene for Our Colonies?'.
101 Critique by military physician Reinhold Ruge on 'Steuber: Über die Verwendbarkeit', 433.
102 This is reflected in expressions such as: 'At all times, a pith helmet as sun protection! Never burden the white soldier with anything but his weapons'; ibid., 432.
103 zur Verth, *Zur Hygiene europäischer Truppen*, 14.
104 Curtin, 'The End of the "White Man's Grave"?'.
105 Livingstone, 'Tropical Climate', 106–10.
106 Jennings, *Curing the Colonizers*, 32–7.
107 Paligot, *République Raciale*, 96–100.
108 Here I refer to Wolfgang Eckart's 'special form of German colonial racism', as contrasted with an internally directed racism in the same time period. Whether this differentiation between the two forms of racism is specific to Germany, however, is doubtful, but it needs further investigation. Eckart, *Medizin und Kolonialimperialismus*, 63.
109 Ibid., 505–40.

Bibliography

Becker, Frank, 'Synthetischer Militarismus. Die Einigungskriege und der Stellenwert des Militärischen in der deutschen Gesellschaft', in *Das Militär und der Aufbruch in die Moderne 1860 bis 1890. Armeen, Marinen und der Wandel von Politik, Gesellschaft und Wirtschaft in Europa, den USA sowie Japan*, ed. Michael Epkenhans and Gerhard P. Gross (München: Oldenbourg, 2003), 125–42.

———, '"Bewaffnetes Volk" oder "Volk in Waffen?" Militärpolitik und Militarismus in Deutschland und Frankreich 1870–1914', in *Der Bürger als Soldat. Die Militarisierung europäischer Gesellschaften im langen 19. Jahrhundert: ein internationaler Vergleich*, ed. Christian Jansen (Essen: Klartext, 2004), 158–74.

Benecke, Werner, *Militär, Reform und Gesellschaft im Zarenreich. Die Wehrpflicht in Russland 1874–1914* (Paderborn: Schöningh, 2006).

Bischoff, Eva, 'Tropenkoller. Male Self-Control and the Loss of Colonial Rule', in *Helpless Imperialists: Imperial Failure, Fear and Radicalization*, ed. Maurus Reinkowski and Gregor Thum (Göttingen: Vandenhoeck & Ruprecht, 2013), 117–36.

Blais, H., I. Surun, C. Fredj, V. Caru, E. Sibeud, P. Singaravélou, M.-A. Suremain, and S. Thénault, *Les sociétés coloniales à l'âge des Empires, 1850–1960* (Neuilly: Atlande, 2012).

Bock, Gisela, 'Geschichte, Frauengeschichte, Geschlechtergeschichte', *Geschichte und Gesellschaft* 14 (1988): 364–91.

Curtin, Philip D., '"The White Man's Grave": Image and Reality, 1780–1850', *The Journal of British Studies* 1 (1961): 94–110.

———, 'The End of the "White Man's Grave"? Nineteenth-Century Mortality in West Africa', *Journal of Interdisciplinary History* 21 (1990): 63–88.

———, *Disease and Empire: The Health of European Troops in the Conquest of Africa* (Cambridge: Cambridge University Press, 1998).

Conrad, Sebastian, *Globalisierung und Nation im Deutschen Kaiserreich* (München: Beck, 2006).

Dienel, Christiane, *Kinderzahl und Staatsräson: Empfängnisverhütung und Bevölkerungspolitik in Deutschland und Frankreich bis 1918* (Münster: Westfälisches Dampfboot, 1995).

Eckart, Wolfgang U., *Medizin und Kolonialimperialismus: Deutschland 1884–1945* (Paderborn: Schöningh, 1997).

Etemad, Bouda, 'Pour une approche démographique de l'expansion coloniale de l'Europe', *Annales de Démographie Historique* 113 (2007): 13–32.

Fabian, Johannes, *Im Tropenfieber. Wissenschaft und Wahn in der Erforschung Zentralafrikas* (München: Beck, 2001).

Ferdinand, Ursula, 'Die Debatte Agrar- vs. Industriestaat und die Bevölkerungsfrage', in *Das Konstrukt Bevölkerung vor, in und nach dem Dritten Reich*, ed. Rainer Mackensen and Jürgen Reulecke (Frankfurt: VS Verlag für Sozialwissenschaften, 2005), 111–49.

Foucrier, Annick, 'Populations coloniales', *Annales de Démographie Historique* 113 (2007): 5–11.

Geggus, David P., 'Yellow Fever in the 1790's: The British Army of Occupation in Saint Domingue, 1793–1798', *Medical History* 23 (1979): 38–58.

———, *Slavery, War, and Revolution: The British Occupation of Saint Domingue 1793–1798* (Oxford: Clarendon Press, 1982).

Gradmann, Christoph, *Krankheit im Labor: Robert Koch und die medizinische Bakteriologie* (Göttingen: Wallstein, 2005).

Hartmann, Heinrich, 'Soldaten in den Tropen, Soldaten aus den Tropen: Neudefinitionen der Wehrkraft im kolonialen Kontext zwischen 1884 und 1914, in *Koloniale Politik und Praktiken Deutschlands und Frankreichs 1880–1962 / Politiques et pratiques coloniales dans les empires allemands 1880–1962*, ed. Alain Chatriot and Dieter Gosewinkel (Stuttgart: Steiner, 2010), 223–46.

———, *Der Volkskörper bei der Musterung. Militärstatistik und Demographie in Europa vor dem Ersten Weltkrieg* (Göttingen: Wallstein, 2011).

———, 'Eine unaufhörliche Schwächung der Wehrkraft unseres Vaterlandes: Rekrutenstatistik und demografischer Diskurs in Europa vor dem Ersten Weltkrieg', in *Die vergangene Zukunft Europas. Kulturwissenschaftliche Analyse von demografischen Prognosen und Wissensordnungen*, ed. Petra Overath (Köln: Böhlau, 2010), 29–56.

Herren, Madeleine, '"Die Erweiterung des Wissens beruht vorzugsweise auf dem Kontakt mit der Aussenwelt", Wissenschaftliche Netzwerke aus historischer Perspektive', *Zeitschrift für Geschichtswissenschaft* 49 (2001): 197–207.

——— and Sascha Zala, *Netzwerk Aussenpolitik. Internationale Kongresse und Organisationen als Instrumente der schweizerischen Aussenpolitik 1914–1950* (Zürich: Chronos, 2002).

Ittmann, Karl, 'Where Nature Dominates Men. Demographic Ideas and Policy in British Colonial Africa, 1890–1970', in *The Demographics of Empire: The Colonial Order and the Creation of Knowledge*, ed. K. Ittmann, D. D. Cordell, and G. H. Maddox (Athens: Ohio University Press, 2010), 89–112.

Jeismann, Michael, *Das Vaterland der Feinde. Studien zum nationalen Feindbegriff und Selbstverständnis in Deutschland und Frankreich 1792–1918* (Stuttgart: Klett-Cotta, 1992).

Jennings, Eric C., *Curing the Colonizers. Hydrotherapy, Climatology, and French Colonial Spas* (London: Duke University Press, 2006).

Koller, Christian, *"Von Wilden aller Rassen niedergemetzelt." Die Diskussion um die Verwendung von Kolonialtruppen in Europa zwischen Rassismus, Kolonial- und Militärpolitik (1914–1930)* (Stuttgart: Steiner, 2001).

———, 'Feind – Bilder. Rassen- und Geschlechterstereotype in der Kolonialtruppendiskussion Deutschlands und Frankreichs, 1914–1923', in *Heimat – Front. Militär und Geschlechterverhältnisse im Zeitalter der Weltkriege*, ed. Karen Hagemann and Stefanie Schüler-Springorum (Frankfurt a.M.: Campus, 2002), 150–67.

———, 'Farbige Truppen', in *Enzyklopädie Erster Weltkrieg*, ed. G. Hirschfeld, G. Krumeich, and I. Renz (Paderborn: Schöningh, 2003), 471–2.

Krumeich, Gerd, 'Zur Entwicklung der "nation armée" in Frankreich bis zum Weltkrieg', in *Die Wehrpflicht. Entstehung, Erscheinungsformen und politisch-militärische Wirkung*, ed. Stig Foerster (München: Oldenbourg, 1994), 133–45.

Lipphardt, Veronika and Mandy Kretzschmar, 'Der Europäer in Übersee', in *Die Erfindung des Europäers*, ed. J. Franke, K. Patel, and F. Stuhl (Berlin: Kreuzberg Museum, 2009), 26–33.

Livingstone, David N., 'Tropical Climate and Moral Hygiene: the Anatomy of a Victorian Debate', *The British Journal of the History of Science* 32 (1999): 93–110.

Mass, Sandra, 'Welcome to the Jungle. Imperial Men, "Inner Africa", and Mental Disorder, 1870–1970', in *Helpless Imperialists. Imperial Failure, Fear and Radicalization*, ed. Maurus Reinkowski and Gregor Thum (Göttingen: Vandenhoeck & Ruprecht, 2012), 91–115.

Maddox, Gregory H., 'Disease and Environment in Africa. Imputed Dynamics and Unresolved Issues', in *The Demographics of Empire. The Colonial Order and the Creation of Knowledge*, ed. K. Ittmann, D. D. Cordell, and G. H. Maddox (Athens: Ohio University Press, 2010), 198–216.

McNeill, John, *Mosquito Empires. Ecology and War in the Greater Caribbean, 1620–1914* (Cambridge: Cambridge University Press, 2010).

Michel, Marc *Les Africains et la Grande Guerre. L'appel à l'Afrique (1914–1918)* (Paris: Karthala, 2003).

Neill, Deborah J., *Networks in Tropical Medicine. Internationalism, Colonialism, and the Rise of Medical Specialty* (Stanford: Stanford University Press, 2012).

Overath, Petra, 'Zwischen Krisendeutung und Kriegsszenarien. Bevölkerungspolitische Vorstellungen in Deutschland und Frankreich (1870–1918)' *Comparativ* 13 (2003): 65–79.

Paligot, Carole Reynaud, *La République Raciale. Paradigme racial et idéologie républicaine 1860–1930* (Paris: PUF, 2009).

Pesek, Michael, *Koloniale Herrschaft in Deutsch-Ostafrika. Expeditionen, Militär und Verwaltung seit 1880* (Frankfurt a.M.: Campus, 2005).

Ronsin, Alain, *La grève des ventres propagande néo-matlhusienne et baisse de la natalité francaise (XIXe-XXe siecles)* (Paris: Aubier, 1980).

Roynette, Odile, *"Bons pour le service." L'expérience de la caserne en France à la fin du XIXe siècle* (Paris: Belin, 2000).

Schubert, Frank, '"Soldiers can get anything free". Idi Amin und das Erbe des Kolonialmilitärs in Afrika', *Historische Anthropologie. Kultur, Gesellschaft, Alltag* 14 (2006): 93–104.

Spörri, Myriam, '"Reines Blut", "gemischtes Blut". Blutgruppen und Rassen zwischen 1900 und 1933', in *Transfusionen. Blut-Bilder und Bio-Politik in der Neuzeit*, ed. Anja Laupe (Berlin: Diaphanes, 2005), 211–25.

Soloway, Richard A., *Demography and Degeneration. Eugenics and the Decline of Birthrate in Twentieth-Century Britain* (Chapel Hill: University of North Carolina Press, 1990).

Streets, Heather, *Martial Races. The Military, Race, and Masculinity in British Imperial Culture, 1857–1914* (Manchester: Manchester University Press, 2004).

Surun, Isabelle, 'Le terrain de l'exploration reconsidéré: les explorateurs en Afrique au XIXe siècle', in *L'Empire des géographes*, ed. Pierre Singaravélou (Paris and Berlin, 2008).

van der Brug, Peter H., 'Unhealthy Batavia and the decline of the VOC in the eigtheenth century', in *Jakarta – Batavia. Socio-cultural essays*, ed. Kees Grijns and Peter J.M. Nas (Leiden: KITLV Press, 2000), 43–74.

van Laak, Dirk, *Imperiale Infrastruktur. Deutsche Planungen für eine Erschliessung Afrikas 1880 bis 1960* (Paderborn: Schöningh, 2004).

———, *Über alles in der Welt. Deutscher Imperialismus im 19. und 20. Jahrhundert* (München: Beck, 2005).

Zimmerman, Andrew, *Anthropology and Antihumanism in Imperial Germany* (Chicago: Chicago University Press, 2001).

Zweiniger-Bargielowska, Ina, *Managing the Body. Beauty, Health and Fitness in Britain, 1880–1939* (Oxford: Oxford University Press, 2010).

Sources

Alsberg, Moritz, *Militäruntauglichkeit und Grossstadt-Einfluss. Hygienisch-volkswirtschaftliche Betrachtungen und Vorschläge* (Leipzig: Teubner, 1909).

Ammon, Otto, *Zur Anthropologie der Badener. Bericht über die von der anthropologischen Kommission des Karlsruher Altertumsvereins an Wehrpflichtigen und Mittelschülern vorgenommenen Untersuchungen* (Jena: G. Fischer, 1899).

Arbo, Carl Oscar Eugen, *Sveriges anthropologi med sammenlignende bemaerkninger til Norges* (Christiania: Dybwad, 1903).

Bircher, Heinrich, *Die Rekrutirung und Ausmusterung der schweizerischen Armee* (Aarau: Sauerländer, 1886).

Braun, Heinrich, *Die Behandlung aussereuropäischer (insbesondere von Tropen-)Risiken in der deutschen Lebensversicherung* (Göttingen: Mittler, 1912).

Bruck, Carl, 'Die biologische Differenzierung von Affenarten und menschlichen Rassen durch spezifische Blutreaktionen', *Berliner klinische Wochenschrift* 26 (1907): 793–7.
Cathoire, 'Relation de deux observations de fièvre méditerranéenne chez des soldats l'un francais, l'autre indigène, de la division d'occupation de Tunisie', *Caducée* 3 (1906): 35–6.
Collignon, René, *Anthropologie de la France. Dordogne, Charente, Corrèze, Creuse, Haute-Vienne* (Paris: Masson, 1894).
Davies, 'How to Judge of the Fitness of Officers and Men for Active Service in Tropical Countries', in *Festschrift, den Teilnehmern am XIV. Internationalen Kongresse für Hygiene und Demographie zu Berlin 1907*, vol. III (Berlin: Puttkammer und Mühlbrecht, 1907), 572.
Deherme, Georges, 'L'alcoolisme dans les colonies', in *Annales antialcooliques* (April 1905).
Francis, Charles Richard, 'On Opium, Narcotics and Alcohol in the Tropics', in *VIIIe Congrès international d'Hygiène et de démographie, tenu à Budapest du 1er au 9 septembre 1894. Comptes-rendus et mémoires* (Budapest: Gerlóczy, 1896), 722–9.
Friedmann, Sigwart, *Niederländisch Ost- und Westindien. Ihre neueste Gestaltung mit besonderer Darstellung der klimatischen und sanitätischen Verhältnisse* (München: Franz, 1860).
Fürst, Carl and Gustav Retzius, *Anthropometria suecica. Beiträge zur Anthropologie der Schweden* (Stockholm: Akademiens Förlag, 1902).
Geneeskundi Tijdschrift voor Nederlandsch Indie 2, no. 3 (1895).
Gould, Benjamin A., *Investigations in the Military and Anthropological Statistics of American Soldiers* (New York: U.S. sanitary commission, 1869).
Gros, H., 'Les enseignements d'une statistiques', *Archives de médecine navale* 78 (1902): 81–113 and (1908): 161–95.
Hey, Fr., *Der Tropenarzt. Ausführlicher Ratgeber für Europäer in den Tropen sowie für Besitzer von Plantagen und Handelshäusern, Kolonial-Behörden und Missions-Verwaltungen* (Wismar: Historff, 1912).
Koch, Robert, *Reiseberichte* (Berlin: Springer, 1898).
Kohlstock, P., *Ratgeber für die Tropen* (Offenbach: Scherz, 1905).
Külz, Ludwig, *Zur Hygiene des Trinkens in den Tropen* (Flensburg, 1905).
———, *Grundzüge der kolonialen Eingeborenenhygiene* (Leipzig: Barth, 1911).
Layet, Alexandre, *La santé des Européens entre les tropiques. Première partie: le climat, le sol les agents vivants d'agressions morbide* (Paris: Félix Alcan, 1906).
Livi, Ridolfo, *Antropometria Militare*, 2 vols (Rome: Giornale medico del Regio Esercito, 1896 and 1905).
Moreira, Juliano and Afriano Peixoto, *Les maladies mentales dans les pays tropicaux* (paper presented at the XVe Congrès International de Médecine, Lisbonne, Portugal, avril 19–26, 1906, Sektion XVII Médecine Coloniale et Navale), 175–92.
Orgéas, J., *Contribution à l'étude du Non-Cosmopolitisme de l'homme. La colonisation de la Guyane par la transportation: étude historique et démographique* (Paris: Octave Doin, 1883).
Pallmann, Reinhold, *Die Bewohnbarkeit der Tropen für Europäer. Eine kulturgeographische Studie aus den Quellen. Vortrag gehalten im Klub der Landwirthe zu Berlin am 21.12.1886* (Berlin: Lehmann, 1887).
Raynal, Guillaume T., *Histoire philosophique et politique des établissements de commerce des Européens dans les deux Indes*, vol. VI (Paris: Berry, 1770).
Reynaud, G., 'Jugement quant à l'aptitude des officiers et des soldats au service dans les pays tropiques', *Archiv für Schiffs- und Tropenhygiene* 12 (1908): 136.

Ruge, Reinhold, 'Review of Steuber: Über die Verwendbarkeit europäischer Truppen in Kolonien vom gesundheitlichen Standpunkte', *Archiv für Schiffs- und Tropenhygiene* 11 (1907): 432–4.

Steudel, Emil, 'Die Beurteilung der Tropendiensttauglichkeit bei Offizieren und Mannschaften', *Archiv für Schiffs- und Tropenhygiene* 12 (1908): S. 73–7.

Süssmilch, Johann Peter, *Die göttliche Ordnung in den Veränderungen des menschlichen Geschlechts aus der Geburt, dem Tode und der Fortpflanzung desselben* (Berlin: Verlag der Buchhandlung der Realschule, 1761), § 216: 418.

Sunder, H., *Kann die weisse Rasse sich in den Tropen akklimatisieren?* (Berlin: Süsserott, 1908).

Théron, Fernand, 'Le récrutement à la Réunion', *Archives de la médecine navale et coloniale* 68 (1897): 5–18.

Tilley, Helen, *Africa as a Living Laboratory: Empire, Development, and the Problem of Scientific Knowledge 1870–1950* (Chicago: University of Chicago Press, 2011).

Vacher de Lapouge, Georges, *L'Aryen. Son rôle social : Cours libre de science politique* (Paris: Fontemoing, 1899).

Verth, Th. zur, *Zur Hygiene europäischer Truppen bei tropischen Feldzügen* (Leipzig: Barth, 1909).

Virchow, Rudolph, 'Akklimatisation', *Verhandlungen der Berliner Gesellschaft für Anthropologie, Ethnologie und Urgeschichte. Zeitschrift für Ethnologie* 17 (1885): 202–4.

Waldeyer, Wilhelm, *Die im Weltkrieg stehenden Völker in anthropologischer Betrachtung* (Berlin: Heymann, 1915).

Weisbach, Augustin, *Körpermessungen verschiedener Menschenrassen* (Berlin: Weidler, 1874).

———, *Die Herzegowiner, verglichen mit Tschechen und Deutschen aus Mähren nach Major Himmels Messungen* (Wien, 1889).

Woodruff, Charles Edward, *The Effects of Tropical Light on White Men* (New York: Rebman, 1905).

Wulffert, Friedrich, *Die Akklimatisation der europäischen und insbesondere der germanischen Rasse in den Tropen und ihre hauptsächlichen Hindernisse* (Leipzig, 1900).

Ziemann, Heinrich, *Wie erobert man Afrika für die weisse und farbige Rasse? Vortrag gehalten auf dem Internationalen Kongress für Hygiene und Demographie zu Berlin 1907* (Leipzig: Barth, 1907).

Zöller, Hugo, *Die deutschen Besitzungen an der westafrikanischen Küste*, vol. II (Berlin: Spemann, 1885).

8 Disease at the confluence of knowledge

Kifafa and epilepsy in Ulanga (Tanzania)

Marcel Dreier

Disease, diagnosis and treatment are knowledge articulated in concrete settings. These concrete settings often have an explicitly translocal character.[1] This chapter looks at the articulation of disease through the lens of an illness called *kifafa*, which is commonly translated as being the Kiswahili word for epilepsy. In Mahenge – a small rural town in the mountains of the Ulanga District in south-central Tanzania, and with a rich tradition as a seat of African polities, colonial administration and Catholic Mission – kifafa acquired particular connotations which go beyond the simple equation which a dictionary translation of the term kifafa as epilepsy insinuates.

Susan Reynolds White has provided an anthropological critique of the disease concepts of epilepsy that biomedicine has come to construct in East Africa.[2] She showed that the fixed characteristics of the disease definition are, under the circumstances of actual lived experiences with epilepsy, very fluent. This chapter adds to this social anthropological reading and looks at how kifafa in Ulanga was constituted in a historical process which Patrick Harries has termed the *confluence of knowledge*.[3] In Ulanga kifafa became a medical syndrome that is positioned at the fringes of scientific knowledge and at the margins of health care systems. At the same time this localised kifafa made it into scientific journals all over the world and connected with illnesses observed in other locations. As we follow the confluence of knowledge, this chapter tries to show how a disease named kifafa became the product of a translocal and transcultural history of health services, science and development.

The history of kifafa in Ulanga is comparably easy to follow, because this history is firmly tied to the scientific career and developmental work of one person, Louise Jilek-Aall, who spent her early career in that region. Louise Jilek-Aall had originally come there in order to do research for Basel pharmaceutical companies on the treatment of schistosomiasis. Eventually she produced an extensive and creative body of knowledge about epilepsy as a particularly prevalent, and kifafa as a peculiar disease. Tellingly, Reynolds White started her chapter on the *Construction of Epilepsy* in East Africa by quoting Jilek-Aall. In this chapter, I look a bit deeper into history of this construction process in order to understand how African, missionary and psychiatric ideas formed the basis from which kifafa was described by Jilek-Aall and to understand how bio-scientific discourses have produced knowledge from it. This is a worthwhile undertaking because Jilek-Aall's work in the early 1960s was indeed groundbreaking. She had to make sense of what she experienced in the clinic without being able to have recourse to firm ground of scientific knowledge. And her epilepsy clinic was probably one of the earliest attempts in that field in Africa.

Although around since the early 1960s, Jilek-Aall's idea that kifafa in Mahenge could be 'a sort of an unknown brain-disease among the people of Mahenge' has recently become more noticed and combined with other strands of neurological research on Nodding Syndrome or on Nakalanga syndrome.[4] The World Health Organization (WHO) has launched research activities on this syndrome in Sudan, and specialists in the WHO agreed on a case definition of the nodding syndrome in August 2012.[5] Since then the nodding syndrome has attracted the selective attention of the public in the West and in Uganda as a sort of freak disease, medical riddle and a 'scourge for the people' to which 'thousands of African children fall victim'.[6] Today the WHO says that nodding syndrome has been first described in Mahenge by Louise Jilek-Aall.[7]

The account of kifafa as the product of a history of knowledge systems shows a process that was complex and to which a number of local and global bodies of knowledge contributed. In this chapter I look at this history by discussing the experiences of kifafa since the colonial period and describing the historical sequence of social configurations in which the disease has been defined, treated and investigated. My narrative starts with kifafa as an illness that found a rather unsystematic treatment response from missionary medicine in Ulanga since the 1930s. This response generated an intensive intellectual pursuit of the affliction by Jilek-Aall in the early 1960s. I will show how Jilek-Aall drew on colonial missionary and local African framings of kifafa and came up with a biomedical theory of the disease. She established a new treatment regime, too. However, in the context of the early post-colonial state's health system, the planners devalued her initiatives – along with her knowledge. Jilek-Aall persisted. She continued to support treatment at a local clinic and I will follow some of her intellectual moves as she drew on a rich array of bodies of scientific knowledge, from transcultural psychiatry to neurology, family genetics or parasitology in order to describe kifafa. Eventually knowledge about kifafa flowed into a pool of strongly localised, "rare" syndromes.

Unlike most of the epidemiologists, parasitologists or neurologists who are interested in the clinical aspects of nodding syndrome, I came to the topic as part of historical research on the local institutional trajectories of health services in that particular area. Jilek-Aall's work contains a lot of ethnographic material, which has not been submitted to a critical discussion. And her overall argument really grew out of a local historical context.[8] This chapter draws on archival material and missionary publications related to Ulanga to discuss publications in scientific disciplines that are foreign to me, and which have been produced in an increasingly transnational context.

Kifafa and missionary care in Ulanga before 1960

In Ulanga, epilepsy has always been far less prevalent than most other diseases, for example common fevers or diseases of the digestive tract. Still, Louise Jilek-Aall was not the first to write about epilepsy and the attention it attracted in

Ulanga. Catholic missionaries, colonial administrators and social anthropologists had reported about epilepsy from the late 1920s. They also noticed that people in Ulanga had a concept of the disease they called kifafa.[9] Knowledge about medicines was substantial, with the Swiss pharmacist Fritz Haerdi reporting at least 21 medical plants used by the people in Ulanga for treating kifafa in the early 1960s.[10]

Local healing traditions for cases of kifafa included dramatic acts by healers involving spirits and fluids, making the missionaries uneasy.[11] It was easy to read African concepts of epilepsy in the light of a Judeo-Christian tradition that considered it as a 'demonic disease'.[12] All the more, the Swiss Catholic missionaries in Ulanga saw epilepsy, much like leprosy, as a chronic disease whose sufferers deserved Christian care and nursing.[13] When they observed that expulsion from family and society was a major treatment option in Ulanga for epileptics in an advanced state of the affliction, the missionaries were truly worried. In 1931 Sisters from Baldegg, who were active in the Swiss-run Catholic Mission in Ulanga, collected money for a nursing home for epileptics in Kwiro, close to the administrative capital of Mahenge. The nursing home for epileptics had reportedly started in response to the plight of a young girl and an elderly woman, both with epilepsy, who had sought asylum at the mission. In the 1930s an average of about 20 patients seem to have stayed at this epilepsy home in Kwiro. Patients who came to the mission were heavily afflicted by the disease or by injuries resulting from fits, and mortality was rather high.[14]

This genealogy of epilepsy care in the hands of the Catholic Mission is different from the way in which the colonial state handled people with psychiatric disorders, which often included people with epilepsy. Central and East African British psychiatric services focused almost exclusively on the control of criminality in the sense of violent and aggressive behaviour of Africans.[15] Spiritual and rebellious individuals were diagnosed with epilepsy and were locked up when they threatened, in the eyes of the colonial state, to induce larger groups of Africans to become collectively unstable. The missionary nursing home tradition was not entirely congruent with the colonial prison-asylum. Missionary institutions could be disciplinary and sometimes abusive in character, but control was as much a question of morality as one of institutional – but not of public – order. Inmates in the Kwiro nursing home were being summoned to work with an underlying moralistic rationale. 'No police is sent after them', when they ran off. The mission was confident that inmates would return when hungry.[16] A major worry for the missionaries however was sexuality and procreation. In the late 1930s and in the 1940s, the mission attempted to gender segregate the nursing home for epileptics. When this seemed not to work and the nursed did not conform to the missionaries' moral behaviour standards, the mission did turn to the powers of the state: Drawing on imageries of heredity and, likely, on the eugenics then common in Switzerland's treatment of the marginalised,[17] they attempted to make sufferers of epilepsy into idiots by law. Probably they did not succeed. Instead, the mission seems to have excluded men from the nursed until the end of the decade.[18]

Louise Jilek-Aall, the missionary tradition and kifafa as a colonial tribal biomedical condition

In the early 1960s kifafa came to be ordered as a truly biomedical condition mainly through the interested gaze and helpful hand of a medical doctor, Louise Jilek-Aall. She added new dimensions and depth to the knowledge about kifafa. She had done her MD in Zürich, and took a post-graduate training course in Tropical Medicine in Basel at the Swiss Tropical Institute, STI.[19] With a grant from a Basel pharmaceutical company she left for Tanganyika in 1959. Her task was to research tropical diseases at the STI's field laboratory in Ifakara. In Ifakara the Swiss Capuchin Mission had just built a large 250-bed hospital. Louise Aall, then a young unmarried woman in her late twenties, was given boarding and accommodation at the hospital. Her research on schistosomiasis quickly came to a standstill. She lacked language skills to interact with patients, suffered from culture shock, and she had to face a medical director of the mission hospital who felt that he himself should have been paid to do Aall's research. Eventually the children who were her probands and whom she treated against parasites were hit by an epidemic of infectious hepatitis, and she struggled to keep them alive. As a result she started a whole new career as a bush-doctor, out of the lab and mostly out of the hospital, the archetype of the missionary doctor at the frontier in Africa.[20] The transfer from STI's field laboratory to mission medicine opened doors for her. An old nun and nurse at the hospital helped her into the local ways of knowing health and medicine: Jilek-Aall was quickly trained in diagnosing the most frequent diseases, in assisting birth, etc. And as she headed out to do mobile medical work in the district she got into the lives of the people in Ulanga (and also into the hospital and home of Albert Schweitzer, with whom she worked in Lambaréné in Gabon, for a number of months, interrupting her time in Tanzania).[21] At that time she seems to have subscribed to the legitimation for (mission) medicine as a tool of enlightenment, modernisation and development.[22]

It was in Kwiro, half a day's journey by car, conditions permitting, from Ifakara, where Jilek-Aall started working with epileptics. If the booming rural township of Ifakara was the place for traders, Kwiro was the seat of the Catholic Mission and it lay less than half an hour on foot from the colonial administrative Centre at Mahenge. The colonial and missionary world in Ulanga was organised in tribes, and this had a strong feedback on Jilek-Aall's ideas about kifafa. The Catholic mission had not only accepted tribal boundaries enforced by the colonial system of administration. It also worked along these lines, often considering itself as the Church of the tribe of the Wapogoro, whose "chiefs" had administrative control over the Mahenge highlands.

The mission context was of great importance for the knowledge produced by Jilek-Aall. Not only because she practised her medicine partly within missionary institutions, but also because her perception of the cultures of ill-health and healing must have been shaped within this context. Many of the patients described in her papers on kifafa and on psychiatric conditions more generally had a direct connection to the institutions of the Catholic Mission Church.[23] The Baldegg sisters and African mission nurses in the clinic acted as interpreters and forwarded information to Jilek-Aall.[24]

Jilek-Aall's first explanation of kifafa as the typical disease of the tribe in her 1965 psychiatric dissertation submitted in Zürich was based on a historical argument centred on the tribal history of the Wapogoro, the people of Mahenge. The Wapogoro were, according to Jilek-Aall,

> a peaceful and timid people of small stature, who chose to retreat from more warlike neighbours into a hardly accessible mountain region where they still lead a rather seclusive existence. . . . Their traditions and customs are better preserved than those of many tribes in Ulanga, partly because the people are shut off by the mountains, . . . and partly because the Wapogoro have a strong clan sense, which results in a disinclination to marry outside their own clan.[25]

This idea of the voluntary seclusion of the Wapogoro strongly undergirds the explanation for kifafa. The Wapogoro had, Jilek-Aall contended, based on a fairly widespread negative image of the Wapogoro in the region, generally an 'un-enterprising outlook' and

> an indifference towards illness. . . . Only as regards one illness does their behaviour follow a different pattern and this is epilepsy. Almost everybody in the tribe knows this disease and fears it. There are countless measures they consider will protect them against it.[26]

While knowledge and usage of plants to treat kifafa in Ulanga were not limited to the Wapogoro, Jilek-All still felt – and I quote from her thesis – 'tempted to call epilepsy their 'tribal sickness'.[27]

A new therapeutic era and the local (re-) arrangements of therapy management

Jilek-Aall's scientific conceptualisation of kifafa was not a renunciation of the missionary tradition. On the contrary she chose the missionary dispensary, rather than the overcrowded small government hospital, as the base for her work and research. But she introduced chemotherapy for the epileptics at the mission.[28] For this beginning of a new therapeutic era there was a narrative of an individual healing breakthrough experience congruent to the 'small girl and old woman' story which the missionaries had offered for the beginning of the nursing home back in the 1930s: The narrative for the advent of chemotherapy was based on the progress of a boy who had come back from being marked as a *maskini*, a crippled, into a productive place in society, becoming an assistant and translator to Jilek-Aall.[29]

Chemotherapeutic treatment was a success and so was (at least partly) social reintegration, and they led Jilek-Aall to hope that families were not only seeking help, but were, reluctantly, ready to labour and care for patients, whom they had expelled and deposited before.[30] Trust was central to the acceptability and success of Jilek-Aall's' chemotherapeutic offer. At the same time this trust paved a way for her understanding of kifafa through a new quality of relationships to knowledgeable people. Jilek-Aall portrays the inputs by local experts, for example by the "native healer" Chava Msolomoka, as central to her grasp of that disease.[31]

Jilek-Aall's own empirical understanding of the disease must have been shaped by the way patients and their family units would manage the disease. Particular versions of local narrative about kifafa, and in particular the issue of family ties, have a strong presence in Jilek-Aall's early descriptions of the disease. The strong link, which was made locally between family and the afflicted individual, plus the concept of exogamous marriage, seemed to combine into a hypothesis that turned into a search for a hereditary disease. Consequently she looked for family links in the community of the diseased.[32]

Seen from a historian's perspective of today, her patient group at the epilepsy clinic did not reflect a representative sample of total society. The problem of the sample was noted later by Jilek-Aall when it came to calculate prevalence of the disease.[33] But the cultural factor seemed to matter only as far as the understanding of kifafa in "traditional" society went. When it came to her own work at the clinic the impact of the cultural factor on treatment arrangements went unseen. How *kifafa treatment* had been negotiated in the exchange with missionary practices of medicine and care seemed to have no importance to the articulation of what is presented as the "Wapogoro" narrative of kifafa.

Treatment management groups,[34] who were "family", were strongly involved in the illness that afflicted one of their members, and once the family had opted for the chemotherapeutic approach, it is no surprise that family members came to the dispensary.[35] Patients highlighted their family connections in a situation where they hoped to receive powerful medicine and general nursing care from the charitable missionaries (who were obsessed with the issue of family life of Africans). This would only be normal for clients who sought access to services and distributive networks by belonging. Family ties were powerful claims to belonging. They were also seen as risks, exposing individuals to threats of contagion and spirit affliction. All this made it likely that patients would highlight how family bonds connected them.

Looking closely at the specific set-up of therapy allows us to understand the power that Jilek-Aall's chemotherapeutic offer had to transform local narratives and practices about kifafa. As Jilek-Aall treated epilepsy with Luminal-pills (Phenobarbital), she turned the simple pill into a more dramatic technology, in tune with local conceptions of modern healing:

> Each patient was given a small bottle for the drugs. He was strictly forbidden to use the bottle for any other purpose. The bottle was to be placed at a certain spot above his sleeping-place in the hut, and where the children could not reach it. Every evening before going to bed he was to take his drug. . . . From its place above the bed the bottle with the medicine became a symbol

> of healing force, protecting the patient and keeping the devil or other evil influences away from him.[36]

During the day, patients would often carry the bottle wrapped into their clothes, as they did with tobacco bags.[37] Patients themselves, claimed Jilek-Aall, did not buy into a purely "mechanical" explanation of the effect of the drug:

> it made no sense to explain to people that Phenobarbital was a common drug against epilepsy in Europe. The Wapogoro believe that it is me alone who has the power to heal epilepsy and that Phenobarbital is my drug that can only be received from me.[38]

Nevertheless, these were modern technologies of healing and they were institutionalised in novel ways. Friday became kifafa clinic day, and three newly trained African assistants and the missionary nurse assisted Jilek-Aall.[39] The African assistants soon knew most of the clients and the conditions under which they lived.[40] Most patients were out-patients, and Jilek-Aall herself rarely saw them having a fit. The treatment results at the clinic were reportedly very good. Out of 110 patients who received medication 61 remained free of fits with even the smallest dose of the drug.[41] The Mission's services for epileptics were overrun by demand and this success outgrew the capacity of the Capuchin Mission station.

Overwhelmed, Jilek-Aall sneaked out of Mahenge, quietly, but with a burden on her: that she wanted to help the treatment to continue.[42] In her absence Jilek-Aall's powers as a healer remained institutionalised in modern technical ways that went beyond the biomedical. Writing was part of the procedure. At the clinic patients were registered and their histories taken as good as possible. This system was considerably enlarged when Jilek-Aall left. Now the missionary nurse in charge of the clinic consulted Jilek-Aall by way of letter writing and it was quite acceptable for patients that this medium of communication could extend the healing relationship they had with the doctor-healer.[43] Knowledge about kifafa was travelling beyond the geographical space of Ulanga, but it had not yet any substance beyond those who belonged to that therapeutic network.

Kifafa in post-independence development: marginal patients, neglected disease

Early in 1961 Jilek-Aall had tried to attract the Government of Tanganyika to support research on and treatment of epilepsy in Ulanga:

> Although until now I have been working in cooperation with the Catholic Mission, I prefer to work on this project in connection with Government, as this is of a general interest, and not only a problem and task of the mission. As the patients with epilepsy generally suffer from many other diseases and form wounds . . . there should be a close cooperation with the local hospital.[44]

Jilek-Aall claimed to have the support of the Member of Parliament for Ulanga District, C.M. Kapilima. But this did not win the Ministry of Health on her side. On the contrary the fact that Jilek-Aall had more than 100 patients under her care 'apparently without the formality of registering as a Medical practitioner' shocked the Chief Medical Officer of newly independent Tanganyika.[45] What was more important was that the Ministry of Health feared the cost of caring for chronically ill patients:

> I very much doubt whether funds of this order would be forthcoming from W.H.O. or any other source. Even if they should be available, if, as at present, they were offset against our Development programme the net effect would be the postponement of other projects of much higher priority. Epilepsy is undoubtedly a very distressing disease, particularly to the patients' relatives, but it is not a disease of economic importance.[46]

Before that, Jilek-Aall had unmistakably pointed out to the Prime Minister that her project was not only medical, but also about development: In 'Community Mental Health Service it was vital that the patients and their families would benefit from the social and agricultural development which such a centre would aim at'. In addition she proposed to integrate the development of modern facilities with the general health care system in Mahenge, allowing 'the hospital to share some of the modern equipment which the centre will necessarily have'.[47]

The fear of an unproductive population and colonial experiences of leprosy care undergirded the Ministry's view on chronic disease. The keeper of the file, whom I could not identify, commented about Jilek-Aall's suggestion to house patients and family in the camp for the beginning of the therapy:

> I foresee that once a family had been removed to such a centre the task of subsequently releasing them . . . would be almost impossible one. Like the leper settlements that have arisen around leprosy hospitals, once the patient and his family has settled down and started cultivating they would be there for thc rcmainder of their lives.[48]

Government medical planners also challenged Jilek-Aall's claims about the prevalence of epilepsy. They believed, somewhat arrogantly, that 'she has fallen into a common error in assessing the incidence of any particular disease, i.e. assuming that if comparatively large numbers are seen the incidence must be high'.[49] Jilek-Aall had learnt about the importance of epidemiological data for securing assistance early and she would keep up the argument about the high incidence all through the decades that followed.[50] The numbers game had the potential for problematic consequences: In the face of a generalising brushing off of numbers as they were experienced on the ground, the argument about the local focus of disease could become attractive to those, like Jilek-Aall who defended the picture – and health services – on the ground.[51]

The clinic in Kwiro continued as a part of the mission health services. Jilek-Aall supported it with her private funds, and with her husband-to-be, also a specialist in epilepsy, she travelled to Tanganyika again in 1963, re-examined the patients and clarified on treatment regimes.[52] Roughly 160 patients were entered into a programme for which Jilek-Aall organised the drugs and advised the missionary sister in charge of the clinic who reported on the development of the cases to her.[53]

Jilek-Aall and the emerging field of transcultural psychiatry

Although Jilek-Aall's suggestion to extend epilepsy care was not unreasonable from a medical systems point at the time, Jilek-Aall's attempts to push the development of treatment and research on epilepsy in Mahenge remained largely unanswered by the mainstream institutions throughout the first three decades of Tanzanian independence.[54] Jilek-Aall lacked professional status and symbolic capital to get such a project moving. When she approached the WHO, her appeal for assistance was turned down with the WHO representative suggesting she should become an expert on epilepsy before support could be granted.[55] She indeed refashioned herself as a psychiatrist.[56]

Another body of knowledge, more directly linked to the psychological than physiological perspectives on epilepsy, should be acknowledged in this context. Jilek-Aall's family history is rich in connections to Vienna and to psychology and Jilek-Aall's interest in culture and "folk medicine" is likely to have been stimulated by this family background. Even her ideas of witchcraft were probably not produced in the concrete colonial situation only. Jilek-Aall's mother, Lily Weiser Aall, was an early student (and the first "habilitated" woman in Vienna) in the field in European Ethnology and highly interested in psychological questions, and in the far-flung history of ritual and witchcraft.[57] The interest in psychology she shared with Louise's father, the philosopher Anathol Aall. This background of a Freudian and Jungian psychology created a rich ontology, which combined well with ideas about "primitive medicine" as well as new ideas in cultural anthropology.[58] While race theories and the images of African societies' tribal basis furthered the idea of genetic factors, it was psychology and psychiatry which provided the pre-eminent episteme in explaining epilepsy before the advent of more biotechnological neurological approaches.[59] At the confluence of these bodies of knowledge, it was easy to conceptualise epilepsy in terms of its tribal psychological aspects.

The international research community of transcultural psychiatrists not only engaged in comparative research, but it also considered the cultural factor in psychology as essential on the level of the individual and of entire societies. In the early 1960s it created a space for new discourses and a range of creative scholars with a footing in the late colonial regions of Africa who came to establish and define this new discipline. While it became a fundamental field from which a new generation of medical anthropologist would draw, transcultural psychiatry is a complex "discipline", transdisciplinary in character, and drawing in knowledge from the fringes of science.[60]

Whilst still in Africa, Jilek-Aall had already connected to the young African community of transcultural psychiatry. She had contacted Thomas Adeoye Lambo and was invited to participate in the 1st Pan-African Psychiatric Conference held 1961 in Abeokuta, in Nigeria.[61] At the conference it was

> suggested that one should build a sort of a research centre in the [Mahenge] district, where one could carry out a proper research work, and where the patients could get a modern treatment, such as is not possible under the conditions now in Mahenge.[62]

Jilek-Aall returned to Zürich to work at the Zürich Psychiatric Hospital Burghölzli where she wrote her thesis with Manfred Bleuler.[63] By 1966, the year in which the proceedings of a fundamental conference in transcultural psychiatry were published, Jilek-Aall was a researcher, together with her husband, W.G. Jilek, at the major hub of the new discipline, McGill University in Canada.[64]

The tribal disease and new traits of epilepsy

Kifafa in Mahenge coagulated for a surprisingly long time, into 'almost a tribal disease'. The definition of kifafa as a 'separate disease entity'[65] in an 'isolated tribe' was reinforced from the end of the 1960s, the moment when the "tribe" left isolation and when Jilek-Aall qualified for a transcultural, rather than a missionary doctor working in one particular tribe.

Even though historical perspectives were not totally absent in local debate, they did not find a way into a refinement of the argument about the tribal disease.[66] When Jilek-Aall had to make sense of the disease as a transcultural psychiatrist, she came up with a description that was built from a colonial way of knowing the African world – a traditional world of witchcraft and incestuous families firmly encapsulated by the culture and psychology of the tribe. At the same time, she noticed a number of traits in kifafa, which seemed new articulations of epilepsy. Locality of kifafa became reinforced as part of the process in which Jilek-Aall tried to make sense of what she now understood, in a comparative perspective, as the specifics of "Epilepsy" she had witnessed. She couldn't even be sure it was epilepsy. Remember that she was no specialist of epilepsy when she came to Ulanga. The mission's epileptics home had brought together patients without a specialist's diagnosis. When Jilek-Aall's chemotherapy worked, it supported the diagnosis.[67] But while she had used the word epilepsy in the title of her dissertation she had then already been cautious to pin down the illness simply as epilepsy as it was known from the medical textbooks, and she had pointed out that she was observing, maybe, a 'genuine epilepsy with a hereditary pathology of its own'.[68] Soon the name of the disease she would use in academic publication titles was "kifafa" and she described it as an 'unusual seizure disorder'.[69]

One of the unusual features Jilek-Aall heard about and observed in Mahenge was a how a 'child suddenly drops its head to the chest, then raises it up again. Sometimes a series of such nods is observed . . . accompanied by saliva drooling'.[70]

This was described in local terms as *amesinzia kichwa*, which Jilek-Aall translated as 'nodding of the head', as in letting your head drop repeatedly in drowsiness. Increasingly this symptom of kifafa made her consider different hypotheses. Kifafa could be a symptomatic epileptic disorder, a hereditary form of idiopathic epilepsy, or an altogether separate disease identity.[71]

In the process of describing the specific characteristics of the illness she witnessed, Jilek-Aall presented the relationship of the Wapogoro with kifafa as almost endowing identity.[72] *The dream of the Wapogoro* popped up as an argument in many of her and her husband's papers. According to the Jileks, the Wapogoro explained this dream, a nightmare about water turning into red colour or into blood as foretelling epilepsy.[73] The presence of the dream argument hints at the way in which psychological thinking informed Jilek-Aall's work. A reading of this dream as informing of sickness is consistent particularly with Jungian analytical knowledge.[74] In any case her knowledge of the dream served to underscore Jilek-Aall's expertise and access to cultural concepts of the disease. In her publications she presented the dream as a very private, and almost a secret knowledge, and how her knowledge of the dream and its meanings was a key that opened the doors to the Wapogoro speaking about their understanding of kifafa. At the same time, Jilek-Aall seems not to have methodologically assessed her claims of ethnic specificity or questioned the mechanics of constructing identity in Ulanga. There is little in the way of cross referencing attributes of an ethnic character and not as much reflection on the context in which she collected statements about epilepsy as historians would want to see. The result was that Jilek-Aall presented kifafa as specific to the Wapogoro and she claimed that even the name for the disease, kifafa, was rarely heard of outside the tribe, and that other tribes were reluctant to marry Wapogoro.[75] Socioeconomic and environmental factors were, to her and her husband, not enough to explain the 'unusual frequency of epileptic manifestations in the tribe' of the Wapogoro.[76] Other "tribes" lived in the same environment, but did not, in the Jilek's perception, suffer as much from kifafa.[77]

This view of the tribal disease lead to kifafa being identified as a localised medical phenomenon: In 1979 the Jileks enumerated 'peculiar features that set many kifafa patients apart from the majority of epileptics we have seen elsewhere' and installed kifafa as 'a hitherto unknown chronic degenerative disease of the central nervous system highly prevalent in this isolated East African tribal population'.[78]

With the notion of kifafa as a distinct disease entity in the Wapogoro tribe, the idea of hereditary kept its momentum. And since the local idea of contagious kifafa (within a tight-knit family context) did not have credit in a biomedical episteme of epilepsy, the patterns of kifafa came to be shaped into a thesis about genetic disposition. In the early 1990s a high-tech genetic analysis of kifafa in Ulangan patients was undertaken.[79] The researchers collected the disease histories of roughly 1,600 relatives of some 22 study participants. On the basis of a disease in a "tribal community", the research transformed a historically configured social identity, with cultural traits like the *dream of the Wapogoro* into a complex epidemiological sample. The research paper found that parents knew the disease phenotype well. For researchers, diagnosis remained difficult, EEG abnormalities were not seen at all and they found that only 2 of the 86 grandparents of the study participants had a kifafa diagnosis, too. Other family members, who entered the study as probands suffering from kifafa, had uncertain diagnoses. The study concluded that family clustering was present, but that simple genetic inheritance was not. Still, the authors 'believe that a genetic component plays a strong role in the occurrence of kifafa in this population' because of an 'unusual . . . inheritance of unstable [genetic] sequences'. With this kifafa entered the knowledge bases of specialists on family genetics. Jilek-Aall's research papers, or papers where she contributed, were now published in journals with a clearly biomedical background. This echoes a shift in scientific views on epilepsy: Jilek-Aall's expert knowledge was changing from a cultural to a more neurological field.

In terms of benefits for the people in Ulanga, the new research in the early 1990s meant a breakthrough for the actual treatment offered in the clinic in Kwiro. Throughout three decades the epilepsy clinic had continued to offer some services, but the provision of drugs was often unstable under the Government health system.[80] Now, the clinic at the hospital was improved to treat additional patients, and it was permanently staffed with a medical assistant and psychiatric nurse as well as two medical doctors, Henry Rwiza and William Matuja, who visited regularly and published a series of scientific articles.[81]

In looking back, the outcome of the ten years of treatment in the 1960s remained unclear when Jilek-Aall and Henry Rwiza made a follow-up study 30 years after regular chemotherapy had started for roughly 160 patients who had been served at the Mahenge Epilepsy Clinic.[82] Treatment had reduced exclusion and marginalisation and had opened new ways to think about kifafa in local society.[83] Services for epileptics now became reattached to networks of health provision, and what Jilek-Aall had suggested in the early 1960s in terms of care for people with epilepsy finally was realised in the early 1990s.

Kifafa cha kusinzia and head nodding beyond Ulanga

But the story of kifafa does not end with the provision of treatment. With the clinic treating some 700 patients, the issue of the prevalence of epilepsy remained a focus of research.[84] The number of epilepsy cases was still very high, and research suggested a series of risk factors such as brain injuries resulting from disease episodes, e.g. from river blindness (onchocerciasis) since childhood, or even from delayed labour.[85] From about 2009, Jilek-Aall joined a debate in parasitology and neurology about the possibility of a connection between "head-nodding", epilepsy and onchocerciasis.[86]

More importantly, kifafa descriptions from Mahenge in the 1960s have helped to produce the definition of a rare nodding syndrome, which reaches beyond Ulanga and might also be present in specific locations of East Africa.[87] Present in South Sudan, Uganda and Tanzania the nodding syndrome has become the object of research that brings observations about medical phenomena into conversation. The combination of Jilek-Aall's notes on kifafa with the concrete idea of a nodding syndrome made it appear that "head nodding" was kifafa. It also gave "head nodding" a historical depth that reached back into the times before spraying with insecticides took place, thus also speaking to rumours about the newly observed epidemics of head nodding in Uganda being a result of chemicals.[88] Not least it made Louise Jilek-Aall a major reference in publications from the most popular to the most specialised. In 2012 *Nature Medicine* talked about head nodding in the following way:

> For years, the disease remained an isolated oddity. But over the past decade, physicians began observing this 'nodding syndrome' in several African countries, where it continues to perplex researchers to this day. 'Nobody has yet been able to find an explanation', says Jilek-Aall, 'and we are not sure that we are always describing the same phenomenon'.[89]

The Mahenge clinic depended no longer only on a legitimation as a specialist treatment centre for epilepsy, it was now also a special institution for a disease specific to Africa. In a report on 50 years of the existence of the epilepsy clinic Jilek-Aall explained:

> Our examinations confirmed the presence of a new type of epileptic seizure in children, the "Head Nodding Syndrome" leading to convulsive attacks in later life. I had already observed and described the Head Nodding Syndrome in the 1960s. As it was unknown in Western countries, the existence of this epilepsy syndrome had then been ignored. However, the Head Nodding Syndrome is today accepted and attracts international attention among experts.[90]

By about 2005, what had started as a largely palliative care offer by the Capuchin Mission in the late 1930s had, after repeated attempts by Jilek-Aall to sponsor both treatment and research, produced a new documentation on a neglected syndrome in an East African setting. In recent research, the nodding syndrome in Mahenge has been renamed into *amesinzia kichwa* by Jilek-Aall and thus constitutes a new field for research discerned from "universal" epilepsy without head nodding.[91] Today, according to van der Weegen, the local word for the disease used locally is *kifafa cha kusinzia*.[92]

Conclusion

The story of kifafa takes us through a rich history of different strands of intellectual traditions of reading society and disease. As they connected, the object of epilepsy became fragmented, and kifafa became an epileptic disorder rather than simply the Kiswahili translation of epilepsy.

For a long time the scientific explanation of kifafa remained grounded in colonial representations about tribal culture that had sprung from African, missionary and ethnopsychiatric ideas (rather than, for example in ecology or ecological changes, or an argument about the traumatic experiences of colonial occupation in Mahenge).[93] The "secluded" tribe of little initiative was quite a convincing host, first for an exceptionally high prevalence of epilepsy and later for a rare disease. When Jilek-Aall's engagement for people who suffered from illness was disqualified by the post-independence government planners, she inserted kifafa into the field of science, drawing creatively on a range of disciplines, some new, but entangled in a post-colonial world – like transcultural psychiatry – and some technological – like genetics – which turned the local, historical context even more into an artefact.

Biomedical research on kifafa and epilepsy does not reflect how this syndrome or disease is historically configured. What today is represented like an early scientific description of a timeless (and "odd") disease is – from the historian's perspective – a historical product of the colonial situation in Ulanga in which it was first described and it is connected to a long process of global entanglements of knowledge exchange, of care and cure. The ways in which knowledge was produced in this situation created a narrative of an "odd" or even a freak disease.[94]

While my account tries to show the depth and importance of transnational entanglements, it also testifies that such transnational connectivity must not necessarily make an affliction relevant and raise it out of "vernacular" knowledge.[95] Moreover, research on kifafa and nodding syndrome, notwithstanding its potential to be popularised, remains marginal, and at the interstices of scientific disciplines.[96]

Notes

1 Prince, 'Situating Health'; Fleischmann et al., *Transnational and Historical Perspectives*.

2 Whyte, 'Constructing Epilepsy'. In this text I have used italics for kifafa only in the first instance, where it is a term in Kiswahili. Otherwise I treat kifafa as the name for an object in a scientific discourse to which "vernacular" knowledge has contributed. Tilley, 'Global Histories, Vernacular Science'.

3 This chapter was meant to feed into a discussion on the production of knowledge systems with Patrick Harries, who had supervised research for my thesis on the history of health services in Ulanga.

4 Diocesan Archives Kwiro (Tanzania): Jilek-Aall, Letter to Bishop Elias. Basel, 08.02.1962; Föger et al., 'Nakalanga Syndrome'. Note that Louise Jilek-Aall is a co-author of this paper, published in early 2017, almost 60 years after her first observations in Ulanga.

5 See www.afro.who.int/news/nodding-syndrome-meeting-researchers-agree-case-definition-and-establish-research-agenda (accessed 16 December 2017); Colebunders, Irani, and Post, 'Nodding Syndrome'.

6 Hochstrasser, 'Medizinisches Rätsel'. See the history of the article on 'nodding disease' in Wikipedia, especially the German one on Nickkrankheit with a first entry on 15.12.2012. Other popularised accounts contain words in their titles like 'haunts', 'baffled scientists', 'mysterious': are Vogel, 'Mystery Disease'; Williams, 'Nodding Syndrome'; Korevaar and Visser, 'Reviewing Evidence'; and most recently Föger et al., 'Nakalanga Syndrome', last sentence.

7 www.who.int/onchocerciasis/symptoms/nodding_syndrome/en/ (accessed 16 December 2017).

8 Dreier, Health, *Welfare and Development*; Larson, 'History of Ulanga'; Green, *Priests, Witches and Power*.

9 Culwick, 'Treatment of Fits'. Tanzanian National Archives (TNA), District Book, District Office Mahenge: Reid, *Notes on Witchcraft*.

10 Haerdi, *Eingeborenen-Heilpflanzen*, 240–4.
11 Jilek-Aall, 'Epilepsy', 72; Widmer, *Geschichte*, 34–5.
12 Jilek-Aall, 'Morbus Sacer'; Temkin, *Falling Sickness*.
13 Hodel, *Lehrbuch der Krankenpflege*, 200; Brentano, *Barmherzige Schwestern*, 8–9.
14 Fischli, 'Gruss Aus Kwiro'; Gürber, 'Heim für die Epileptischen'; Anon., 'Bericht des Schwestern-Institutes'; Wolfisberg, 'Aus den Missionen'. This was still the same in the 1960s: Jilek-Aall, Jilek and Miller, 'Clinical and Genetic Aspects', 617. The story about the two women in Widmer, *Geschichte*, 35. It is probably based on information gathered from Sr. Valentina Wolfisberg.
15 Vaughan, *Curing Their Ills*, chapter 5; Mahone, 'East African Psychiatry'.
16 Gürber, 'Heim für die Epileptischen', 9.
17 Huonker, *Diagnose: moralisch defekt*.
18 Anon., 'Eine Missionsschwester erzählt'; Jilek-Aall, *Call Mama Doctor*, 193; Gürber, 'Heim für die Epileptischen'; Institutsarchiv Baldegg: Chronik Kwiro. Häne, 'Vom Dienst an den Kranken'; TNA 61/129H, 1940T: Maclean, Notes on Ulanga District, 18.01.1940.
19 Krahl, 'Zum 80. Geburtstag'.
20 Jilek-Aall, *Call Mama Doctor*, 7–8, 39–46; Provinzarchiv der Schweizer Kapuziner, Luzern (PAL), Sch 1061.6: Schöpf, Letter to E. Maranta. Ifakara, 15.04.1959.
21 Jilek-Aall, *Call Mama Doctor*; Jilek-Aall, *Working with Dr. Schweitzer*.
22 Jilek-Aall, 'Epilepsy', 59.
23 See for example patient stories presented in Jilek and Jilek-Aall, 'Psychiatric Concepts and Conditions'.
24 Ibid., 206; Jilek-Aall, 'Epilepsy'.
25 Jilek-Aall, 'Epilepsy', 61.
26 Ibid., 63.
27 Ibid.
28 Jilek-Aall and Jilek, 'Problem of Epilepsy'; Jilek-Aall, 'Kifafa'; Jilek-Aall, *Call Mama Doctor*, 191.
29 Jilek-Aall, *Call Mama Doctor*, 183–4, 90. In Jilek-Aall, 'Forty Years', this individual story is absent and it is more a history of trial of a new medication.
30 Research undertaken at a later time showed that seizures were stopped or reduced significantly. Survival of epileptics was still 'poor', however, showing that neglect, poverty and a lack of psychosocial support was not fully offset by chemotherapy. Jilek-Aall and Rwiza, 'Prognosis of Epilepsy', 648–9. Jilek-Aall, *Call Mama Doctor*; Jilek-Aall, Jilek and Miller, 'Clinical and Genetic Aspects'. Jilek-Aall read this as a reduction of stigma. There is good reason to remain cautious about the extent of stigmatisation towards epilepsy patients in Tanzania. Denisenko, *Perceiving Epilepsy*; Whyte, 'Constructing Epilepsy', 238.
31 Jilek and Jilek-Aall, 'Psychiatric Concepts and Conditions', 206 mentions in particular: Chava Msolomoka and 'the famous mbui Linkono'.
32 Jilek-Aall, 'Geisteskrankheiten'. This argument is particularly strong in Jilek-Aall, Jilek and Miller, 'Clinical and Genetic Aspects'.
33 Jilek-Aall, Jilek and Miller, 'Clinical and Genetic Aspects', 616.
34 Janzen, 'Therapy Management'.
35 Whyte observed in a late 1970s social anthropological study of epilepsy in Tanzania that families would tend to 'overprotect' their epileptic members. WHO, Programme for Mental Health; Whyte, 'Constructing Epilepsy', 227, 236, 238.
36 Jilek-Aall, 'Epilepsy', 74. The bottle was still a major element of treatment in the 1990. Jilek-Aall et al., 'Psychosocial Study', 786.
37 Jilek-Aall, *Call Mama Doctor*, 201.
38 Jilek-Aall, 'Geisteskrankheiten', 247.

39 Jilek-Aall, *Call Mama Doctor*, 201–3. The missionary nurse was also training African staff, see Jilek-Aall, 'Kifafa', 38.
40 Jilek-Aall, 'Epilepsy', 61.
41 Ibid., 74–5.
42 Ibid., 59–60.
43 Jilek-Aall, 'Geisteskrankheiten', 202.
44 TNA 450/HL 777/22: [Jilek-]Aall, letter to W.J.M. Evans, Oslo probably 03.01.1961.
45 TNA 450/HL 777/22: note by the Chief Medical Officer of Tanganyika, dated 01.06.1962.
46 TNA 450/HL 777/22: Note by unidentified author, dated 21.05.1962.
47 TNA 450/HL 777/22: Letter to Prime Minister Rashid Kawawa, Zürich 07.05.1962.
48 TNA 450/HL 777/22: Note dated 21.05.1962.
49 TNA 450/HL 777/22: Note by C.M.O, 01.06.1962. Today, research suggests, that in Africa, 'contrary to developed countries, the most frequent neurological disorder amongst hospital inpatients was seizures. Febrile convulsions and Epilepsy were major causes'. Mosser, 'Pattern of Epileptic Seizures'.
50 Diocesan Archives Kwiro (Tanzania): Jilek-Aall, Letter to Bishop Elias, Basel, 08.02.1962; TNA 450/HL 777/22: C.V. Mtawali, letter to L. Aall, Dar es Salaam, 06.06.1962.
51 In the face of neglect the ethnic dimension of a disease can become highlighted, as in the Ugandan case of nodding syndrome. Van Bemmel, Derluyn, and Stroeke, 'Nodding Syndrome or Disease'.
52 During this trip Jilek-Aaal collected a specific bark that a healer under her treatment had given her as the drug he applied against Epilepsy. In laboratory tests in Switzerland the bark was found to have anti-convulsive effect, but development into a drug was deemed inefficient. Jilek-Aall, *Call Mama Doctor*, 198–9, 214–19.
53 Jilek-Aall and Rwiza, 'Prognosis of Epilepsy', 646. Later Jilek-Aall received support from Canadian pharmacy company Elliott-Marion. In 1973 Government took responsibility for the Epilepsy clinic in Kwiro albeit on a low level. State mental health services in independent Tanzania were seriously underdeveloped and underfunded. The total number of psychiatrists in the country at the end of that decade was still the same as in the early 1960s: two specialists for the entire country. Tanzania then started two pilot projects for Epilepsy care with the support of the Danish government development service DANIDA and WHO. But they were not concentrating on Mahenge at all. WHO, Programme for Mental Health; Kilonzo and Simmons, 'Development of Mental Health Services'.
54 African Medical and Research Foundation, *The Health Services of Tanganyika*, 136.
55 Jilek-Aall, *Call Mama Doctor*, 206.
56 Unlike the 'East African School' of psychiatry Jilek-Aall did not push 'deculturation' and culture conflict as drivers of disease. Rather she expanded the idea of the tribal disease, bridging the worlds of ethnopsychiatry and transcultural psychiatry in a peculiar way. Mahone, 'East African Psychiatry'; Sadowsky, *Imperial Bedlam*, 104–10; McCulloch, *Colonial Psychiatry*.
57 Keintzel, *Wissenschafterinnen*, 799–800.
58 Harries and Dreier, 'Medizin und Magie'. Benedict, *Patterns of Culture*.
59 On the issue of psychiatry versus neurology as the leading discipline in the treatment of Epilepsy see Reynolds and Trimble, 'Epilepsy, Psychiatry, and Neurology'.
60 Bains, 'Race, Culture and Psychiatry'; Opler, *Culture and Mental Health*, see especially the introduction by Opler and the last chapter by Margaret Mead.
61 Jilek, 'The Early History'; Heaton, *Black Skin, White Coats*. Lambo, one of the first African psychiatrists, had started a model village at Aro in Nigeria in 1954: Lambo, 'The Village of Aro'. He joined the WHO in the 1970s and eventually became Deputy Director General.

62 TNA 450/HL 777/22: Aall, letter to W.J.M. Evans, Oslo probably 03.01.1961.
63 Jilek-Aall, 'Epilepsy'.
64 Krahl, 'Laudatio'. De Reuck and Porter, *Transcultural Psychiatry*. The initiator of this conference, E. D. Wittkower, had been a supervisor of Louise Jilek-Aall at McGill.
65 Jilek-Aall, 'Kifafa', 59.
66 Jilek and Jilek-Aall, 'Psychiatric Concepts and Conditions', 206; Jilek-Aall and Jilek, 'Problem of Epilepsy', 44.
67 Jilek-Aall, 'Kifafa', 58; Jilek-Aall, Jilek and Miller, 'Clinical and Genetic Aspects', 614.
68 Jilek-Aall, 'Epilepsy', 84.
69 Jilek-Aall and Jilek, 'Problem of Epilepsy'; Jilek-Aall, 'Kifafa'; Jilek-Aall, Jilek and Miller, 'Clinical and Genetic Aspects'.
70 Jilek-Aall and Jilek, 'Problem of Epilepsy', 46.
71 Witkowski et al., *Lexikon der Syndrome*, 686; OMIM (Online Mendelian Inheritance in Man), Entry 245180 Kifafa Seizure Disorder, www.omim.org/entry/245180 (accessed 16 December 2017).
72 Jilek-Aall, 'Kifafa', 58.
73 For example in Jilek-Aall, 'Epilepsy', 71; Jilek-Aall and Jilek, 'Problem of Epilepsy', 44; Jilek-Aall, 'Kifafa', 58.
74 Verbal information by Denise Pfander, a psychologist trained at CG Jung Institute in Zürich, August 2016.
75 Jilek-Aall, 'Kifafa', 57–8; Jilek-Aall, Jilek and Miller, 'Clinical and Genetic Aspects', 618.
76 Jilek-Aal, 'Problem of Epilepsy', 45.
77 Jilek-Aall, Jilek and Miller, 'Clinical and Genetic Aspects', 618.
78 Ibid., 617, 621.
79 This paragraph based on Neuman et al., 'Genetic Analysis of Kifafa'. See also: Jilek-Aall, Jilek and Miller, 'Clinical and Genetic Aspects', 618.
80 Jilek-Aall, 'Fifty Years Epilepsy Clinic'; Lischer, *50 Jahre Baldeggerschwestern*, 39–43; Jilek-Aall and Rwiza, 'Prognosis of Epilepsy'. Treating roughly 200 patients meant that the mission had to dispense more than 100,000 pills annually. Jilek-Aall, *Call Mama Doctor*, 203.
81 Rwiza et al., 'Knowledge, Attitude, and Practice'.
82 Jilek-Aall and Rwiza, 'Prognosis of Epilepsy'.
83 Jilek-Aall et al., *Psychosocial Study*, 789.
84 Matuja et al., 'Risk Factors for Epilepsy'.
85 According to Jilek-Aall, the Wapogoro seem to have made a connection between adultery on the part of the pregnant mother and epilepsy in the child. Jilek-Aall, 'Morbus Sacer', 383; Green, *Priests, Witches and Power*, 85–6.
86 Jilek-Aall, Jilek and Miller, 'Clinical and Genetic Aspects'; Jilek-Aall, 'Fifty Years Epilepsy Clinic', 618; Marin et al., 'Onchocerciasis-related Epilepsy'; Kaiser et al., 'Association'; Wamala et al., 'Nodding Syndrome'; Winkler et al., 'Head Nodding Syndrome'; König, 'Role of Onchocerca Volvulus'. The mission doctor at Mahenge, Alois Gabathuler, had argued repeatedly in the 1940s that Ulanga, and particularly the area around Mahenge, was a region of particularly high infestation with the parasite onchocerca volvulus. TNA 26367: Gabathuler, Letter to Director of Medical Services, Mahenge 12.07.1945; Maria Gabathuler and Alois Gabathuler, 'Report of Onchocerciasis'. Also Geigy, Colas and Fernex, 'Endemic Onchocerciasis'.
87 With a map: Vogel, 'Mystery Disease'; Föger et al., 'Nakalanga Syndrome'.
88 Spencer, Palmer and Jilek-Aall, 'Nodding Syndrome'.
89 Williams, 'Nodding Syndrome'.
90 Jilek-Aall, 'Fifty Years Epilepsy Clinic'.
91 Spencer, Schmuthard and Winkler, 'Nodding Syndrome'.

92 Van der Weegen, *Shifting Meanings of Illness*, chapter 4.
93 Heaton, *Black Skin, White Coats*, conclusion; Sadowsky, *Imperial Bedlam*; McCulloch, *Colonial Psychiatry*.
94 Williams, 'Nodding Syndrome'.
95 Tilley. 'Global Histories, Vernacular Science'.
96 Colebunders, Irani and Post, 'Nodding Syndrome'; Årdal and Røttingen, 'Financing'.

Bibliography

African Medical and Research Foundation, *The Health Services of Tanganyika; A Report to the Government* (London: Pitman Medical Publishers, 1964).

Anon., 'Bericht des Schwestern-Institutes und Töchterpensionates Baldegg', *Jahrbuch des Akademischen Missionsbundes Freiburg* 12 (1931): 120–2.

Anon., 'Eine Missionsschwester erzählt', *Jahresbericht der Schweizer Kapuziner in Afrika* (1951): 24–8.

Årdal, Christine and J.-A. Røttingen, 'Financing and Collaboration on Research and Development for Nodding Syndrome', *Health Research Policy and Systems* 14 (2016): 19.

Bains, Jatinder, 'Race, Culture and Psychiatry: A History of Transcultural Psychiatry', *History of Psychiatry* 16 (2005): 139–54.

Benedict, Ruth, *Patterns of Culture* (New York: Penguin Books, 1952 [1st ed. 1934]).

Brentano, Clemens, *Die Barmherzigen Schwestern in Bezug auf Armen- und Krankenpflege*, Gesamtausgabe, vol. 14 (München and Leipzig: Georg Müller, 1912).

Colebunders, R., J. Irani, and R. Post, 'Nodding Syndrome – We Can Now Prevent It', *International Journal of Infectious Diseases* 44 (2016): 61–3.

A. T. and G. M. Culwick, 'Treatment of Fits By the Wambunga', *Man* 34 (1934): 136.

de Reuck, Anthony and Ruth Porter, *Transcultural Psychiatry* (London: Churchill, 1965).

Denisenko, Marina, *Perceiving Epilepsy: Interpretations of Kifafa in Kigamboni, Tanzania* (master's thesis, Leiden University, 2013).

Dreier, Marcel, *Health, Welfare and Development in Rural Africa* (PhD diss., University of Basel, 2015).

Fischli, Fridolin, 'Gruss Aus Kwiro', *Missionsbote der Schweizer Kapuziner in Afrika* 12 (1932): 43–4.

Fleischmann, E., S. Grypma, Mi. Marten, and I. M. Okkenhaug, eds., *Transnational and Historical Perspectives on Global Health, Welfare and Humanitarianism* (Kristiansand: Portal, 2013).

Föger, K., G. Gora-Stahlber, J. Sejvar, E. Ovuga, L. Jilek-Aall, E. Schmutzhard, C. Kaiser, and A. S. Winkler, 'Nakalanga Syndrome: Clinical Characteristics, Potential Causes, and Its Relationship with Recently Described Nodding Syndrome', *PLOS Neglected Tropical Diseases* 11 (2017): e0005201.

Gabathuler, Maria and Alois Gabathuler, 'Report of Onchocerciasis in the Ulanga District (Eastern Province, T.T.)', *East African Medical Journal* XXIV (1947): 188–95.

Geigy, Rudolf, J. Colas, and M. Fernex, 'Endemic Onchocerciasis in the Ulanga Area, Tanzania', *Acta Tropica* XXII (1965): 70–3.

Green, Maia, *Priests, Witches and Power: Popular Christianity After Mission in Southern Tanzania* (Cambridge: Cambridge University Press, 2009).

Gürber, Valentina, 'Das Heim für die Epileptischen bei der Missionsstation Kwiro', *Missionsbote der Schweizer Kapuziner in Afrika* 15 & 16 (1935 & 1936): 185–8 & 9–10.

Haerdi, Fritz, *Die Eingeborenen-Heilpflanzen des Ulanga-Distriktes Tanganjikas* (Basel: Verl. für Recht und Gesellschaft, 1964).

Häne, Ansgar, 'Vom Dienst an den Kranken in der Mission der Schweizer Kapuziner im Vikariat Daressalaam', *Missionsärztliche Caritas* (1939): 30–2.

Harries, Patrick and Marcel Dreier, 'Medizin und Magie in Afrika', in *Nach Feierabend: Gesundheit* (Zürich: Diaphanes, 2012), 85–104.

Heaton, Matthew M., *Black Skin, White Coats: Nigerian psychiatrists, decolonization, and the Globalization of Psychiatry* (Athens, Ohio: Ohio University Press, 2013).

Hochstrasser, Xaver, 'Medizinisches Rätsel und Geissel für das Volk der Acholi', *Neue Zürcher Zeitung (NZZ)*, 12 December 2013, 54–5.

Hodel, Angelina, *Lehrbuch Der Krankenpflege* (Baldegg: Selbstverlag der Pflegerinnenschule Baldegg, 1927 [1st ed. 1916]).

Huonker, Thomas, *Diagnose: moralisch defekt* (Zürich: Orell Füssli, 2003).

Janzen, John M., 'Therapy Management: Concept, Reality, Process', *Medical Anthropology Quarterly* 1 (1987): 68–84.

Jilek-Aall, Louise, 'Geisteskrankheiten und Epilepsie im tropischen Afrika', *Fortschritte der Neurologie, Psychiatrie und ihrer Grenzgebiete* 32 (1964): 213–59.

———, 'Epilepsy in the Wapogoro Tribe in Tanganyika', *Acta Psychiatrica Scandinavica* 41 (1965): 57–86.

——— and W.G. Jilek, 'The Problem of Epilepsy in Rural Africa: 'Kifafa' in a Tanzanian Tribe', *Transcultural Psychiatry* 7 (1970): 43–8.

———, 'Kifafa: A Tribal Disease in an East African Bantu Population', in *Anthropology and Mental Health: Setting a New Course*, ed. Joseph Westermeyer (The Hague, Paris: Mouton, 1976), 57–67.

———, *Call Mama Doctor. African Notes of a Young Woman Doctor* (Seattle: Hancock House, 1979).

———, W.G. Jilek, and J. R. Miller, 'Clinical and Genetic Aspects of Seizure Disorders Prevalent in an Isolated African Population', *Epilepsia* 20 (1979): 613–22.

———, *Working with Dr. Schweitzer: Sharing His Reverence for Life* (Seattle: Hancock House, 1990).

——— and Henry T. Rwiza, 'Prognosis of Epilepsy in a Rural African Community: A 30-Year Follow-up of 164 Patients in an Outpatient Clinic in Rural Tanzania', *Epilepsia* 33 (1992): 645–50.

———, M. Jilek, J. Kaaya, L. Mkombachepa, and K. Hillary, 'Psychosocial Study of Epilepsy in Africa', *Social Science & Medicine* 45 (1997): 783–95.

———, 'Morbus Sacer in Africa: Some Religious Aspects of Epilepsy in Traditional Cultures', *Epilepsia* 40 (1999): 382–6.

———, 'Forty Years of Experience with Epilepsy in Africa', in *Epilepsy in Our World: Stories of Living with Seizures from Around the World*, ed. Steven C. Schachter and Lisa F. Andermann (New York: Oxford University Press, 2008), 33–47.

———, 'Fifty Years Epilepsy Clinic and Research in Mahenge, Tanzania. Summarizing Report, December 2010', http://mahenge.wordpress.com/about/mahenge-Epilepsy-clinic/ (accessed 2010).

Jilek, W.G. and Louise Jilek-Aall. 'Psychiatric Concepts and Conditions in the Wapogoro Tribe of Tanganyika', *Beiträge zur vergleichenden Psychiatrie* 5 (1967): 205–28.

———, 'The Image of the African Medicine-Man', *Bibliotheca Psychiatrica Et Neurologica* 133 (1967): 165–78.

———, 'The Early History of Cultural Psychiatry (1820–1980)', *World Cultural Psychiatry Research Review* 9 (2014): 3–15.

Kaiser, C., T. Rubaale, E. Tukesiga, W. Kipp, G. Kabagambe, J. O. Ojony, and G. Asaba, 'Association Between Onchocerciasis and Epilepsy in the Itwara Hyperendemic Focus,

West Uganda', *The American Journal of Tropical Medicine and Hygiene* 85, no. 2 (2011): 225–8.

Keintzel, Brigitta, *Wissenschafterinnen in und aus Österreich: Leben – Werk – Wirken* (Wien: Böhlau, 2002).

Kilonzo, G. P. and N. Simmons, 'Development of Mental Health Services in Tanzania: A Reappraisal for the Future', *Social Science & Medicine* 47 (1998): 419–28.

König, R., A. Nassri, M. Meindl, W. Matuja, AR. Kidunda, V. Siegmund, G. Bretzel, T. Löscher, L. Jilek-Aall, E. Schmutzhard, and AS. Winkler, 'The Role of Onchocerca Volvulus in the Development of Epilepsy in a Rural Area of Tanzania', *Parasitology* 137, no. 10 (2010): 1559–68.

Korevaar, Daniël A. and B.J. Visser, 'Reviewing the Evidence on Nodding Syndrome, a Mysterious Tropical Disorder', *International Journal of Infectious Diseases* 17 (2013): e149–52.

Krahl, Wolfgang, 'Laudatio auf Professor Wolfang Jilek zur Ernennung als Ehrenmitglied der Agem', *Curare* (2003): 168–71.

———, 'Zum 80. Geburtstag von Louise Jilek-Aall', *Curare* 35 (2012): 324–5.

Lambo, T.A., 'The Village of Aro', *Lancet* 284 (1964): 513–14.

Larson, Lorne, 'A History of the Mahenge (Ulanga) district, ca. 1860–1957' (PhD diss., University of Dar es Salaam, 1976).

Lischer, Erika, *50 Jahre Baldeggerschwestern in Tansania 1921–1971* (Baldegg and Dar es Salaam: Institut Baldegg, 1971).

Mahone, Sloan, 'East African Psychiatry and the Practical Problems of Empire', in *Psychiatry and Empire*, ed. Sloan Mahone and Megan Vaughan (London: Palgrave Macmillan, 2007), 41–66.

Marin, B., M. Boussinesq, M. Druet-Cabanac, J. Kamgno, B. Bouteille, P.-M. Preux, 'Onchocerciasis-Related Epilepsy? Prospects at a Time of Uncertainty', *Trends in Parasitology* 22 (2006): 17–20.

Matuja, W.B.P., G. Kilonzo, P. Mbena, RL Mwango'mbola, P. Wong, P. Goodfellow, L. Jilek-Aall, 'Risk Factors for Epilepsy in a Rural Area in Tanzania', *Neuroepidemiology* 20 (2001): 242–7.

McCulloch, Jock, *Colonial Psychiatry and the 'African Mind'* (Cambridge: Cambridge University Press, 1995).

Mosser, P., 'The Pattern of Epileptic Seizures in Rural Tanzania', *Journal of Neurological Sciences* 258 (2007): 33–8.

Neuman, R.J., J.M. Kwon, L. Jilek-Aall, H.T. Rwiza, J.P. Rice, and P.J. Goodfellow, 'Genetic Analysis of Kifafa, a Complex Familial Seizure Disorder', *American Journal of Human Genetics* 57 (1995): 902–10.

Opler, Marvin K., *Culture and Mental Health: Cross-Cultural Studies* (New York: Palgrave MacMillan, 1959).

Prince, Ruth, 'Situating Health and the Public in Africa', in *Making and Unmaking Public Health in Africa*, ed. Ruth J. Prince and Rebecca Marsland (Athens: Ohio University Press, 2014), 1–51.

Reynolds, Edward H. and M. R. Trimble, 'Epilepsy, Psychiatry, and Neurology', *Epilepsia* 50 (2009): 50–5.

Rwiza, H.T., W.B. Matuja, G.P. Kilonzo, J. Haule, P. Mbena, R. Mwang'ombola, L. Jilek-Aall, 'Knowledge, Attitude, and Practice toward Epilepsy among Rural Tanzanian Residents', *Epilepsia* 34 (1993): 1017–23.

Sadowsky, Jonathan, *Imperial Bedlam* (Berkeley: University of California Press, 1999).

Spencer, Peter S., V.S. Palmer, and Louise Jilek-Aall, 'Nodding Syndrome: Origins and Natural History of a Longstanding Epileptic Disorder in Sub-Saharan Africa', *African Health Sciences* 13 (2013): 176–82.

Spencer, Peter S., E. Schmutzhard, and A. S. Winkler, 'Nodding Syndrome in the Spotlight – Placing Recent Findings in Perspective', *Trends in Parasitology* 33 (2017): 490–92.

Temkin, Owsei, *The Falling Sickness*, 2nd ed. (Baltimore: Johns Hopkins Press, 1971).

Tilley, Helen, 'Global Histories, Vernacular Science, and African Genealogies; Or, Is the History of Science Ready for the World?', *Isis* 101 (2010): 110–19.

van Bemmel, Karin, I. Derluyn, and K. Stroeken, 'Nodding Syndrome or Disease? On the Conceptualization of an Illness-in-the-Making', *Ethnicity & Health* 19 (2014): 100–18.

van der Weegen, Kim, *Shifting Meanings of Illness: An Anthropological Study of Nodding Syndrome in Tanzania* (master's thesis, Utrecht University, 2014).

Vaughan, Megan, *Curing Their Ills: Colonial Power and African Illness* (Stanford, CA: Stanford University Press, 1991).

Vogel, Gretchen, 'Mystery Disease Haunts Region', *Science* 336, no. 6078 (2012): 144–6.

Wamala, J.F., M. Malimbo, F. Tepage, L. Lukwago, C.L. Okot, R. Cannon, A. Laudisoit, and R. Colebunders, 'Nodding Syndrome May Be Only the Ears of the Hippo', *PLOS Neglected Tropical Diseases* 9 (2015): e0003880.

Whyte, Susan Reynolds, 'Constructing Epilepsy: Images and Contexts', in *Disability and Culture*, ed. Benedicte Instad and Susan Reynolds Whyte (Berkeley: University of California Press, 1995), 226–45.

Widmer, Edgar, *Zur Geschichte der Schweizerischen ärztlichen Mission in Afrika* (Basel: Benno Schwabe, 1963).

Williams, Sarah C. P., 'Nodding Syndrome Leaves Baffled Scientists Shaking Their Heads', *Nature Medicine* 18, no. 3 (2012): 312–34.

Winkler, A.S., K. Friedrich, R. König, M. Meindl, R. Helbok, I. Unterberger, T. Gotwald, J. Dharsee, S. Velicheti, A. Kidunda, L. Jilek-Aall, W. Matuja, and E. Schmutzhard, 'The Head Nodding Syndrome – Clinical Classification and Possible Causes', *Epilepsia* 49 (2008): 2008–15.

———, B. Wallner, K. Friedrich, B. Pfausler, I. Unterberger, W. Matuja, L. Jilek-Aall, and E. Schmutzhard, 'A Longitudinal Study on Nodding Syndrome – A New African Epilepsy Disorder', *Epilepsia* 55 (2014): 86–93.

Witkowski, R., O. Prokop, E. Ullrich, and G. Thiel, *Lexikon der Syndrome und Fehlbildungen: Ursachen, Genetik and Risiken* (Berlin and Heidelberg: Springer-Verlag, 2013).

Wolfisberg, Nikolata, 'Aus den Missionen', *Providentia* 8 (1934): 43.

World Health Organization Division of Mental Health, *A Programme for Mental Health in a Country Struggling for Development: The Tanzanian Experience* (Geneva: WHO, 1987).

9 Standards and standardisations

The history of a malaria vaccine candidate (SPf66) in Tanzania

Lukas Meier

Introduction

In a review on six malaria vaccine trials using the vaccine candidate SPf66, malaria experts Patricia Graves and Hellen Gelband concluded that there is neither reason to support the introduction of SPf66 as a public health measure for malaria prevention globally, nor is there any justification for further trials with SPf66 in Africa.[1] This verdict marked the endpoint of two decades of heated debates about this vaccine that once again raised the hopes of having a powerful technological tool to combat the deadly disease.

SPf66 was produced by the Colombian scientist Manuel Elkin Patarroyo working at the *Instituto National de Inmunologia* in Bogotá. It was the first product in a series of malaria vaccines that has been synthesised chemically rather than made from genetically altered or dead pathogens (viruses or bacteria).[2] Patarroyo's vaccine also differed further from earlier efforts in that it was designed, not to prevent blood-stage infection, but to reduce the number of parasites in the blood – preventing life-threatening malaria while allowing natural immunity to develop. Thus, the vaccine's major advantage was that it was not tailored to the needs of tourists or business travellers, but to those of people living in malaria-endemic areas.

SPf66 has become history. In fact, it would be easy to regard the appearance and disappearance of SPf66 as just one more episode in the whole series of unsuccessful attempts to come up with a malaria vaccine.[3] The reasons of this general failure are multi-faceted and historians of science and medicine tirelessly added major or minor points to the picture.[4]

This paper is not an investigation about *why* Manuel Patarroyo's vaccine candidate did not match the high hopes of Western malariologists. Rather, the history of SPf66 aims to understand *how* science is organised and how the attempts to test the vaccine on a global scale questions the construction of a "living laboratory" in one specific African community.[5] The major argument here is that even though the standardisation of trial protocols and trial populations were seen as basis for the successful execution of the SPf66 trial, the social reality of African villages would not allow for a complete control of all the internal and external factors. It is only in juxtaposing the local strategies of standardisation with trial arrangements on a global scale that those local characteristics of trial sites became evident.

Drawing extensively on the historical material of the Swiss Tropical Institute (STI), one of the major players in the SPf66 trial in Tanzania, as well as on oral evidence from the trial population in Idete/Tanzania, the narrative is organised chronologically and switches constantly between a global and a local perspective. The story follows the scientific process ranging from the production of the vaccine to the trial population in rural Tanzania, the interpretations of the results among the "scientific community" and ends with the failed attempts to reproduce the results in other places. A first part then reiterates the political implications surrounding the fact that the vaccine candidate was produced in the Third World rather than in Europe or the USA and the question of how the trial in Idete was organised. A second part takes a closer look on the impact of the vaccine trial on the local population and explores the local dynamics unleashed by the announcement of hosting a vaccine trial in the village as well as the local understandings of this new technology.[6] The third part is concerned with the interpretations of the results in international scientific circles while the last part investigates the failed attempts to standardise field trials on a global scale.

SPf66 – a vaccine candidate from the South

When, in the late 1980s, Manuel Patarroyo proved SPf66 to be safe and immunogenic in *Aotus* monkeys and subsequently also in humans, hardly anybody in the scientific community would have questioned the need for what was seen as a scientific breakthrough. Soon after the publication of his results in *Nature*, however, several malariologists expressed scepticism as to the reproducibility of the results. Though criticisms were also voiced by Colombian scientists, by far the loudest objections came from the West, where such a success from the Third World 'represents a blow to the scientific establishment that believes itself to be the trustee of malaria vaccine research'.[7] Nevertheless, an ad hoc committee of the World Health Organization (WHO) and the Pan American Health Organization visiting Bogotá in 1990 concluded that 'SPf66 merits further study' and recommended that 'randomized, placebo-controlled trials should be carried out urgently among children living in areas of high transmission, particularly in Africa'.[8]

Setting up such trials was, however, highly contentious. The UK Medical Research Council (MRC), which operated large laboratories in the Gambia, twice refused to test the product on the grounds that the required technical information about the formulation of the vaccine was still lacking. On the basis of such limited information, the argument went, it would never allow a trial to be conducted in Britain and therefore it refused to do so in Africa.[9]

The MRC's hesitation opened up possibilities for other research groups: Pedro Alonso – a Spanish malariologist collaborating with Patarroyo in Colombia and Brian Greenwood, the principal investigator engaged for the Gambia trials – approached Marcel Tanner, director of the Swiss Tropical Institute (STI), to explore whether the institute in Tanzania was willing and able to provide the infrastructure required for vaccine trials.[10]

When asked why the STI stepped into the breach and agreed to test the vaccine in rural Tanzania, Marcel Tanner replied that, from his point of view, there had – at that time – been no reason to hesitate. The primary objective of the first Phase III trial with SPf66 outside Latin America was to determine the efficacy of the vaccine in preventing malaria episodes in a hyperendemic area, highly representative of large parts of Africa.[11] To him and his collaborators, it was important to test the product independently of Manuel Patarroyo's group once all the ethical clearances were at hand.[12] Alonso's proposal also matched the overall aims of the STI's Kilombero Malaria Project, which prioritised research on possible vaccine candidates.[13] Naturally enough, with the Ifakara Center taking the lead in the SPf66 trial, the political tensions surrounding the vaccine did not vanish altogether. As it turned out, there were strains in the relationship between the principal co-investigators of the Tanzania trial – Marcel Tanner (STI), Pedro Alonso (Hospital Clinic I Provincial) and Thomas Teuscher (Ifakara Center) – and the chair of the WHO TDR/IMMAL Steering Committee, Howard Engers. Although funds were released by the WHO, the investigators suspected the organisation of giving priority to another SPf66 trial, led by Jerald Sadoff and the Walter Reed Army Institute of Research in Washington. The US Army researchers, collaborating closely with Patarroyo, used a vaccine developed in California (rather than Bogotá).[14] The impression that WHO was eager to downplay the Tanzanian trial was also reflected in the media. In an article in the *New Scientist*, for example, Phyllida Brown gave a detailed account of the development of a new malaria vaccine without even mentioning the Tanzanian trial.[15] In a letter to Odile Puijalon, chair of the IMMAL/TDR subcommittee, Pedro Alonso expressed his concerns as follows:

> I would like to draw your attention to a recent news report appeared in Science. This plus other circulating comments seem to be creating the distinct impression that the IMMAL supported Tanzanian trial is not the adequate trial. This is, to say the least, scientifically debatable. However, IMMAL may be seen as contributing to create this impression by recommending the urgent execution of another trial, which as we all know, intends to use the SPf66 molecule synthesized in the US. I am worried that IMMAL might be seen as promoting the use of a US synthesized product rather than one, at least equally good, coming from a developing country and which is already being used in a WHO supported trial.[16]

Against this political backdrop, it was clear to Marcel Tanner and his collaborators that their trial should at least not be 'criticized on design and operational grounds', and that their aim should be 'to generate an efficacy figure that will be accepted by the scientific establishment'.[17]

Standards and standardisations

The trial was a joint venture involving the Ifakara Centre, the Tanzanian National Institute for Medical Research (NIMR), the Swiss Tropical Institute (STI), the

London School of Hygiene and Tropical Medicine (LSHTM), the Instituto de Parasitologia (Spain) and the Foundation for Biomedical Research (Hospital Clinic i Provincial de Barcelona, Spain); it received funds from Spain and the UNDP/World Bank/WHO Special Program for Research and Training in Tropical Diseases (TDR).[18] This programme was launched in the 1970s in an attempt to raise the profile of parasitological research and to focus on six diseases particular prevalent in Africa: malaria, filariasis, schistosomiasis, trypanosomiasis, leishmaniasis and leprosy.[19] Like the (cancelled) trial in the Gambia, the Tanzanian project was a placebo-controlled, double-blind vaccine trial. Important for the later history of SPf66 in Idete was the definition of malaria and the methods to be used for recording cases of the disease. It was clear that in order to assess the impact of the vaccine on clinical malaria, a suitable definition had to be found for this endpoint. In areas where malaria is not highly endemic, case definition is relatively unproblematic: diagnosis is usually based on symptoms of fever and detection of malaria parasites in the blood. However, in most parts of sub-Saharan Africa, where malaria is endemic or hyperendemic, this case definition would be inadequate. There, 'to assume that a child who presents with fever and who has parasitaemia is ill from malaria is not valid, and will result in over-diagnosis'.[20] The researchers therefore defined clinical malaria as a combination of fever (37.5 °C or higher) and high parasite density – the latter being uncommon in asymptomatic cases of malaria. The organisation of a vaccine trial thus entailed adaptation to the new setting.

The work of the anthropologist Adriana Petryna investigates the pharmaceutical industry's growing demand of 'treatment-naïve' human subjects and the offshoring of clinical trials to mid- and low-income countries. In the logic of global research, this is legitimised by the fact that the trials constitute a social good in themselves, providing extensive health care for the people involved. Moreover, these experimental terrains are depicted in the language of 'humanitarian crises', creating spaces of emergency where research easily transgresses what might be ethically justifiable.[21] The context of the SPf66 trial in Tanzania was different. On the one hand, the logistics of clinical trials and field trials are not readily comparable. On the other, the scientific product in question was developed in a laboratory in Colombia and was therefore seen as a successful example of 'South-South collaboration', with Western assistance.[22] Petryna's argument is, however, compelling with regard to the different standards concerning risk-benefit ratios applied to medical research in Europe and Africa. The transfer of the vaccine from Latin America to rural Africa involved not merely the movement of a scientific object produced in Bogotá and formulated and bottled in Spain for injection into the right upper arm of 580 children aged 1–5 in rural Tanzania. It also involved the acceptance of certain risk associated with the administration of a substance tested on adults in countries of low malaria transmission to semi-immune children in an area of much higher disease prevalence.[23] While scientific objects and arrangements can be transferred from one place to another, health insurance schemes to cover the adverse effects of a vaccine seem to be more bound to specific locations. In this regard, it is worth noting the response given by a Swiss legal firm when it was asked whether the villagers involved in the trial could be included in a comprehensive health insurance scheme:

> From a legal point of view, the insurance coverage of the vaccination test program has to be in conformity with the requirements in the country in which the program is executed. I assume that legislation and relevant guidelines in Tanzania reflect the absence of general health insurance in that country and do therefore not require health insurance in connection with a vaccination test program. Guidelines existing in Switzerland are designed to protect Swiss population along the standards of the Swiss social security and health care system cannot be transferred to a developing country and applied to its inhabitants.[24]

In medical research, Africa thus evokes a double discourse and double standards. In the official discourse, the continent is an ideal site for testing new medications because diseases are widespread and the new products should reach those places where sufferers are living. At the same time, Africa constitutes a good testing ground because of the availability of more treatment-naïve subjects than in European countries and because the urgency of health interventions overshadows the lack of a legal framework that would protect individuals from adverse effects of the interventions in question. In stark contrast to the considerations about where research is to be carried out, how diseases are to be defined and how trials are to be organised, are the collective anxieties and individual decisions concerning participation in this form of research at the local level.[25] It is to these aspects that we now turn.

Idete

Idete was established as a ujamaa village in 1974, as part of the socialist villagisation programme, and buoyed by its location on both the road and the TAZARA railway line to Zambia. Most of the village's older dwellers had been forcibly relocated and given a plot where they could grow maize and naturally irrigated rice. There were no health facilities in the 1970s, and a single bicycle was the only means of transportation in an emergency.[26] My own research conducted during 2010 revealed contradictory perceptions of malaria. The differences were due not just to varying individual statements but also to the different sources consulted. Not surprisingly, in an area of high malaria transmission, the record book kept at the village dispensary registered 600–800 cases of malaria each month in the early 1990s.[27] Almost 20 years later, many villagers reported that malaria was no longer much of a problem because of the widespread use of mosquito nets. Others, again, reported that the disease still takes a steady toll among the village community. There is no doubt, however, that Idete has been visited frequently over the years by researchers working for the Tanzanian government and for the Ifakara Centre (renamed the Ifakara Health Institute/IHI in 2008). The village was included as a research site in the Kilombero Malaria Project, and today fieldworkers from the IHI still meticulously record the health situation, patterns of migration, evidence of prosperity and destitution, and birth and death rates for all households and at regular intervals. The IHI and Idete's dispensary staff were also the main disseminators of biomedical messages to villagers in Idete. They explained the connection between mosquitoes and malaria, advocated appropriate prevention measures and urged sufferers to seek medical advice as soon as the first symptoms appeared. In her study of community understanding of malaria, the anthropologist Susanna Hausmann Muela showed that the population living in and around Ifakara understood the biomedical message very well, but that this information coexisted and sometimes merged with pre-existing ideas about the disease – a phenomenon termed "*medical syncretism*".[28]

Vaccines for the prevention of childhood diseases were a familiar biomedical tool, and the Expanded Program on Immunization (EPI) – launched by the WHO and UNICEF in the mid-1980s to improve vaccine coverage among children – is still considered highly effective by the villagers. Nevertheless, the initial reaction to the announcement of the SPf66 trial included both relief and widespread unease.

Here, it is important to mention the methodological difficulties and constraints that I faced during my discussion with the trial population. Jensen Charles – the son of Charles Leutel, who was in charge of Idete's dispensary at the time of the SPf66 trial – still knew many of the parents who had allowed their children to be included in the vaccine trial and could easily arrange meetings with them. His close relationship with the participants could well explain a certain bias in their answers. More methodologically challenging, however, was the fact that our purposive sample included only those who had ultimately decided to participate in the trial and who were therefore assumed to have had a positive attitude towards the vaccination project. To counter methodological bias, I took an indirect approach, asking participants about the arguments of those who had refused to participate in the trial. I thus learned that the village had been very much divided on this issue: participants had been accused of rashly making their children available for a biomedical trial, the intention of which was to decimate the population – 'quite similar to the family planning initiatives', as one of the interviewees remarked.[29] People had also been accused of agreeing to participate purely because of the material benefits involved, and it had been claimed that malaria was too severe a disease to be successfully controlled by a vaccine.[30] One mother recalled:

> First, we were very much afraid on that case and why were we afraid? It was because people said that they take the blood and the blood is for business so they take a lot of blood from the bodies of the children so they are going to die. Secondly, other mothers around us, they said, ah, why are you taking your children to the vaccination station, because of the soap, you don't have money to buy soap, so you bring your children because of the soap. . . . But I have taken the child to the hospital because there is malaria in the household. Because of Skola (her daughter, LM). She was always suffering from malaria, Skola, she was always suffering. I said, ah, I don't try to hear this

words around so I will take my child – if she dies ok, if she's not dying, ok. If I don't take my child to the vaccination station she is going to die anyway. She is always suffering in the house. It is better to bring her to the hospital. If she dies, no problem. But around, some mothers insisted that I am going to take the child to the vaccination for the soap and they asked: Why? Don't you have money to buy soap? Only 100 Shillings for one piece of soap? But my child was still suffering from malaria, so I decided to take her, but there were some discussions around.[31]

The stories circulating within the village after the announcement of the vaccine trial stoked villagers' anxieties about having their children injected with an unknown substance.[32] The villagers' answers to the question of why they had participated in the trial fell into two broad categories. In the first of these, the severity of the disease was weighed against the risks of the vaccine – for example: 'We did not know what the vaccine was about, but we are very aware what malaria is all about, and that is why we were happy to take our children to the vaccination station'.[33] The second set of answers invoked social aspects and the vital role played by the village leadership in the villagers' decision-making: 'The government would never be able to sacrifice their people'.[34] Others again decided to participate because others were doing so, or were persuaded by villagers working at the dispensary.[35]

As it was not possible for villagers to assess the likely effects of SPf66 on the basis of their past experiences with vaccinations, their decisions on whether or not to participate in the trial were determined by perceptions of the threat of malaria, considerations about risks and potential benefits, and social and economic concerns. It is important to note that the uncertainties, concerns and mutual accusations arising at the local level were echoed among biomedical experts around the world; as we will see below, the scientific community was also deeply divided about the effects of SPf66.

The interpretation of truth

On 9 October 1994, the principal investigators and their collaborators published the data from the Idete trial in the *Lancet*. While they reported that the SPf66 vaccine 'reduces the risk of malaria among children highly exposed to natural infection', the estimated efficacy of 31% (95% confidence interval 0–52%) warranted only cautious optimism.[36] The authors concluded: 'The estimated efficacy of SPf66 is lower than that of most vaccines in use for other infections. However, since the burden of malaria morbidity and mortality is vast, measures with a moderate efficacy merit development'.[37] But not everyone within the scientific community agreed that the efficacy figure justified consideration of the use of SPf66 as a public health measure. Worse still, critical voices were raised concerning the reliability of the data presented. After publication of the *Lancet* article, Pierre Druilhe of the Institut Pasteur in Paris complained: 'We have no clear evidence about the efficacy or the inefficacy of SPf66'. Echoing concerns expressed by

Jean-François Trapé (ORSTOM) and Christophe Rogier (Institut Pasteur, Dakar), he claimed that 'the ways . . . the data were collected on the spot were deficient' and that 'the number of fever cases measured in nonvaccinated children was inexplicably low'.[38]

For the purpose of my argument, several aspects of the ensuing scientific debate are of interest, as they indicate the instability of experimental terrains and the gap between scientific assumptions and local realities. In a draft reply to the criticisms from Dakar and Paris, the scientists involved in the Idete trial considered possible reasons why the number of episodes of clinical malaria was lower than would have been expected in an area of high malaria transmission.[39] It would appear that their idea of establishing standardised trial conditions, where all external factors could be brought under control, had been too optimistic. The first factor that could have explained the low number of cases reported was the free medical services provided at the dispensary during the trial – a finding of major public health relevance.[40] Secondly, the incidence was higher among children living closer to the dispensary, suggesting that reporting rates were lower for more remote parts of the village. Thirdly, the main data analysis depended on passive case detection (PCD), which meant that it included all cases of clinical malaria recorded at the dispensary. This approach was likely to exclude all cases of asymptomatic parasitemia. Fourthly, the frequency of clinical malaria decreased as trial subjects got older.[41] Thus, not only were the results dependent on an appropriate definition of malaria in a hyperendemic area, but the findings were influenced by the nature of the intervention itself (improved health service delivery, ageing of the trial population). But if, as the scientists argued, the social fabric of the population is fundamentally changed by the performance of a trial, then the very notion of the medical trial is called into question.

Mechanisms of critical self-reflection are inherent in all scientific activities. The application of scientific results requires a certain level of approval and acceptance among the wider scientific community. The less one's results are accepted by outside experts, the more critical one tends to be of one's own products. The Idete scientists, at least, were preoccupied by the question whether or not the reactions to the *Lancet* study were justified. In a letter to a collaborator at the LSHTM headed "What are we trying to do?", an STI investigator reflected on how science progresses:

> I thought that I was trying to find out from the data as much as possible about what SPf66 is likely to be doing in these children. This involves making hypotheses, and testing them against the alternatives using not just the evidence from the trial but also what we already know from other sources. Since the primary effect (published in the Lancet) that we saw is so small, and subject to considerable uncertainty, nothing relating to clinical episodes can be expected to be free of reasonable doubt, and so we must hedge all conclusions with words like "likely", "suggest", "appear to indicate". You would not want to send anyone to prison on the strength of evidence as weak as this. However, this evidence is the best that we have. If someone comes

> along next week and shows that we are wrong, I would not be surprised. This is how science progresses.[42]

The collaborator at the LSHTM, however, was less inclined to accept this Popperian view of science's progress and argued that SPf66 should not yet be abandoned: research should now focus on large-scale application and on the question of whether SPf66 would have any effect on severe disease and death. In this, she shared the views of most public health experts pursuing effective means of protecting people from malaria. However, the future of SPf66 would depend on the outcomes achieved in other endemic settings. Two trials – in the Gambia and Thailand – sealed the fate of SPf66, ultimately making it just one more episode on the rocky road to a malaria vaccine.

The global blow

The Gambian trial, conducted in collaboration with Patarroyo's group, included the vaccine produced in Colombia as well as the US formulation. But the results obtained with SPf66 were far from encouraging. In a letter to Patarroyo, Brain Greenwood wrote:

> At the beginning of May, the MRC Trial Monitoring Committee, chaired by Malcolm Molyneux met to review the side effects and serological data that we had obtained after administration of the third dose of American or Colombian SPf66 to the children in our pilot trial. There were no major differences in the incidence of local or systemic side effects among children who had received SPf66 or IVP (Polio-Vaccine, LM). However, the committee noted that more children in the malaria vaccine group had had a positive blood film with or without accompanying fever than had children given polio. The tendency was present with both the American and Colombian vaccines and was most prominent in the group that had received the high dose Colombian vaccine.[43]

The efficacy figure reported for the Gambian trial was a mere 3%, and the SPf66 and placebo groups did not differ significantly in parasitic rates or in any other index of malaria.[44] However, Marcel Tanner believed that the results from this parallel study should not affect the decision to assess the efficacy of SPf66 in infants under one year of age: 'Given the results from the Tanzanian trial for an area of high and perennial transmission, trials with immunization of infants and with severe malaria as endpoint represent a logical evolution'. Pointing to major differences between the two trials in terms of the protocol and population, he argued that it would be premature to abandon SPf66 without having assessed its efficacy in infants living in Kilombero valley.

The argument based on differences in the local characteristics of study sites became more compelling after the publication of sobering news from Thailand. Here, the efficacy of SPf66 was assessed in a double-blind placebo-controlled trial involving Karen children at a military-controlled refugee camp. Principal investigator François Nosten (Mahidol University) and his co-workers concluded that 'there appears to be little justification for further trials with this vaccine'.[45] As mentioned above, this trial had long been a thorn in the side of the Swiss-Spanish-British-Tanzanian team because it used a version of SPf66 manufactured in the US rather than in Colombia. While these results probably dealt the final blow to the vaccine, questions about the comparability of populations, sites and products remained. While Marcel Tanner and Pedro Alonso accepted the conclusions drawn by Nosten's group, they raised a number of questions:

> Firstly, we have no evidence as to the possible role that the genetic make up of the population of parasites or volunteers may play in modulating efficacy. Secondly, there are a number of differences between the Thai trial and all other trials, including the type of placebo used, the very intense active daily case detection, and perhaps the specificity of the case definition used. Finally, and probably most important, the authors acknowledge that there is strong evidence that the US manufactured product is not identical to the Colombian manufacture SPf66. Not only are there differences in the proportion of monomer to polymer between the two products, but these differences have been shown to imply significant differences in immunogenicity both in mice and in humans, with the US manufactured version having been shown to be consistently less immunogenic than the Colombian SPf66.[46]

Besides the ethical question of why the persons in charge for the Thai trial had chosen a vaccine shown in previous studies to be less immunogenic than Colombian-manufactured SPf66, what the whole matter revealed was:

> once again the difficulty and importance of standardizing products for field trials . . . as well as the need of standardizing protocols and procedures in order to allow for adequate comparison of results. The Thai trial has also highlighted the need to choose the trial sites and subsequently interpret the results in relation to the relevance to the different populations of the malaria endemic areas. In other words, how relevant are the results obtained in northern Thailand among refugees living in a camp under daily medical surveillance and the protection of the Thai army, to the malaria endemic populations of sub-Saharan Africa?[47]

Conclusion

Looking back, the history of malaria vaccine development has much to say about the difficulties encountered in translating medical research into sustained public health action. In the case of SPf66, two points should be mentioned in particular. Firstly, it suffered from a birth defect – rather than being "merely" a chemical substance, its life story was shaped by political and economic interests. The fact that SPf66 was produced in Colombia, and that the patent rights were not sold to the pharmaceutical industry but assigned to the WHO, was more than enough to provoke serious doubts about its efficacy and safety. Secondly, there is the difficulty of standardising scientific products, trial sites and trial protocols. The SPf66 compound manufactured in the US laboratories was not identical to that produced in Colombia, and the differences between the results of the Gambian and Tanzanian trials were partly attributable to the different study protocols. However, quite apart from methodological differences, the attempt to establish laboratory conditions in African villages was impeded by "reality effects": the presence of Western experts in the village, the desires aroused by the objects (nets, sodas, soaps, etc.) they introduced, the renovation of the dispensary and provision of essential drugs, the ageing of the trial population – all these elements were not only beyond the scientists' control but influenced the results of the study.

Notes

1 Graves and Gelband, 'Vaccines for Preventing Malaria', 11. I am very much indebted to Jensen Charles Leutel (Ifakara Health Institute) who translated the interviews held with the population of Idete/Tanzania. The article is based on my PhD-study financed by the R. Geigy Foundation in Basel.
2 Spurgeon, *Southern Lights*, 41; a more scientific explanation would be: 'SPf66 is a synthetic peptide consisting of aminoacid sequences derived from three asexual stage proteins (85, 55 and 35 kD) linked by NANP sequences derived from the circumsporozoite protein of P. falciparum', see: Alonso et al., 'A Trial of SPf66', 3.
3 See for example Desowitz, *The Malaria Capers*.
4 See for example Packard and Gadelha, *Land Filled with Mosquitoes*, 197–213; Packard, *Making of a Tropical Disease*; and Litsios, 'Malaria Control', 255–78.
5 Tilley, *Africa as Living Laboratory*.
6 Fairhead, 'Public Engagement with Science?', 103–16.
7 MacLeod, 'Creation of First Malaria Vaccine', 1321.
8 Alonso, Teuscher and Tanner, 'Research for Development', 99.
9 Brown, 'Colombia's Malaria Vaccine Approved', 1.
10 Correspondence Pedro Alonso (Archiv Schweizerisches Tropeninstitut, ASTI), 1–2.
11 Secondary objectives were (2) to measure any immediate or delayed side-effects associated with the administration of SPf66 in a semi-immune population and (2) to assess the immunogenicity of each dose of SPf66, see: Alonso et al., 'A Trial of SPf66', 9.
12 Personal communication.
13 Kilombero Malaria Project, Summary (ASTI), 1.
14 Brown, 'Colombia's Malaria Vaccine Approved', 1.
15 Ibid.; see also: Brown, 'Malaria Vaccine Trials Stalled', 1.
16 Pedro Alonso (ASTI), 2.
17 Marcel Tanner (ASTI), 1.
18 Alonso et al., 'Synthetic Malaria Vaccine', 182.
19 Charlesworth et al., *Life among the Scientists*, 230.
20 Alonso et al., 'Synthetic Malaria Vaccine', 184.
21 Petryna, 'Globalizing Human Subjects Research', 51; Petryna, 'Clinical Trials Offshored', 21–40; Petryna, *When Experiments Travel*, 1–272.
22 Alonso, Teuscher and Tanner, (ASTI).
23 Pedro Alonso noted: 'The parameters of toxicity, safety and immunogeneicity may vary from population to population due to the different genetic background of those populations and to the pressure of the disease. For that reason it is of the outmost importance to carefully determine those parameters before the main trial starts in order

to avoid potential unwanted risks. Thus, it is recommended that the main trial should not be carried out until these parameters are clearly defined for the present situation', see: Alonso, Ethical Evaluation of the Study Protocol on the 'Trial of a Potential Synthetic Malaria Vaccine SPf66 in Tanzania (ASTI)', 1; unpublished letter, archives of the Swiss TPH (Swiss Trophical and Public Health Institute).

24 Letter of Christian Brückner on the Malaria Vaccination Program in Tanzania, 06.07.1992, 2; unpublished letter, archives of the Swiss TPH (Swiss Trophical and Public Health Institute).
25 Fairhead, 'Public Engagement with Science?', 103–16.
26 Interview with LR in Idete, 03.04.2010.
27 Mpanju and Molyneux, SPf66 Candidate Malaria Vaccine Trial Ifakara Centre, (ASTI), 2.
28 Hausmann Muela, 'Community Understanding of Malaria'.
29 Interview with JM, 03.04.2010.
30 Interview with AL, 17.04.2010.
31 Interview with GN in Idete, 22.03.2010.
32 Leach and Fairhead, *Vaccine Anxieties*.
33 Interview with MK, 22.03.2010.
34 Interview with AD, 03.04.2010.
35 Ibid.
36 Alonso et al., 'Randomized Trial', 1180 (After a preliminary study in 1992, the trial started in Idete in February 1993).
37 Ibid., 1181.
38 Interrogations sur Le Vaccine Colombien Contre le Paludisme, (ASTI).
39 Draft Reply to Lancet Letters (ASTI).
40 See also: Alonso et al., 'Randomized Trial', 1180.
41 Ibid.
42 What Are We Trying To Do?, (ASTI), 1–6.
43 Brian Greenwood to Manuel Patarroyo, (ASTI), 1.
44 The Lancet: Press Release, Malaria Vaccine not so Efficacious.
45 Nosten et al., 'Placebo-Controlled Trial', 707.
46 Pedro Alonso and Marcel Tanner (ASTI), 1.
47 Ibid., 2.

Bibliography

Alonso, Pedro, Thomas Teuscher, and Marcel Tanner, 'Research for Development or Development for Research', *Vaccine* 12 (1994): 99–101.

———, Thomas A. Smith, Joanna Armstrong Schellenberg, and Clara Menéndez, 'Randomized Trial of Efficacy of SPf66 Vaccine Against Plasmodium Falciparum Malaria in Children in Southern Tanzania', *The Lancet* 344 (1994): 1175–81.

———, Marcel Tanner, Thomas A. Smith, Richard J. Hayes, Joanna Armstrong Schellenberg, Myriam C. López, Inez Bastos de Azevedo, Clara Menéndez, Edith Lyimo, and Niklaus Weiss, 'A Trial of the Synthetic Malaria Vaccine SPf66 in Tanzania: Rationale and Design', *Vaccine* 12, no. 2 (1995): 181–6.

Brown, Phyllida, 'Malaria Vaccine Trials Stalled a Second Time', *New Scientist*, 21/28 December 1991.

———, 'Colombia's Malaria Vaccine Approved for Trials', *New Scientist*, 26 September 1992.

Charlesworth, Max, Lyndsay Farrall, Terry Stokes, and David Turnbull, *Life among the Scientists: An Anthropological Study of an Australian Scientific Community* (Melbourne, New York: Oxford University Press, 1989).

Desowitz, Robert S., *The Malaria Capers: More Tales of Parasites and People, Research and Reality* (New York: W. W. Norton, 1991).

Fairhead, James, 'Public Engagement with Science? Local Understandings of a Vaccine Trial in the Gambia', *Journal of Biosocial Science* 38 (2006): 103–16.

Graves, Patricia M. and Hellen Gelband, 'Vaccines for Preventing Malaria (SPf66)', *The Cochrane Library* 2 (2009, reprint): 1–30.

Leach, Melissa and James Fairhead, *Vaccine Anxieties: Global Science, Child Health and Society* (London: Routledge, 2008).

Litsios, Socrates, 'Malaria Control, the Cold War, and the Postwar Reorganization of International Assistance', *Medical Anthropology* 17 (1997): 255–78.

MacLeod, Kirsten, 'Creation of First Malaria Vaccine Raises Troubling Questions about "Intellectual Racism"', *Canadian Medical Association Journal* 153, no. 9 (1995): 1319–21.

Muela, Susanne Hausmann, 'Community Understanding of Malaria, and Treatment-Seeking Behavior in a Holoendemic Area of Southeastern Tanzania' (PhD diss., University of Basel, 2000).

Nosten, François, Christine Luxemburger, Dennis E. Kyle, W. Ripley Ballou, Janet Wittes, Eh Wah, Tan Chongsuphajaisiddhi, Daniel M. Gordon, Nicholas J. White, Jerald C. Sadoff, and D. Gray Heppner, 'Randomized Double-Blind Placebo-Controlled Trial of SPf66 Malaria Vaccine in Children in Northwestern Thailand', *The Lancet* 348 (1996): 701–7.

Packard, Randall and Paulo Gadelha, 'A Land Filled with Mosquitoes: Fred. L. Soper, the Rockefeller Foundation, and the Anopheles Gambiae Invasion of Brazil', *Parassitologia* 36 (1994): 197–213.

———, *The Making of a Tropical Disease: A Short History of Malaria* (Baltimore: Johns Hopkins University Press, 2007).

Petryna, Adriana, 'Globalizing Human Subjects Research', in *Global Pharmaceuticals: Ethics, Markets, Practices* (Durham and London: Duke University Press Books, 2006), 33–60.

———, 'Clinical Trials Offshored: On Private Sector Science and Public Health', *BioSocieties* 2 (2007): 21–40.

———, *When Experiments Travel: Clinical Trials and the Global Search for Human Subjects* (Princeton: Princeton University Press, 2009).

Spurgeon, David, *Southern Lights: Celebrating the Scientific Achievements of the Developing World* (Ottawa: International Development Research Centre, 1995).

Tilley, Helen, *Africa as Living Laboratory: Empire, Development, and the Problem of Scientific Knowledge, 1870–1950* (Chicago: University of Chicago Press, 2011).

Part III

International discourses, transnational circulations of knowledge

10 The politics and production of history on the birth of archaeology at the Cape (1827–2015)

Tanja Hammel

The 1820 settler community of the rural area around Grahamstown (population c. 8,000 in 1865), Albany, at the eastern frontier of the Cape Colony, produced a remarkable group of antiquarians during the second half of the 19th century. These contributed to the then inextricable, yet slowly differentiating disciplines of geology, palaeontology and archaeology.

Archaeology from the very beginning was 'a science of the past – expressed in collection – to articulate a history and geography of possession'.[1] It was inextricably linked with the annexing of land. Settler nations – more than other imagined communities – have a particular need and urgency in imagining their origins. In the dispossessory take over of land, the land's past and its past inhabitants both have to be aligned with the new reality of the incoming settler communities in order to provide them with legitimacy and a sense of continuity.

In this chapter, I will analyse the contributions of Thomas Holden Bowker and James Henry Bowker to the early settler endeavours to create archaeological collections and knowledge. Diverse people at different times have acknowledged both of them as the first antiquarian in South Africa and one of the first in the world.[2] Both of them actively fought against the African inhabitants of the lands they wished to conquer for the settler farmers. I use the two brothers to examine the discourses around the travelling artefacts[3] and thereby contribute to the critical history of archaeology.

Much has been published on the professionalisation of the discipline and the corresponding establishment of the first chair of archaeology in South Africa in 1923, the "South Africanization" of the discipline under Astley John Hilary Goodwin (1900–1959) as well as the importance of Africans in archaeological excavations.[4] Yet the discipline's birth has been neglected.

Besides the Bowkers' private and scientific correspondence, contemporary publications and 20th-century archaeologists' discussions of their contributions, I examine stone implements in South African and British collections and how they have been displayed in museums. The time period spans from the allegedly first archaeological finds by a white settler to my last visit to museums where stone artefacts have been on display. I first elaborate on the beginning of archaeological research at the Cape and contextualise it within the settlers' claims for land. I focus on the prime interpretation of the artefacts among the Bowker siblings and then trace the history of narrating the story of the first finds in a local museum over time, which provides insights into how curators have dealt with the colonial past. I also explore what happened to the artefacts which had been sent to British museums over time.

This chapter, thus, aims to contribute to the social history of archaeology, to an entangled history of the micro-politics of knowledge as well as the emerging field of South African Empire Studies, all part of a wider endeavour to rethink South Africa's past.

Between struggles for compensation and carving out a career

Since 1835 Thomas Holden Bowker had been concerned with 1820 settlers' compensation claims for Frontier War losses and had been known in Albany under the cognomen of "Compensation Bowker".[5] He acted a member of the House of Assembly (the lower house) and the legislative council (the upper house) of the Parliament of the Cape of Good Hope in Cape Town (founded in 1853) for Albany, Victoria East and Queenstown from 1854 to 1863. After the Eighth Frontier War (1850–3), he was part of a commission which distributed farms to supposedly deserving burghers. Bowker is remembered in family traditions as having selflessly waited for the last piece of available land to be distributed before claiming his share, but that this would ultimately prove unsuitable for farming.

On experiencing subsequent financial difficulties, he wrote a memorandum which was discussed in parliament during the 1857–8 sessions, when he was in Cape Town. In the memorandum he listed his achievements and declared it his aim to achieve recompense for his deeds as an overlooked national hero.[6] Due to the government's lack of a frontier defence plan, he argued, he had had to sacrifice his career and fortune to protect his compatriots.

He also demanded land, but not just any land: he aimed for Theopolis, which had been a long-established London Missionary Society (LMS) station. It was situated to the east of Grahamstown. This site was of symbolic importance, since Bowker was strongly opposed to missionaries in general and especially those who had lived and worked at Theopolis. The Scottish missionary Dr. John Philip (1775–1851), the superintendent of the LMS stations in South Africa, had frequently criticised the colonists and the colonial government for their treatment of Africans and Bowker was particulalry opposed to him.[7] Had Bowker received former LMS land, this would have been a significant victory for the settlers in this ideological battle as well as a measure of revenge for Bowker.

Bowker's memorandum was dicussed in Government House on 31 March 1858 by Sir George Grey, who was the governor of the Cape Colony from 1854 to 1861.[8] A committee questioned witnesses and resolved to report on the matter to the House before adjourning. Ultimately, the government did not recognise any of Bowker's claims, but Grey offered to lend him 100 pounds.[9]

At the same time, he turned to a quest for archaeological artefacts. During the same parliamentary session, in March 1858, he visited the curator Edgar Leopold Layard at the South African Museum in Cape Town. Bowker entered Layard's office while he was unpacking a collection of flint artefacts from Copenhagen.

According to *The Lower Albany Chronicle*, the 20-year-old Thomas Holden Bowker who had then been living at the Cape for seven years, had found flake points which he used as arrowheads on hunting expeditions in December 1827.[10] He had promised to send these to Layard, as the latter reported, if the barn, where Bowker had stored them on the farm Tharfield, 'had remained undisturbed, and had escaped the ravages and burnings of [Bowker's] foes, the Kaffirs' in the intervening decades.[11] After his return from Cape Town in 1858, Bowker searched at the mouths of the Kowie and the Kleinemonde, two small rivers which opened on to the beach on his land, and found further stone implements.

Public recognition continued to elude him. He was not elected president of the Orange Free State in 1863, even though he had been asked to be a candidate and supported by the press.[12] Bowker thus searched for fame in other ways. For example, he sent a letter on his finds to Layard, who in turn forwarded it to professor Richard Owen at the British Museum in London. In this letter, Bowker maintained that the stone implements which he had found were produced by the same people who had made those which he had seen from Copenhagen.

Some of the perforated implements he had gathered bore the marks of strikes from a hammer or long, hard pebble, an act of which he did not believe "a stalwart savage" was capable; as Bowker argued, Africans had no knowledge of any skills besides shooting. He claimed to have also found other, less complex implements which bore the mark 'of the savage whose ideas went no further in the art of stone cutting than is necessary for chipping a flint'.[13]

Debates on the producers of stone artefacts

In search of support for his theory that the implements he had found were made by white original people populating Southern Africa, Bowker first turned to his sister Mary Elizabeth Barber. By 1865, she had become a recognised botanist, entomologist and ornithologist.[14] Her reply brought to mind that she did not believe that the original inhabitants of the region had been white; if they had been, as she argued in a staunch Social Darwinist manner, they would not have subsequently vanished.[15]

Bowker then looked for support by the more influential former governor of the Cape Colony and then governor of New Zealand, George Grey. While governor at the Cape, Grey's schemes for colonial expansion had 'equated metropolitan utilitarianism with colonial despotism'.[16] Eager to eradicate tribalism and superstition and to spread Christianity and civilisation, Grey had aimed to integrate and assimilate Africans into a Cape Colony, which he had imagined as a union of states ready to take its place in 'the expanding universe of western civilization'.[17]

Grey held a wide interest in prehistory and ethnology.[18] In 1838, he had been the first to document rock art in the Kimberley region of Australia, arguing that it was hardly probable that a self-taught Aborigine could have created these paintings and that their origins were open to conjecture. The Aborigines whom he had asked, meanwhile had not claimed their ancestors to be the original artists of these paintings, but that 'the moon, who was a man' had created them, something which

Grey took as a reference to a white man. Like Bowker, Grey had also argued for two distinct styles of cultural artefacts. The rock art near the coast was 'nothing but the rudest scratches'. The more complex drawings, which culminated at the furthest point from the sea, however, pointed towards a lost tribe of whites in the interior.[19]

In a letter to Grey, Bowker argued that because the first European visitors had not seen the arrow and spear heads in use and the contemporary Khoekhoen did not use them, they had to be the only remaining evidence of the original inhabitants of the Cape, who would have been 'far anterior to the advent or immigration of' the Xhosa, Khoekhoen or San.[20] In line with Jean-Jacques Rousseau, who had claimed that "savages" had never used tools, and the Scottish anatomist, ethnologist and medic Robert Knox's conviction that the Cape had remained unaltered over time,[21] Bowker recorded that he had not seen any Africans using tools to make fire or obtain honey as he himself had always done.[22] As he considered Africans to be "living fossils" who had not developed since the pre-historic period, he assumed that the San would still use such Stone Age tools if they had once made them.

At the same time, settlers also perceived indigenous people as irritants who intruded on and disturbed their peace.[23] Settlers suffered from "mnemonic myopia", disregarding the histories that preceded their arrival and considering themselves to be the first actual inhabitants of the land on which they settled.[24] As the literary scholars Anna Johnston and Alan Lawson argue, a settler narrative follows two goals: it aims to suppress and efface the indigene, while, conversely, also seeking to indigenise the settler.[25] Archaeological discourse in Southern Africa is a clear example of how these two aims supplemented each other.[26]

Given that this genealogy of Thomas Bowker's archaeological contributions and land claims has never been traced, it is apt to show how archaeologists and museum curators in South Africa have remembered him.

"Screen memories", "anachronistic space" and visible concealment

In 1947 – the year before the Afrikaner National Party was able to gain a strong majority in the South African parliament and begin implementation of its apartheid ideology – a new display case was arranged in the second oldest museum in South Africa, the Albany Museum in Grahamstown, established in 1855. The museum had burnt down in September 1941, but had been rebuilt in its original style before its re-opening in 1945. New displays were needed, however, and the descendants of the 1820 settlers decided to celebrate their ancestors' achievements accordingly. One of the new exhibits was to prominently feature Thomas Holden Bowker's archaeological finds.

The zoologist, archaeologist and director of the museum, John Hewitt, had written to the Royal Artillery Museum at Woolwich, in London, where some of Bowker's stone implements had been stored for almost a century. He convinced the curators to return these to Grahamstown because of their local historical

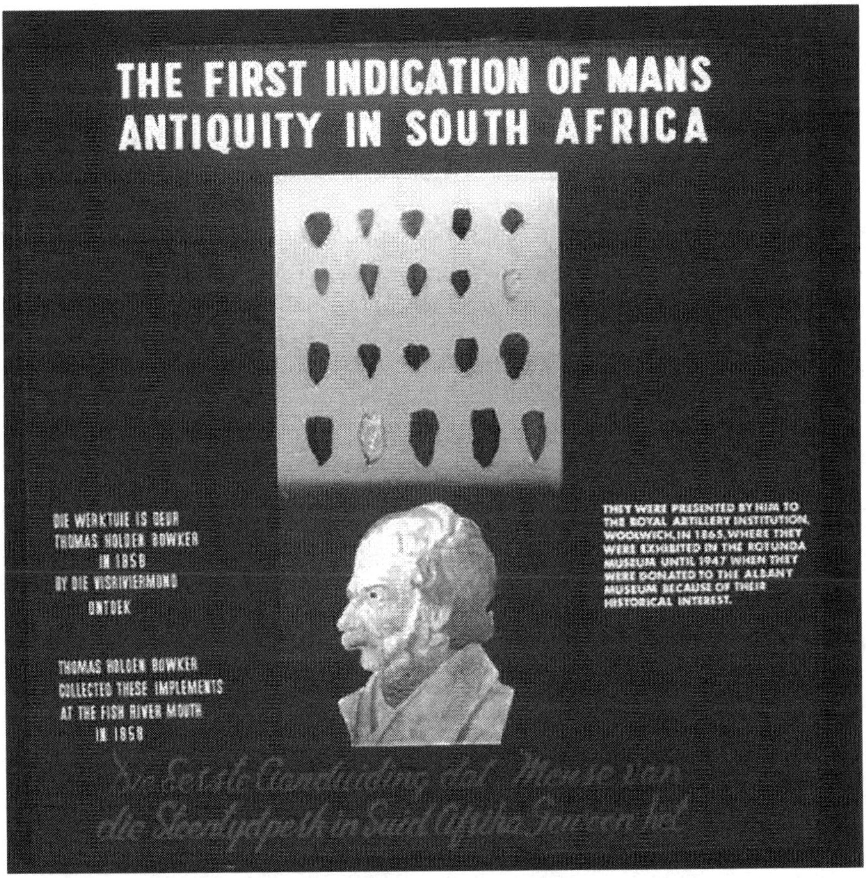

Figure 10.1 Former display at the Albany Museum (Archaeology Section). Upon seeing the title and picture of the display at first glance, the white original inhabitant's hypothesis immediately comes to mind. Display photographed by Hilary John Deacon, the archaeologist and geologist who was Hewitt's successor from 1963 to 1971. He used this photograph in his 2009 lecture on '150 years of Archaeology in South Africa – Discovering the Past', which he gave to the Western Cape Branch of the South African Archaeological Society.

Source: www.docfoc.com/150-years-of-archaeology-in-sa-hilary-john-deacon-hjdeaconiafricacom-discovering, Slide 3 (accessed 6 July 2016).

importance.[27] Twenty-one arrow points, an oil painting of Bowker and a short description in Afrikaans and English of where he found them and how they reached the museum were included in the display (see Figure 10.1).[28]

This display requires contextualisation within archaeological research in South Africa which suffered during the economic depression of the 1930s. The teaching of archaeology at South African institutions decreased during the Second

World War. However, excavations such as at Mapungubwe continued, and Europeans such as the French archaeologist Abbé Henri Breuil carried on conducting research in the country.[29] In reaction to the financial difficulties of the period, Astley John Hilary Goodwin, the first professor of archaeology at the University of Cape Town,[30] strengthened the role of amateurs in the discipline. In his survey of archaeology in South Africa, which he composed in 1935, he included references to Bowker's finds between 1855 and 1858 as an example of what amateur archaeologists could achieve.[31]

Goodwin "South Africanized" archaeology by introducing a local typology and nomenclature for the Stone Age and the subsequent stages of prehistory.[32] In 1945, he co-founded the South African Archaeological Society along with its journal, the *South African Archaeological Bulletin*. He did so to assert the importance of the discipline in the country as well as its independence from British traditions. It also helped to continuously encourage amateur archaeologists to contribute their time and effort to archaeological endeavours.

South Africa's recurrent prime minister Jan Christiaan Smuts (1919–24, 1939–48) acted as a patron to archaeology as he was interested in its power to prove South Africa the "cradle of mankind".[33] This, he hoped, would play an important role in nation-building and the formation of national pride and identity.[34] In 1947, the first Pan-African Congress on Prehistory took place in Nairobi, Kenya, with one-third of its delegates coming from South Africa. Smuts had arranged for them to be flown in by the military and requested that they invite participants to come to South Africa for the next congress, which he hoped would be held in Johannesburg in 1952.[35] It was more than anything else a means of putting South Africa on the map.[36]

After the National Party's general election victory in 1948, Afrikaner nationalists began to actively intervene in the area of archaeological knowledge creation. They disseminated a propagandist history which stressed Bantu-speaking agriculturalists' recent migration to South Africa and highlighted instability and conflict among African ethnic groups.[37] In doing so, they glorified the early Afrikaner settlers, and turned them into one of many contemporary migrant groups who were represented as having civilised a "wild, nearly empty land".[38] Artefacts from excavations, however, did not support this narrative.

Consequently and unsurprisingly, the National Party government announced that it was inopportune to host the planned congress in Johannesburg at the last moment, as 1952 was to be reserved for the tercentenary celebrations of the arrival of Jan and Maria van Riebeeck at the Cape. This was to be commemorated with a series of events geared towards the media: a staged re-enactment of the landing of Van Riebeeck at Granger Bay, a pageant in Adderley Street in Cape Town, and a series of historical exhibitions devised to celebrate the commonality of all whites in the country.[39] Meanwhile, funding and public support for archaeological research stagnated in the 1950s.

To counter this marginalisation of archaeology, Goodwin chose 1955 as the year to commemorate the centenary of Bowker's finds as the centenary of the discipline in South Africa. Bowker, Goodwin underlined, had been the first to find stone implements which humans had prepared to use as arrowheads in South Africa. Predating finds in France and Britain made Bowker's finding of human evidence in a geological deposit all the more significant.[40] What remained ignored during this commemoration were details of the artefacts themselves, who had helped Bowker find the implements and most offensively the people who had created and used them.

The British archaeologist John Desmond Clark followed with his own commemoration of Bowker, but chose 1858 as the starting date of archaeological research in South Africa – probably to promote his book written around the centenary. He argued that Bowker's finds of the Middle-Stone-Age arrow points occurred one year before the French archaeologist and antiquary Jacques Boucher de Crèvecoeur de Perthes had convinced his colleagues of the antiquity of the Palaeolithic artefacts he had found in Abbeville and St. Acheul, France.[41]

Goodwin's student Berry Malan celebrated Bowker in 1970 as the first archaeologist to have excavated pre-historic artefacts – 'stone spear or arrowheads' – from a depth of 18 feet near the mouth of the Fish River in 1855, and highlighted that Bowker's finds had been simultaneous to those of de Perthes in France, a fact of which Bowker himself had been unaware. Malan acknowledged that de Perthes had found his first stone implements in 1841 and published his claims in 1846 and 1853. Yet he was concerned to emphasise that it was not until 1859 that the British geologist Joseph Prestwich, the archaeologist and geologist John Evans and the archaeologist John Lubbock visited de Perthes and confirmed these finds. Consequently, Malan argued, Bowker's discoveries pre-dated the validation of de Perthes's claims by a wider scientific community. Malan ended by lamenting that he felt that Bowker, whom Goodwin had called "our first antiquary", had not been given appropriate recognition.[42]

During the demanding years which archaeologists experienced under apartheid, Goodwin, Malan and Clark utilised Bowker's legacy to highlight that the archaeology which had emerged in South Africa was a truly South African discipline, and that South Africans should consequently retain a leading role in the field.

Bowker was furthermore invoked to promote an Anglophone South African nationalism and to emphasise the unique abilities of the 1820 settlers and their descendants. Since the opening of the 1820 Settlers Memorial Museum (today known as the History Museum) in Grahamstown in 1965 next to the Albany Museum (now the Natural Science Museum), Bowker and Barber have been commemorated locally for their scientific achievements. The museum was built to celebrate settler history during a period of concern that the Afrikaner nationalist government would neglect it, and was one of a series of institutions honouring the history of British settlers opened in Grahamstown within a decade of each other, such as the National English Literary Museum (1971) and the 1820 Settlers National Monument (1974). After the 1820 Settlers Memorial Museum's inauguration, the permanent exhibition remained unaltered presenting the history of a number of 1820 settler parties and their achievements for more than three decades.

In 1996, the only recently elected post-apartheid government's White Paper on Arts, Culture and Heritage policy requested museums to transform their displays.[43] As a provincial museum funded by the new Eastern Cape Department of Sport, Recreation, Arts and Culture, the 1820 Settlers Memorial Museum was consequently required to introduce displays that would reflect these new policies. But the museum also had to satisfy the demands of the 1820 Settlers' Association, a non-profit organisation with 1,200 members who have a keen interest in the museum and finance the genealogist's position there.

As the curator and her predecessors had shown little interest in Xhosa history and culture before the end of apartheid, the museum held no corresponding collection of objects. Combined with scarce funding, this may further explain why it would take ten years for the museum to implement any transformative measures.[44]

In 2007, 13 years after the introduction of democracy, the institution was renamed 'History Museum'. The permanent 'Settler Gallery' exhibition was retitled the 'Nineteenth-Century Lifestyles Gallery'. It was announced that it would be 'renovated to reflect the interaction and changing lifestyles of various groups on the Eastern Cape frontier' at the time.[45] The gallery walls were repainted in a vivid dark clay- or salmon-colored earthy backdrop. The glass cases devoted to settler parties and their achievements were rearranged by topic, such as the development of the press in Grahamstown and agricultural industry in Albany. Settler photographs on the pillars were removed in order to create space for new displays that were added on the south side. The museum became part of the Albany Museum Complex that includes five museums.

During this transformation, which was completed in 2009, the display 'The First Indication of Man's Antiquity in South Africa' (Figure 10.1) at the neighbouring Natural Sciences Museum seems to have also been removed. At the History Museum, however, Bowker has remained an integral part of the exhibition. His nine arrowheads glued on card can still be viewed, with a label announcing them as 'DISCOVERIES: The first record of Palaeolithic stone implements in S. A. and apparently in the Southern Hemisphere found at the Fish River mouth by Mr TH Bowker in 1858'.[46]

It is unclear whether these stone implements were part of the family estate before being given to the museum and whether these artefacts were indeed the ones he had found in 1857.[47] They are in the Bowker Case that includes objects belonging to the 1820 settler family of Miles and Anna Maria Bowker (née Mitford) – Thomas Bowker's parents – which have ever since the museum's inception been prominently placed in the gallery.[48] The grandson of Bowker's brother William Monkhouse Bowker, genealogist, big game hunter and farmer Frank Bowker (1871–1942) of Thorn Kloof, and his son Francis donated the Bowker Case to the museum, presumably long before the museum was opened. Ever since the stone arrowheads have been placed at the centre as the highlight of the display.[49]

Near the Bowker Case, new displays provide information on the history of the Xhosa living in the area, while immediately next to the Bowker Case are seven black-and-white photographs taken by an anthropologist at the end of the 20th century (see Figure 10.2).

This display is a classic example of the post-apartheid add-on mode of representation, an "extend-rather-than-revise" approach to redress 'the imbalance of

Figure 10.2 Bowker Case and Display at the19th-Century Life-Styles Gallery, History Museum, Albany Museum Complex, Grahamstown. The photographs depict a woman sewing, an older woman sitting and smoking, a young man milking a cow, two women fetching water, two children herding two oxen and three women talking alongside a child. They are idealised, romanticised depictions of agriculture, manual labour and household chores.

Source: Photography by Tanja Hammel in April 2014.

the apartheid-era symbolic landscape'.[50] Instead of transforming the entire exhibition, the curators attempted to insert the local African population into the settler narrative. The overall effect of the exhibition was to historicise settlers, while traditionalising and ethicising the Xhosa. With the settler past contained in cases and the African present displayed on clay-colored walls, this reinvigorated ethnic separation created the impression the two peoples shared no common past. The gallery's name, perhaps unconsciously, implies that 20th-century Xhosa lived in the same way as their ancestors had done in the 19th century, thereby depriving them of their historical context and denying them their social evolution.

A static view is thus presented which is reminiscent of earlier "noble savage" representations and of "anachronistic space", a space in which people are presented as contiguous with modernity yet are figured as temporarily out of place. The colonial social hierarchy based on race, which the settlers believed themselves to head and which the museum's curator had been used to portraying throughout apartheid, was projected on the axis of time and thereby naturalised in this display.[51] The 20th-century photographs of the Xhosa individuals – that come to represent the Xhosa – give the impression of predating the 1820 settlers even though the photographs were taken a century later.

The 20th-century archaeologists' discussion of Bowker's contributions to archaeology as well as the two displays in Figures 10.1 and 10.2 are what Sigmund Freud called 'screen memories'.[52] Screen memories are inaccurate reconstructions that obscure what really happened or depict compromises between 'an unconscious recognition of the importance of an experience and an equally unconscious desire not to recognize the experience at all'.[53] It highlights the importance of colonisation, while at the same time suppressing the destruction of the indigenous population. The complete neglect of histories which preceded the arrival of the "first settlers" is a feature of the politics of memory generally observed in settler colonial contexts, and evidence of settlers' own mnemonic myopia. It is somewhat ironic that archaeological finds – objects that predate European settlement – are the central theme, and that these objects then are placed within the settler display, not the sections reserved for ever-stagnant indigenous peoples.

The displays are thus revealing in terms of what they conceal and, similarly to other marked sites of initial exploration such as monuments, are typical of the attitude which the Australian historian Inga Clendinnen identifies as what she terms the '"smoke rising from slab huts" narratives' – namely feel-good stories which settlers told their descendants and have subsequently been passed down from generation to generation.[54]

The display in Figure 8.2 therefore does not inform about Xhosa history, but instead is informative on the ways in which history has been produced in this curatorial setting. Visitors gain little insight into hitherto hidden histories through the old and barely modified narratives presented in the museum. Historians have shown that such displays embody the "history frictions", tensions, debates and conflicts between different communities, each with their own range of interests, perspectives and aims, in places such as Grahamstown.[55] The curatorial practices were not intended to display these frictions but to leave them concealed from the majority of the visitors' untrained eyes.

They are nevertheless visible in the displays (e.g. Figure 8.2), and as such serve as an example of *visible concealment*, in analogy to what Rebekka Habermas has called "eloquent silence". Eloquent silence, according to Habermas, is the process by which a scandal, for instance, serves to hide more than it admits to see. Through the circulation of stereotypes amid the simultaneous silencing of nuanced insights, images were created which have persisted to this very day and served time and again to reinforce convictions of racial superiority and the benefit of European modernity which have been juxtaposed in stark opposition to supposed African primitiveness and lack.[56] In the case of the displays discussed at length, I was interested in how the mechanisms of heroising (as opposed to scandalising) also contributed to the process of silencing. As such, the display makes the concealing of Xhosa life stories and the importance these stone artefacts had in their daily practices as visible as the history frictions within the local community both past and present.

The anglicising of South African archaeology

In Great Britain and Ireland, in the last two decades of the 19th century, ethnologists and anthropologists were trying to piece together a global history of humanity and to discover South Africa's place within it. To do so, they catalogued archaeological material from South Africa in much the same way as they classified plant specimens. The artefacts seemingly merely convinced Eurocentric European scholars that the stone implements were of antiquity and that Africa had been inhabited much earlier than expected, while contributing to Britons' need to know about different parts of the Empire and demonstrating that archaeology was a discipline with universal appeal. More and more collections of similar artefacts reached British institutions, which soon became overwhelmed with so much material they could not exhibit or even unpack it. Many collections which the Bowker siblings had sent to British institutions were lost over time.[57]

One who sent stone implements from the Cape Colony to the metropolis was James Henry Bowker (1822–1900). He was born on the farm Tharfield, on the Kleinemonde River, north of Port Kowie. In 1855, he was appointed inspector of the frontier armed and mounted police, later promoted as commandant. He also acted as high commissioner's agent in Basutoland. During the Seventh and Eighth Xhosa Wars he was promoted to colonel and he was chief commissioner on the diamond fields of Griqualand West. He was mainly interested in Lepidoptera and over the years became probably the leading collector of butterflies in South Africa. He was almost never seen without his net and became known as "Butterfly Bowker". It was even reported that, on at least one occasion, he had downed weapons in the midst of a fierce battle in order to capture an unusual butterfly.[58]

He was much more open to Africans and their knowledge than his older brother Thomas Holden Bowker. There are traces in the family archive that show him as sympathetic to the Zulu from whom he learned much about nature.[59] In 1880, Bowker collected stone artefacts in KwaZulu-Natal after the Anglo-Zulu War and reported that he saw Maluti San using stone arrowheads.[60] Yet, in an article in the *Natal Witness* on 17 April 1880, he argued that the artefacts which he had found at Rorke's Drift and Isandhlwana did not differ from those found at the Cape, in Griqualand West or in the Free State. He thus argued for a white original people populating the entire region.[61]

In 1884, after Basutoland became a British Crown Colony, he published an article, 'Other Days in South Africa'. In it, he attempted to justify the British annexation of more than half of the territory's arable land in 1871, when the boundaries were fixed. He spurred imperialists into reconquering further agriculturally promising land near the sources of the Caledon River.[62] He also reported on how Tswana refugees from the Caledon River valley had been employed on his father's farm and that a boy who had learnt English had told him about the barbarian practices of his people. Their supposedly cannibalistic and uncivilised practices allowed Bowker to justify the colonisation of their cultivatable land.

In 1867, James Henry Bowker had sent his archaeological finds from East London to the director of the Royal Botanic Gardens Joseph Dalton Hooker in

London.[63] Bowker also collected stone implements during excavations in rock shelters in what is today western Lesotho, presumably in 1868, which he sent to Hooker in 1870.[64] Hooker then donated all of these to the British Museum a few months later.[65]

Apart from a rolled dolerite hand-axe from the Early Stone Age and a spokeshave of a Later Stone Age industry dating back to the second half of the Holocene, the 42 artefacts were of mostly Middle Stone Age origin.[66] According to archaeologist Peter Mitchell, the 'hornfels artefacts are variably patinated and rolled, which may indicate that not all of them have had the same depositional history and/or that they are not all of the same age'.[67]

When they arrived at the British Museum, August Franks, keeper of British and Medieval Antiquities (1866–96), most likely catalogued them. In his descriptions it becomes clear that he did not know much about where they had come from and how they had been used, but, according to the archaeologist Jill Cook, Franks was the first to "neatly" mark 'with its provenance, a practice so uncommon at the time that he recommended it should be generally adopted'.[68] He numbered each and recorded details in three columns: the left-hand column for details on who had collected from whom and at what price an object had been obtained, the middle contained a description with a drawing and the right-hand side indicated its accession number. An illustrator whom he paid out of his own pocket produced the drawings.[69] The slips were then bound together to form a catalogue.

Franks was a close friend of the archaeologists John Lubbock and John Evans, member and president of the Society of Antiquaries, Trustee of the Christy Collection. He was well informed and, through his connections, was able to add significantly to the British Museum's Stone Age collections. But Franks was not enthusiastic about stone tools. He even questioned 'how could people collect such odious things?'[70]

The stone tools are today stored in a dilapidated brick stone warehouse – presumably an old factory building – in the east of London that serves as an outpost for the British Museum. The artefacts have fallen into a deep slumber and there is little chance of their awakening through exhibition or a change of location. Embedded in polystyrene, they are hidden, decontextualised and their biographies silenced.

These specimens form part of the Christy Collection, which is named after businessman Henry Christy (1810–65), who came to wealth through manufacturing hats and cotton goods as well as pursuing stock investments. Interested in botany and particularly archaeology, he made trips to numerous countries to acquire an archaeological and ethnographic collection. He offered the collection to the British Museum and left money with which the Christy Fund was established that allowed the purchasing of important objects or collections.[71] Bowker's collection seems to have subsequently been incorporated into the Christy collection.

Besides the collection at the British Museum, there are nine stone arrowheads at the Pitt Rivers Museum, Oxford, which have been similarly neglected and decontextualised.[72] James Henry Bowker had sent them to the archaeologist and geologist John Evans (1823–1908), whose son donated them to the museum in 1928.[73] After the British Museum, the Pitt Rivers Museum is – with 5,010 artefacts – (one of) the largest and most diverse collections of Stone Age archaeological finds from South Africa.[74] On its homepage, the museum provides a photograph of nine stone tools that Bowker had collected in KwaZulu-Natal in 1880. The photo seems to have been made for publication.[75]

Neither the artefacts at the British Museum nor at the Pitt Rivers Museum ever seem to have been exhibited. It can be assumed that the Bowkers' artefacts archived in these institutions had previously been shown at the Royal Anthropological Institute or the Society of Antiquaries, at least when Edgar Layard and George Grey's papers were read in the early 1870s.[76] Before the Bowkers' stone implements arrived there had already been similar artefacts collected by C. J. Busk and the Superintendent-general of education, Langham Dale, from the Cape Flats.[77] Those had been exhibited at the Third International Congress of Archeology in Norwich in 1868, which explains why Bowker's collected items were not received with keen interest as novelties.[78] 'There has not been a significant proportion of southern African collections on exhibition in the British Museum since' the last display, writes Peter Mitchell, 'had to be disbanded during the Second World War'.[79] The stone tools, such as those collected by Bowker, remained stored away, but are available for research and have been praised as important in documenting the beginning of 'Stone Age archaeology in southern Africa'.[80]

Repatriating the stone tools to Southern African institutions seems no alternative. The Iziko South African Museum in Cape Town, for instance, already has 'a number of designated archaeological repositories' and 'about 10,000 boxes of material',[81] many of which are most likely very similar. These are "locked up" in museums and collections and resemble cemeteries. Catalogues and projects such as online exhibitions act as tombstones indicating in which archives, boxes and store rooms the artefacts are buried. While repatriation cases of human remains from European museums to South Africa have attracted much scholarly attention,[82] the South African state has not repatriated to other areas within the former South African Empire.[83] There has not been space for this in the attempts to reconcile and create a common national identity in post-apartheid South Africa.

Provenance research in the UK has led to a number of articles on the Bowkers' collections in the late 1990s, following the end of apartheid.[84] These articles and catalogues provide information on who collected the stone artefacts from where and how they came into the British museum collections. A context is created, particularly by stressing the network of non-African collectors. However, the explanation of how they were interpreted that I provided in the first section of this article is missing, be it as the authors had no access to the sources in South Africa or because this critical analysis did not serve the various authors' purposes to write themselves into the tradition of British archaeology in South Africa.

While British-South African collaboration is stressed, this occurs in a one-sided manner, which takes into account the importance of British education and funding for South African archaeologists, but ignores the impact of particularly black South Africans on their British colleagues.[85] This is typical of a center-periphery approach to knowledge making, in which the South is seen to provide the material while the North provides the evaluation.[86] In the process, a massive body of research on the politics of archaeology in South Africa is elided. Particularly dangerous is the neglect of meticulous research on African archaeologists.[87]

Conclusion

Archaeology was not founded in Europe with its standards subsequently diffused to the South. The American sociologist of science Robert K. Merton coined the concept of "multiple discovery" or "simultaneous invention" arguing that multiple people make most scientific innovations more or less simultaneously.[88] The birth of the discipline coincided in many areas of the world at the same time, one of which was the Cape Colony.

Archaeological knowledge grew and formed within time-bound values; it built on collecting and selecting certain objects, and telling particular stories of origin and cultural use. The Bowkers' staunch imperialism made them imagine a massive empire, which they would actively help to constitute. One form of finding legitimisation for their activities, particularly in regions with valuable mineral resources, was archaeological research.

The Bowkers' making sense of the artefacts became entangled with Africans' even though they did not properly acknowledge their collaborators. Their interpretations were also interlinked with similar contemporary reasoning, such as George Grey's in Kimberley, Northern Australia. This indicates that knowledge circulated widely among European colonial powers and across other colonial empires.

In the History Museum of the Albany Museum, the complex multi-layered display visualises the colonial legacy in curators' and scholars' thinking about the past, and their production of histories today. There is a need, I argue, for more revisionist history writing on archaeology that would in turn influence museum displays. The history of the origins of South African archaeology examined in this chapter follows on from Karen Cereso's argument that the 'reluctance to acknowledge the role of Eurocentrism and hidden colonialism implies that 21st century anthropologists and historians may be inadvertently justifying their continuity'.[89] The need and urgency that fueled the Bowkers' imaginings of human origins in southern Africa is still a task for a changing and evolving South Africa today.

Notes

1 Griffiths, *Hunters and Collectors*, 85.
2 Hewitt, 'Further Light on the Bowker Implements', 94–5; Walker, *The Cambridge History of the British Empire*, 912; Mitchell, 'The South African Stone Age', 32; Cohen, 'Barber, the Bowkers and South African Prehistory', 126.
3 In analogy to Said, 'Travelling Theory'. My approach is similar to Witz, 'The Making of an Animal Biography'.
4 See e.g. Shepherd, 'The Politics of Archaeology in Africa', 189–209; Shepherd, 'Disciplining Archaeology', 127–45.
5 Mitford-Barberton and Mitford-Barberton, *The Bowkers of Tharfield*, 175.
6 Message from His Excellency the Governor to the Honorable House of Assembly, transmitting Copy of a Memorial from Mr. Thomas Holden Bowker. Printed by Order of the House of Assembly, 4 May 1858. CL: MS 18 641.

7 The British Immigrants of 1820, *Some Reasons for Our Opposing*, 208. See: Ross, *John Philip 1775–1851*, 59.
8 It was one among 400 applications for land, mainly from British settlers, discussed at the time. Lester, *Imperial Networks*, 86.
9 Mitford-Barberton and Mitford-Barberton, *The Bowkers of Tharfield*, 177.
10 Mitford-Barberton, *Comdt. Holden Bowker*, 28.
11 Layard, 'Note on the Stone Implements of South Africa', xcviii.
12 Mitford-Barberton and Mitford-Barberton, *The Bowkers of Tharfield*, 178–9.
13 Layard to Owen, South African Museum, 14 April 1864, NHM, Letter 240, General Library: Owen Collection Layard, Edgar Leopold (1825–1900), England and South Africa. 14, 1 reply, and printed, February 1854 – June 1884 and undated. 17/226–59; Suppl. 3.
14 Hammel, 'Thinking with Birds'; Hammel, 'Mary Barber's Travel Journal'; Hammel, 'Racial Difference in Mary Elizabeth Barber's Knowledge on Insects'.
15 Barber to T.H. Bowker, *Highlands*, 14 June 1865, History Museum, Albany Museum Complex, Grahamstown, S.M.D. No 932.
16 Lester, *Imperial Networks*, 180.
17 Dubow, *A Commonwealth of Knowledge*, 66.
18 Ibid., 65.
19 Grey, *Journals of Two Expeditions of Discovery in North-West and Western Australia*, 261, 263–4.
20 Bowker to Grey, Tharfield (near Bathurst), 8 February 1869, National Library South Africa, Cape Town: Sir George Grey (Auckland) Collection, MSB 223, 1 (22) (75).
21 Delaney, *Starting with Rousseau*, 45; Brantlinger, *Dark Vanishings*, 44.
22 Barber to T.H. Bowker, *Highlands*, 14 June 1865, History Museum, Albany Museum Complex, Grahamstown, S.M.D. No 932.
23 Veracini, *Settler Colonialism*, 89.
24 Ibid., 90, 93.
25 Johnston and Lawson, 'Settler Colonies', 369.
26 See e.g. Kuklick, 'Contested Monuments'; Schütte, 'Die "Entdeckung der Ruinen von Zimbaoe" '.
27 Deacon and Deacon, *Human Beginnings in South Africa*, 2–3.
28 These were exhibited in 1977. Thorpe, *Tharfield*, 77. No additional information could be gleaned on how long they were on display. The portrait was painted by W.G. Bevington and was taken from a pencil sketch by a Mr. Edwards of the Cradock Commando. Mitford-Barberton and Mitford-Barberton, *The Bowkers of Tharfield*, 174–5 and Mitford-Barberton, *Comdt. Holden Bowker*, 274.
29 Saul Dubow is currently working on Breuil for a forthcoming article.
30 See e.g. Malan, 'Astley John Hilary Goodwin', 123–5; Inskeep, 'Curriculum Vitae', 111–13.
31 Goodwin, 'A Commentary', 295.
32 See: Goodwin and van Riet Lowe, 'The Stone Age Cultures of South Africa', 1–289; Goodwin, *Method and Prehistory*.
33 See e.g.: Schlanger, 'Making the Past for South Africa's Future', 200–9.
34 Ouzman, 'Imprints', 18.
35 Zipkin, 'Archaeology under Apartheid', 22–3.
36 Ibid., 18–24. Also see Shepherd, 'State of the Discipline', 832–5.
37 See e.g. Cobbing, 'The Mfecane as Alibi', 487–519; Hamilton, *The Mfecane Aftermath*.
38 On the empty land myth, see: Marks, 'South Africa', 7–8; Deacon, 'Fabric of Stone Age Research', 48; Crais, 'The Vacant Land', 225–75; Bonner, 'Myth of the Vacant Land', 141–3.
39 Zipkin, 'Archaeology under Apartheid', 24. See e.g.: Witz, 'Solly Sachs', 45–62; Witz, Apartheid's Festival; Rassool and Witz, 'The 1952 Jan Van Riebeeck Tercentenary Festival', 447–68.

40 Goodwin, 'Thomas Holden Bowker', 2.
41 Clark, *The Prehistory of Southern Africa*, 24.
42 Malan, 'Remarks and Reminiscences', 88–92.
43 Department of Arts, Culture, Science and Technology, 'White Paper on Arts, Culture and Heritage'.
44 In the annual report 2009–2010 the Albany Museum manager warned that 'until somebody begins to listen to [their] plights [the] museum will continue to go the way of the dodo bird'. The state subsidy for 2010–11 (R798 500) would 'go straight into the payment of municipal services leaving nothing for programs let alone transformation'. The manager further claimed that to 'function properly a museum the size of the Albany Museum needs an operational budget of at least R 10 million a year'. Albany Museum 2009–2010, 9.
45 Figure 5 in: Witz, 'Museums, Histories and the Dilemmas of Change', 17. According to curator emeritus Fleur Way-Jones, these displays were changed by herself and her assistants, Zene Schwaiba and Vovo Mabutya, in 2009. Way-Jones to Hammel, 2 August 2014.
46 Bowker Case (Number 15), Albany Museum History Museum, labels by curator Fleur Way-Jones.
47 If so, their journey may have led them to a coffee plantation in Kenya with Thomas Holden Bowker's daughter Mary Layard, to her son Raymond Mitford-Barberton in Australia or Gareth Mitford-Barberton in England and then back to Grahamstown. They may also have been part of a collection returned from an English institution, such as the Rotunda Museum.
48 Mitford-Barberton, *Comdt. Holden Bowker*, 267.
49 Plug, 'Bowker, Mr Francis William Monkhouse'.
50 Marschall, 'Symbols of Reconciliation or Instruments of Division?', 159.
51 McClintock, *Double Crossings*, 19–20.
52 See e.g. Freud, 'Über Kindheits- und Deckerinnerungen', 285–310; Freud, 'Über Deckerinnerun-gen', 531–54.
53 Veracini, *Settler Colonialism*, 90.
54 Ibid., 90; Clendinnen, 'The History Question', 3–4.
55 Karp et al., *Museum Frictions*, 2.
56 Habermas, *Skandal in Togo*, 17.
57 Such as the one at the Dublin Museum. Jones, 'Some Evidences of Primitive Man'.
58 Cohen, 'Barber, the Bowkers and South African Prehistory', 124.
59 Barber, 'Wanderings', vol. 3, MS 10560 (c), 116–18.
60 Bowker, 'Note on Perforated Stones', 55–7.
61 Mitchell, 'Archaeological Collections from the Anglo-Zulu War', 16.
62 Eldredge, 'Drought, Famine and Disease', 68.
63 Gooch, 'Stone Age of South Africa', 153.
64 Ibid.
65 Royal Botanic Gardens, Kew, Director's Correspondence, South Africa, vol. 189, nos. 400–12.
66 Mitchell, 'Archaeological Collections from the Anglo-Zulu War', 13. Christy Collection, ex Bowker, Franks House, The British Museum, +7539 to +7570, + 7571 to +7580.
67 Mitchell, 'Archaeological Collections from the Anglo-Zulu War', 16.
68 Cook, 'A Curator's Curator', 121.
69 Ibid., 121.
70 Ibid., 125.
71 www.britishmuseum.org/research/search_the_collection_database/term_details.aspx?bioId=40853 (accessed 6 August 2015).
72 PRM (Pitt Rivers Museum) Accession Number 1928.68.279–81, 1928.68.285–86 and 1928.68.381–3.

73 Mitchell, 'Stone Age Sub-Saharan Africa', 24.
74 Ibid., 16–17.
75 http://databases.prm.ox.ac.uk/fmi/webd#objects_online (accessed 6 August 2015); Mitchell, 'Stone Age Sub-Saharan Africa', 23.
76 See Layard, 'Note on the Stone Implements of South Africa'; Grey, 'On Quartzite Implements from the Cape of Good Hope'.
77 Dale, 'On a Collection of Stone Implements'.
78 Discussed in Busk, 'Stone Antiquities found in Africa'; Mitchell, *Catalogue of Stone Age Artefacts*, 10.
79 Mitchell, *Catalogue of Stone Age Artefacts*, 27.
80 Ibid., 28.
81 Shepherd, 'The Humility of Sarah Baartman'.
82 Legassick and Rassool, Skeleton in the Cupboard; Rassool, 'Human Remains', 133–56.
83 Shepherd, 'The Humility of Sarah Baartman'; Henrichsen et al., 'Rethinking Empire in Southern Africa'.
84 See e.g. publications by Cohen and Mitchell quoted in this chapter.
85 See e.g. Mitchell, *Catalogue of Stone Age Artefacts*, 9–16.
86 See Connell, *Southern Theory*, ix, 89–110.
87 See e.g. Shepherd, 'When the Hand that Holds the Ttrowel is Black. . .', 334–52.
88 See Merton, 'Resistance to the Systematic Study of Multiple Discoveries in Science', 237–82; Merton, *The Sociology of Science*.
89 Cereso, review of *Inside African Anthropology*.

Bibliography

Albany Museum 2009–2010, www.ru.ac.za/media/rhodesuniversity/content/albanymuseum/documents/AM_ARsml.pdf (accessed 2 February 2015).

Barber, Mary Elizabeth, 'Wanderings in South Africa by Sea and Land', 3 vols., Cory Library, Rhodes University Library, Grahamstown: MS 10560 (a – c).

Bonner, Philip, 'The Myth of the Vacant Land', in *A Search for Origins: Silence, History and South Africa's 'Cradle of Humankind'*, ed. P. Bonner, A. Esterhuysen, and T. Jenkins (Johannesburg: Wits University Press, 2007), 141–7.

———, Amanda Esterhuysen, and Trefor Jenkins, eds., *A Search for Origins: Science, History, and South Africa's "Cradle of Humankind"* (Johannesburg: Wits University Press, 2007).

Bowker, James H., 'Note on Perforated Stones', *Cape Monthly Magazine* 25 (1872): 55–7.

Brantlinger, Patrick, *Dark Vanishings: Discourse on the Extinction of Primitive Races, 1800–1930* (Ithaca: Cornell University Press, 2003).

Busk, G., 'Stone Antiquities found in Africa', in *Transactions of the International Congress of Prehistoric Archaeology* (Third Session, London: Norwich, 1868).

Cereso, Karen, review of *Inside African Anthropology: Monica Wilson and Her Interpreters*, by Andrew and Leslie Bank, *African Studies Bulletin* 75 (Winter 2013/14): 129–32,http://lucas.leeds.ac.uk/review/inside-african-anthropology-monica-wilson-and-her-interpreters/ (accessed 18 August 2017).

Clark, John Desmond, *The Prehistory of Southern Africa: The Latest Discoveries about the Origins and Cultural History of Primitive Man in Southern Africa* (Harmondsworth: Penguin Books, 1959).

Clendinnen, Inga, 'The History Question: Who Owns the Past?', *Quarterly Essay* 23 (2006): 1–72.

Cobbing, Julian, 'The Mfecane as Alibi: Thoughts on Dithakong and Mbolompo', *Journal of African History* 29 (1988): 487–519.

Cohen, Alan, 'Mary Elizabeth Barber, the Bowkers and South African Prehistory', *The South African Archaeological Bulletin* 54 (1999): 120–7.

Connell, Raewyn, *Southern Theory: The Global Dynamics of Knowledge in Social Science* (Cambridge, Malden, MA: Polity Press, 2007).

Cook, Jill, 'A Curator's Curator: Franks and the Stone Age Collection', in *A. W. Franks: Nineteenth Century Collecting and the British Museum*, ed. Marjorie Caygill and John Cherry (London: British Museum Press, 1997), 115–29.

Crais, Clifton C., 'The Vacant Land: The Mythology of British Expansion in the Eastern Cape, South Africa', *Journal of Social History* XXV (1991): 255–75.

Dale, Langham, 'On a Collection of Stone Implements and Pottery from the Cape of Good Hope', *The Journal of the Anthropological Institute of Great Britain and Ireland* 1 (1872): 345–8.

Deacon, Hillary J., 'Weaving the Fabric of Stone Age Research in Southern Africa', in *A History of African Archaeology*, ed. P.T. Robertshaw (London: James Currey, 1990), 39–58.

——— and Janette Deacon, *Human Beginnings in South Africa: Uncovering the Secrets of the Stone Age* (Claremont: Hirt & Carter, 1999).

Delaney, James J., *Starting with Rousseau* (London and New York: Continuum, 2009).

Department of Arts, Culture, Science and Technology (DACST), 'White Paper on Arts, Culture and Heritage: All Our Legacies, Our Common Future' (Pretoria, 1996), www.dac.gov.za/content/white-paper-arts-culture-and-heritage-0 (accessed 13 March 2015).

Dubow, Saul, *A Commonwealth of Knowledge: Science, Sensibility and White South Africa 1820–2000* (Oxford: Oxford University Press, 2006).

Eldredge, Elizabeth A., 'Drought, Famine and Disease in Nineteenth-Century Lesotho', *African Economic History* 16 (1987): 61–93.

Freud, Sigmund, 'Über Kindheits- und Deckerinnerungen', *Monatsschrift für Psychiatrie und Neurologie* 18 (1899): 285–310.

———, 'Über Deckerinnerungen', in *Gesammelte Werke*, vol. 1 (Frankfurt a.M.: S. Fischer Verlag, 1953), 531–54.

Gooch, W. D., 'Stone Age of South Africa', *Journal of the Anthropological Institute* 11 (1881): 124–82.

Goodwin, Astley John Hilary, 'A Commentary on the History and Present Position of South African Prehistory with Full Bibliography', *Bantu Studies: A Journal Devoted to the Scientific Study of Bantu, Hottentot and Bushmen* 9 (1935): 292–417.

———, *Method and Prehistory: An Introduction to the Discipline of Prehistoric Archaeology with Special Reference to South African Conditions* (Claremont: South African Archaeological Society, 1953).

———, 'Thomas Holden Bowker: Our Archaeological Centenary', *The South African Archaeological Bulletin* 10, no. 37 (1955): 2.

——— and Clarence van Riet Lowe, 'The Stone Age Cultures of South Africa', *Annals of the South African Museum* 27 (1929): 1–289.

Grey, George, *Journals of Two Expeditions of Discovery in North-West and Western Australia, During the Years 1837, 38, and 39, in 2 vols.*, vol. 1 (London: T. and W. Boone, 1841).

———, 'On Quartzite Implements from the Cape of Good Hope', *The Journal of the Ethnological Society of London* 2, no. 1 (1870): 39–43.

Griffiths, Tom, *Hunters and Collectors: The Antiquarian Imagination in Australia* (Cambridge: Cambridge University Press, 1996).

Habermas, Rebekka, *Skandal in Togo: Ein Kapitel Deutscher Kolonialherrschaft* (Frankfurt a.M.: S. Fischer Verlag, 2016).

Hamilton, Carolyn, *The Mfecane Aftermath: Reconstructive Debates in Southern African History* (Bloomington: Indiana University Press, 1995).

Hammel, Tanja, 'Thinking with Birds: Mary Elizabeth Barber's Advocacy for Gender Equality in Ornithology', *Kronos* 41 (2015): 85–111.

———, 'Mary Barber's Expedition Journal: An Experimental Space to Voice Social Concerns', in *Expedition as Experiments: Practising Observation and Documentation*, ed. Marianne Klemun and Ulrike Spring (Basingstoke, New York: Palgrave MacMillan, 2016), 121–40.

———, 'Racial Difference in Mary Elizabeth Barber's Knowledge on Insects', in *The Politics of Nature and Science in Southern Africa*, ed. Maano Ramutsindela, Giorgio Miescher, and Melanie Boehi (Basel: Basler Afrika Bibliographien, 2016), 39–58.

Henrichsen, Dag, Giorgio Miescher, Ciraj Rassool, and Lorena Rizzo, 'Rethinking Empire in Southern Africa', *Journal of Southern African Studies* 41, no. 3 (2015): 431–5.

Hewitt, John, 'Further Light on the Bowker Implements', *The South African Archaeological Bulletin* 10, no. 39 (1955): 94–5.

Inskeep, Ray [based on Goodwin's original manuscript], 'Curriculum Vitae: Astley John Hilary Goodwin', *South African Archaeological Bulletin* 51 (1996): 111–13.

Johnston, Anna and Alan Lawson, 'Settler Colonies', in *A Companion to Postcolonial Studies*, ed. Henry Schwarz and Sangeeta Ray (Malden, MA: Blackwell, 2000), 360–76.

Jones, Justice, 'Some Evidences of Primitive Man', *The E. P. Magazine* 2, no. 1 (1893): 15.

Karp, Ivan, Corinne A. Kratz, Lynn Szwaja, and Tomas Ybarra-Frausto, eds., *Museum Frictions: Public Cultures/Global Transformations* (Durham and London: Duke University Press, 2006).

Kuklick, Henrika, 'Contested Monuments: The Politics of Archaeology in Southern Africa', in *Colonial Situations: Essays on the Contextualization of Ethnographic Knowledge*, ed. George W. Stocking (Madison: University of Wisconsin Press, 1991), 135–69.

Layard, Edgar L., 'Note on the Stone Implements of South Africa', *The Journal of the Anthropological Institute of Great Britain and Ireland* 1 (1872): xcvii–c.

Legassick, Martin and Ciraj Rassool, *Skeletons in the Cupboard: South African Museums and the Trade in Human Remains, 1907–1917* (Cape Town: Iziko Museums of SA, 2015).

Lester, Alan, *Imperial Networks: Creating Identities in Nineteenth-Century South Africa and Britain* (London: Routledge, 2001).

Malan, Berry D., 'Astley John Hilary Goodwin, 1900–1959', *The South African Archaeological Bulletin* 14, no. 56 (1959): 123–5.

———, 'Remarks and Reminiscences on the History of Archaeology in South Africa', *The South African Archaeological Bulletin* 25, no. 99/110 (1970): 88–92.

Marks, Shula, 'South Africa: "The Myth of the Empty Land"', *History Today* 30, no. 1 (1980): 7–8.

Marschall, Sabine, 'Symbols of Reconciliation or Instruments of Division? A Critical Look at New Monuments in South Africa', in *Culture and Belonging in Divided Societies: Contestation and Symbolic Landscapes*, ed. Marc Howard Ross (Philadelphia: University of Pennsylvania Press, 2009), 151–75.

McClintock, Anne, *Double Crossings: Madness, Sexuality and Imperialism, The 2000 Garnett Sedgewick Memorial Lecture* (Vancouver: Ronsdale Press, 2001).

Merton, Robert K., 'Resistance to the Systematic Study of Multiple Discoveries in Science', *European Journal of Sociology* 4, no. 2 (1963): 237–82.

———, *The Sociology of Science: Theoretical and Empirical Investigations* (Chicago: The University of Chicago Press, 1973).

Mitchell, Peter, 'Archaeological Collections from the Anglo-Zulu War in the Collections of the British Museum', *Southern African Field Archaeology* 7 (1998): 12–19.

———, 'The South African Stone Age in the Collections of the British Museum: Content, History and Significance', *The South African Archaeological Bulletin* 53, no. 167 (1998): 26–36.

———, *Catalogue of Stone Age Artefacts from Southern Africa in The British Museum* with contributions by Alison Roberts, Alan Cohen, and Karen Perkins, Occasional Paper no. 108 (London: British Museum, 2002).

———, 'Stone Age Sub-Saharan Africa', in *World Archaeology at the Pitt Rivers Museum: A Characterization*, ed. Dan Hicks, Alice Stevenson (Oxford: Archaeopress, 2013), 16–34.

Mitford-Barberton, Ivan and Raymond Mitford-Barberton, *The Bowkers of Tharfield* (Oxford: Oxford University Press, 1952).

———, *Comdt. Holden Bowker. An 1820 Settler Book Including Unpublished Records of the Frontier Wars* (Cape Town and Pretoria: Human and Rousseau, 1970).

Ouzman, Sven, 'Imprints: An Archaeology of Identity in Post-Apartheid Southern Africa' (PhD diss. in Anthropology, University of California, Berkeley, 2008).

Plug, Cornelius, 'Bowker, Mr Francis William Monkhouse (Frank) (mammal collection)', www.s2a3.org.za/bio/Biograph_final.php?serial=320 (accessed 29 December 2016).

Rassool, Ciraj, 'Human Remains, the Disciplines of the Dead, and the South African Memorial Complex', in *The Politics of Heritage in Africa: Economies, Histories, and Infrastructures*, ed. Derek Peterson, Kodzo Gavua, and Ciraj Rassool (Cambridge: Cambridge University Press, 2015), 133–56.

———, 'Re-Storing the Skeletons of Empire: Return, Reburial and Rehumanisation in Southern Africa', *Journal of Southern African Studies* 41, no. 3 (2015): 653–70.

——— and Leslie Witz, 'The 1952 Jan Van Riebeeck Tercentenary Festival: Constructing and Contesting Public National History in South Africa', *The Journal of African History* 34, no. 3 (1993): 447–68.

Robertshaw, Peter, ed., *A History of African Archaeology* (Portsmouth: Heinemann, 1990).

Ross, Andrew, *John Philip 1775–1851: Missions, Race and Politics in South Africa* (Aberdeen: University Press, 1986).

Said, Edward, 'Travelling Theory', in *The World, the Text, and the Critic* (Cambridge, MA: Harvard University Press, 1983), 226–47.

Schlanger, Nathan, 'Making the Past for South Africa's Future: The Prehistory of Field-Marshal Smuts (1920s–40s)', *Antiquity* 76 (2002): 200–9.

Schütte, Michael, 'Die "Entdeckung der Ruinen von Zimbaoe": Zur Erfindung "weisser Geschichte" in "Schwarzafrika" 1871–1872' (Magisterarbeit, Seminar für Mittlere und Neuere Geschichte, Philosophische Fakultät, Georg-August-Universität Göttingen, 2013).

Shepherd, Nick, 'Disciplining Archaeology: The Invention of South African Prehistory, 1923–1953', *Kronos* 28 (2002): 127–45.

———, 'The Politics of Archaeology in Africa', *Annual Review of Anthropology* 31 (2002): 189–209.

———, 'State of the Discipline: Science, Culture and Identity in Southern African Archaeology, 1870–2003', *Journal of Southern African Studies* 29 (2003): 823–44.

———, '"When the Hand that Holds the Trowel is Black. . . ": Disciplinary Practices of Self-Representation and the Issue of "Native" Labour in Archaeology', *Journal of Social Archaeology* 3, no. 3 (2003): 334–52.

———, 'The Humility of Sarah Baartman', *Monday Monthly* 33, no. 10 (1 December 2014), www.uct.ac.za/mondaypaper/?id=9923, (accessed 18 May 2015).

The British Immigrants of 1820, *Some Reasons for Our Opposing the Author of the "South African Researches" The Rev. John Philip, D. D. (Aberdeen), Superintendent of the London Society's Mission in South Africa* (Cape Town: A. S. Robertson; Graham's Town: Godlonton; London: J. M. Richardson, 1836).

Thorpe, C., *Tharfield: An Eastern Cape Farm* (Port Alfred: Private Publication, 1978).

Veracini, Lorenzo, *Settler Colonialism: A Theoretical Overview* (Basingstoke and New York: Palgrave Macmillan, 2010).

Walker, Eric Anderson, ed., *The Cambridge History of the British Empire* (Cambridge: Cambridge University Press, 1963).

Witz, Leslie, *Apartheid's Festival: Contesting South Africa's National Pasts* (Bloomington: Indiana University Press, 2003).

———, 'The Making of an Animal Biography: Huberta's Journey Into South African Natural History, 1928–1932', *Kronos* 30 (2004): 138–66.

———, 'Solly Sachs, the Great Trek and Jan van Riebeeck: Settler Pasts and Racial Identities in the Garment Workers' Union, 1938–1952', in *Rethinking Settler Colonialism: History and Memory in Australia, Canada, Aotearoa New Zealand and South Africa*, ed. Annie E. Coombes (Manchester, New York: Manchester University Press, 2006), 45–62.

———, 'Museums, Histories and the Dilemmas of Change in Post-Apartheid South Africa', *University of Michigan Working Papers in Museum Studies* 3 (2010): 1–23.

Zipkin, Andrew Michael, 'Archaeology under Apartheid: A Preliminary Investigation into the Potential Politicization of Science in South Africa' (Honors Thesis, College of Agriculture and Life Sciences, Biology and Society Program, of Cornell University, May 2009).

11 Davos of Ghana?

Local, national and international perspectives on tuberculosis treatment and control (ca. 1920–1965)[1]

Pascal Schmid

Tuberculosis was one of the major causes of death in the Gold Coast throughout the later colonial period. Measures against tuberculosis were introduced in the 1920s and 1930s. They were, however, limited in scope and focused almost exclusively on the mining areas and towns. Isolation and treatment of cases was limited to a few facilities. Only from the 1950s onwards a comprehensive approach to controlling the disease followed. At the same time, at the mission hospital in Agogo, a rural town in Ashanti, tuberculosis was seen as a possible field in which the hospital could specialise in order to secure and justify its existence. Soon, plans for a new tuberculosis centre attached to the hospital were elaborated. However, in the context of the changing predominant anti-tuberculosis strategies and the development of tuberculosis control in the decolonising Gold Coast and Ghana respectively, such a project did not find the external support it needed from government and other sources.

This chapter engages with the strategies and measures in tuberculosis treatment developed at a hospital founded and run (until the early 1960s) by the Basel Mission. It outlines how this development evolved in a rapidly changing socio-political environment, in the frame of a decolonising health care system and in conjunction with the tuberculosis control policies of international organisations. The changing concepts and practices were shaped by varying figurations of interests, values and ideologies within and around the hospital. Based on missionary and government sources, the chapter aims to demonstrate this by concentrating on the actors on the ground and tracing their interaction with and relations to institutions and networks at local, national and international levels. It shows how tuberculosis treatment at Agogo Hospital including surgical interventions and drug therapy was promoted as a contribution not only to the goals of the medical mission but also the colonial and national health care system, but eventually failed to connect to the changing national and international policies.

A mission hospital in a decolonising health care system

The Basel Mission had been present in the Gold Coast from the first half of the 19th century. By the beginning of the 20th century, it assumed a considerable importance for civil society in the colony. During the First World War, the Basel Mission as a bi-national, Swiss-German institution was expelled from the territory. The Free Church of Scotland took over its activities, and the Presbyterian Church grew out of this arrangement. When the Basel missionaries returned to the Gold Coast after the war, the resumption of medical work was one of their first major projects. In 1928, the building of a hospital started in the small town of Agogo in Ashanti.

Agogo Hospital was inaugurated in 1931 and operated at first by one and, later, by two doctors at a time, along with some three or four European "sisters" and around ten African nurses. It was at that time the only mission hospital in the Gold Coast and, for decades to come, remained one of the biggest hospitals in the Gold Coast (and Ghana) and by far the largest medical institution in Ashanti outside Kumasi. In a time when the porous colonial health care system was quasi-absent in rural areas, the Basel Mission established a big hospital delivering high-standard services in a small town, of 4,000 inhabitants, off the major traffic routes.

The expansion of the hospital's infrastructure, staff, services, capacity and geographical area during the 1930s led to conflicts between the doctors managing the hospital and the mission leadership on the ground (i.e. in Kumasi) and in Basel. The doctors saw the increasing number of patients, including European and well-off African patients from all over the colony, as a means to secure the hospital's existence. The missionaries accused the hospital management of establishing supremacy of the "technical" over the "spiritual", and were anxious to put a halt to any expansion and reinforce the missionary character of the hospital. They felt compelled to do so because the Mission financially supported the hospital, even though it was meant to be self-supporting.

When the Second World War started, even though the Basel Mission could maintain its overseas activities, as it became an entirely Swiss organisation.[2] The hospital, however, had to be closed, as the entire German staff (all but one nurse) was detained. The legacy that remained consisted of the self-perception and reputation of a hospital with high standards, a mission leadership preoccupied to secure its own, namely missionary, interests on the ground and a neglected local health care system with a "peripheral centre".

When the hospital re-opened in 1947, its staff faced a socio-political environment of rapid change. The Second World War had changed the character of colonialism. Development and welfare became key themes of the refurbished colonial state.[3] In the Gold Coast, the dynamic and turbulent post-war years laid the ground for a powerful demand for self-government and for the new constitution of 1951.[4] The 1940s were also a period of major change in colonial health care in general, and the colonial health care system in the Gold Coast saw the initiation of major reforms. Manpower, rural health and preventive medicine were the keywords. At the same time, the government started to promote and support the engagement of mission societies in health care. In contrast to the 1930s and 1940s, the government started to promote missionary medical work and to collaborate with missions in order to expand the health care system. In 1951, there were, apart from Agogo Hospital with its 100 beds, only two other mission hospitals: one at Worawora (Bremen Mission/Evangelical Presbyterian Church) with 18 beds and a maternity clinic at Jirapa (White Fathers/Catholic Church) with 14 beds.[5] Between 1951 and 1960, the number of mission hospitals increased from three to 27 (with almost 1,500 beds) while the number of government hospitals only went up slightly from 31 to 33.[6] The Basel Mission also expanded its health care services during the 1950s, first into Brong Ahafo and, in the course of the Presbyterian Church's expansion into the North, into what is today the Upper East region.

While the new mission clinics and hospitals were built into emerging local and regional health care structures, Agogo Hospital, the mission's long-established flagship, had to find a place for itself in a decolonising health care system, which struggled between technical requirements for public health and political necessities of curative services. The hospital's institutional legacies of the pre-war period remained – the conception and perception of the institution as a centre of hospital medicine with high technical standards; the rural location without direct connection to traffic axes and political or commercial centres; and the commitment to Christian values and missionary aims. Now, these legacies were to be consolidated with a changing socio-political environment and dynamic configurations of old and new actors in and around the hospital. A new generation of doctors questioned the hospital's justification as a general hospital and felt obliged to redefine its strategic orientation. Tuberculosis was seen as one possible field in which the hospital could specialise in order to justify its maintenance.

Early tuberculosis measures in the Gold Coast

Until the Second World War, although the colonial medical authorities did consider tuberculosis in the Gold Coast as an increasing threat to public health, measures against the disease remained limited. The Gold Coast was no exception in this. Harrison and Worboys argue that (British) colonial medical services 'had other disease-control priorities and programmes', while the perception of tuberculosis as a disease of civilisation and the "virgin soil" theory led to the idea that tuberculosis 'was a problem to be settled in historical time, not by immediate medical or sanitary intervention'.[7] In the Gold Coast, as a first attempt, a "sanatorium" for tuberculosis patients was started in Accra in 1922, with the idea that '[p]atients were housed in a small isolation camp where they received treatment and were encouraged to live in the open air'. But the incidence of tuberculosis was not considered sufficiently high to justify the costs to consolidate this "experiment".[8]

Apart from the urban centres, the disease was commonly associated with the mining industry, both by the contemporary authorities and in Ghana's medical historiography.[9] It was to the mining areas that measures against tuberculosis were almost entirely restricted before the Second World War. Mining had become a major commercial activity by the beginning of the 20th century, and, as Dumett notes in his study on health among gold miners, in the first two decades of the 20th century, the miners' shanty towns 'became the natural dissemination grounds for a wide range of respiratory and faecal tract diseases'.[10] Until 1925, the health of the Africans in the mining areas was of no concern to the government (or to the mining companies). After an embarrassing report on the sanitary and health conditions in the mines, the Mining Health Ordinance of 1925 introduced several measures to be implemented by the mining companies. Dumett, in his study, concludes that the slow adoption of the measures by the mining companies including prevention, early detection and improved hospital treatment had a positive effect on the death rates for tuberculosis and other respiratory diseases of the African miners.[11]

Outside the mines (where the companies had borne the cost of the "investments" in African health), specific measures against tuberculosis were virtually absent. During the 1930s, however, the awareness of tuberculosis as a major threat to public health grew in the Gold Coast. In 1934, tuberculosis was the second important cause of death in Ashanti (after the group of "pneumonia, bronchopneumonia and bronchitis"), and over 90% of all fatal tuberculosis cases were due to pulmonary tuberculosis. The regional Senior Health Officer noted that: 'An infection once instituted almost invariably ends in death in a comparatively short period'.[12] In 1936 the Medical Department considered tuberculosis 'the most direct threat to the future of the people of the Gold Coast', as Addae notes.[13] The number of reported deaths due to tuberculosis of the respiratory tract continuously increased from 1924, when the disease was made notifiable by government doctors, up to the mid-1940s.[14] And, most notably, tuberculosis increasingly became a major cause of death. While in 1920 it accounted for less than 6% of all deaths, it accounted for almost double that percentage towards the end of the decade and into the 1930s.[15]

At the same time the number of cases treated increased, but tuberculosis outpatients at clinics and hospitals never accounted for more than 0.65% between 1918 and 1955.[16] By the end of the 1930s, the authorities countered the disease almost exclusively with public health measures in the urban centres, mainly housing schemes (in order to prevent "overcrowding") and health education. Not the least for financial reasons, medical measures in terms of care, treatment and/or isolation were absent with a few exceptions.[17] In 1940, the Colonial Advisory Medical Committee recommended to isolate sputum-positive cases. In the Gold Coast, however, only the Gold Coast Hospital in Accra in its isolation block and the Contagious Diseases Hospital in Kumasi had the infrastructure to isolate cases. A tuberculosis ward was envisaged at the new hospital in Sekondi and an '[e]xtension of the principle will be considered as funds become available'.[18] The absence of a substantial infrastructure of sanatoria, asylums and tuberculosis hospitals was characteristic for (British) colonial Africa. As Harrison notes, '[b]y the end of the 1930s . . . little or nothing had been done in most colonies with regard to tuberculosis and other non-tropical diseases'.[19]

Tuberculosis in Agogo

At Agogo Hospital, an increase in tuberculosis cases was reported throughout the 1930s.[20] As one of the sisters recounted in 1934, the prospect of cure was 'much more hopeless than at home'. Thus, the case was rather for good (because it was "civilised" and Christian) care than for cure: 'Yet, at least these poor [people] may experience the relief of good care in our hospital and are spared the torments of indigenous treatment'.[21]

In 1938, the idea of converting the leper settlement into a tuberculosis station came up for the first time. The Medical Department had turned down a renewed application for support of the leper work in Agogo. The hospital conference discussed whether the financially deficient leper work should be continued. The superintendent suggested giving up the leper settlement and using the buildings for a "tuberculosis station". He argued that the buildings were suitable and the treatment 'significantly more satisfying, at least as soon as the X-ray apparatus is here'. Moreover, he expected that the tuberculosis patients, especially those from a wealthier background, would cover the costs of their treatment – in contrast to the lepers currently in the settlement.[22] In Basel, however, the suggestion to create a tuberculosis station at Agogo did not gather support. The Mission House indicated interest but warned against taking any swift decisions, as a higher workload and a higher risk for the European Sisters was expected.[23] The Africa Inspector at the Mission House in Basel later rejected the idea arguing that the leper settlement should not be given up after such a short time and pointing to the infection of one of the doctors with tuberculosis and the risk tuberculosis patients bore for the European staff.[24]

Tuberculosis became a prominent issue soon after the hospital's re-opening in 1947. At first, however, the disease did not receive any special attention in the reports of the first post-War Basel Mission doctor, Otto Golder. The fact that a Swiss nurse contracted pleurisy in May 1947 (which was later related to tuberculosis)[25] did not trigger any particular medical consideration or activities. The first time Golder reported about tuberculosis was when he had to write in a telegram that he himself was 'bazillus koch positive'.[26] Within three weeks, Golder, who had an open tuberculosis, was first brought to the "Europeans' hospital" in Accra, then flown to Geneva, and finally admitted to the Hospital in the Swiss town Frauenfeld. Later he was transferred to the *Thurgauisch-Schaffhausische Heilstätte* in Davos in the Swiss Alps.[27] There, he learned that another Swiss nurse had contracted a "Pleuropneumonie", which was first also associated with tuberculosis (even though later "no serious findings" were made at the Tropical Institute in Basel).[28] Alarmed by the accumulation of respiratory diseases among its staff, the Mission House asked Golder to report on the tuberculosis situation in the Gold Coast and the handling of tuberculosis at Agogo Hospital. The Mission Committee in the first place feared the infection of more of its staff, but also asked questions about the risk for other patients – and considered the exclusion of tuberculosis patients from the hospital.[29]

Golder reported back that tuberculosis was obviously on the increase and referred to a survey conducted in 1947.[30] For the hospital in Agogo, Golder reported 81 "clinically manifest cases" among 3,800 African patients (the non-manifest cases could not be determined because the X-ray equipment was broken down since the war), and two cases (the nurse and himself) among the 50 Europeans and "Levantines" (a discriminatory term that mainly marked Syrians) between

April 1947 and March 1948. Even though 18 of the 35 African staff members had more or less close contact with tuberculosis patients, none of them got infected. Golder insisted that the African staff were tested "as soon as they coughed", and that for the Swiss nursing sisters the same principal counted (though for them an examination was "at least" not compulsory). The higher vulnerability of the European staff surprised Golder given the 'African's accepted higher predisposition'.[31]

When Golder was recruited by the Basel Mission to re-establish medical work in Agogo, he had just closed his office in the Swiss town of Frauenfeld. Golder was recommended by a former mission doctor who advised the mission leadership regarding the recruitment of medical staff and certified Golder's good training and surgical skill, but also his Pietist attitude.[32] With no experience in "tropical medicine", Golder took a five-month course at the London School of Hygiene and Tropical Medicine.[33] Though he had initially planned to spend some time working at the government hospital in Accra in order to gain some "tropical experience", he would later drop this idea, 'because I would get to see enough of all those things and find my way on the black skin without difficulty'.[34]

It is no surprise that Golder did not have much knowledge about the handling and treatment of tuberculosis at government hospitals. But he did note that not one single tuberculosis patient at Agogo was ready to seek treatment at a government hospital, even though they would have received free treatment there.[35] Golder's observation supports the assumption that, before the introduction of effective drugs, the attractiveness of hospital treatment without prospect of cure was rather low. The "poor reputation" of the hospitals, according to Patterson, constituted a major factor in the spread of tuberculosis in the Gold Coast. As '[s]ufferers could not be detained in hospitals or in open air "sanatoria" by compulsion; they could leave at will, even if they were in a highly infectious state'.[36]

According to Golder, treatment at Agogo Hospital was more attractive to patients than treatment in government hospitals. It is hard to assess the validity of this claim, and thus any positive effect of tuberculosis treatment in Agogo on the spread of the disease. There are no reasons to believe that at that time the prospective for cure at Agogo Hospital was higher than in government hospitals. But it can be assumed that hospitalisation at Agogo, whereas not free, was more agreeable, because treatment of symptoms stood before any public health idea. "Detention" was not an aim, neither "voluntarily" nor compulsively. One of the sisters gave an account of a tuberculosis patient and her treatment at Agogo in one of her reports for a wider public at home. The "young girl" came to the hospital with her sister and was diagnosed with an open pulmonary tuberculosis. She was admitted to the "isolation hut" ("Isolierhäuschen") and treated with

> pills for the cough, drops for the already weakened heart, injections in order to strengthen the entire organism, and then first of all something in order to sleep, when the painful body did not allow any rest to the tired body.

The food, which the patient's sister and her cousin cooked for her, was complemented with 'bread, milk, oat flakes and eggs with red wine'. In the course of time the young woman got skinnier and skinnier. Finally, when it was evident that she would not survive the disease, her sister brought her home where she passed away a week later.[37]

Golder recounts that hospitalisation rarely took longer than one month, because of "financial reasons" in the first place, and that he mainly treated symptoms, for which the patients were thankful. The tuberculosis patients were, "if possible", accommodated in the isolation building. But until constructional measures were taken by the end of 1947, only either male or female patients were accommodated there. Until then, a small adjoining room in the respective ward was used in addition. Still, Golder claimed, a case of patient-to-patient contagion had never occurred.[38]

Golder and his staff had a casual attitude concerning the handling of tuberculosis, did not follow any "public health idea", and lacked medical means to effectively cure the disease. Still, the doctor vigorously defended the treatment of tuberculosis patients arguing that it would be 'a great and by Christian thinking inexcusable mistake to exclude the tuberculous [in German: "die Tuberkulösen"] from Agogo because of own exposure'. Gold admitted that cure could not be expected, but rhetorically asked: 'Are we at Agogo only in order to cure diseases?'[39] The vehement defence of tuberculosis treatment with the reference to the risk of own-exposure, can be linked – as Golder's successor Hans Meister would do (see section ' "Davos" of the Gold Coast') – to the role leprosy with its biblical associations played for medical missionaries in the 19th and into the 20th century. The malady was seen as an opportunity to express highest dedication for the missionary ideals and the stigmatised patients as the perfect objects of proselytisation.[40] Indeed, the importance of tuberculosis treatment, and its connection to the ethical imperatives of a mission hospital was also recognised by Golder's successors, but coupled with more sophisticated and systematic approaches to deal with the disease.

Golder was replaced by Arnold Brack, a young Swiss doctor who was at the time working at Albert Schweitzer's mission hospital in Lambarene. He had applied to the Basel Mission for a position in a hospital in China. When Golder fell sick and had to quit his engagement in Agogo, the Mission persuaded Brack to travel to Agogo directly from Lambarene instead of returning to Europe and preparing for service in China.[41]

Under Brack, tuberculosis was handled and treated in a medically sounder manner. Now, as an expedient, the leper settlement was converted into a tuberculosis settlement (see Figures 11.1 and 11.2). By the end of the 1940s, there were, as Brack noted, "only a few leprosy cases", and thus the leper settlement was not used to the full.[42] Thus five huts of the settlement started to host long-term tuberculosis patients, some staying there with their families.[43] The settlement was considered as a temporary solution. Brack judged its location too far from the other hospital buildings for the intensive care of severe cases and, as it was situated in a hollow, as climatically not suitable for pulmonary patients. The treatment was, as Brack stated, 'not always as hopeless as it is frequently assumed'. He reported that many patients 'even reacted extremely well to the drugs also used in Europe'

Figure 11.1 The leper settlement in Agogo built in 1935

Figure 11.2 The leper settlement in Agogo built in 1935

Source: 'The leprosy settlement. The newest house under construction'. Unknown, ca. 1935 (Basel Mission, QD-30.018.0083); 'The leprosy settlement. Dedicating a new house'. Unknown, ca. 1935 (Basel Mission, QD-30.018.0084).

and were discharged after a short period.[44] The time Brack wrote this report, the use of anti-tuberculosis chemotherapy in Africa (for Africans) was not very common. The drugs he used – probably Streptomycin and PAS – were introduced to the world in the late 1940s and only during the 1950s and in combination with Isoniazid (discovered in 1951) increasingly used in Africa.[45]

For the European medical staff, Brack postulated systematic BCG vaccination – which was subsequently introduced by the Mission Committee in Basel[46] – and, referring to the previous tuberculosis cases, a restriction of the period of duty to two years, as doctors and nurses were in constant contact with tuberculosis, and the organism's resistance to the disease decreased "considerably".[47]

' "Davos" of the Gold Coast'

In 1950, Agogo Hospital was working to capacity, but hardly covered its expenses. In contrast to many of his successors (before and after the War) and colleagues, Brack did not see the solution in the expansion and improvement of the offer in order to increase income. He was convinced that the hospital work as it was could not be continued for more than a few years. Generally, he pointed out, the government services had improved their technical standards and Agogo was not able any more to match up to the hospitals in Accra and Kumasi in terms of equipment and specialised staff, and would thus fail to attract well-off patients from further away as it did before the war. For a mission hospital, Brack argued, this was not necessarily a problem, as it could assume the function of a more basic district ("Bezirksspital") or maternity hospital. However, Agogo's location was too isolated for such a role.[48]

In the mid-1940s, Agogo had 4,500 inhabitants, which was 'fairly typical of the larger centres of the cocoa areas in Ashanti', as the Ashanti Survey points out. The 18-mile road from Kumasi was judged a "second class road". The Afram plains north of Agogo – which would, decades later, develop into a relevant catchment area – were at the time still declared 'almost uninhabited'.[49] In the early 1950s, the district was judged relatively well supplied medically – also, but not only because of Agogo Hospital. In Konongo, a mining hospital with more than 40 beds had been built during the War. The area was, according to a commission enquiring the state of the health care system in the Gold Coast, sufficiently covered with hospital services. The commission's report points to the mining hospital in Konongo and to Agogo Hospital, but also to two projected mission hospitals in Mampong and Mpraeso.[50]

In this situation, Brack saw the only solution in order to maintain a hospital in Agogo in the specialisation in a specific medical field. He suggested a tuberculosis station, an ophthalmologic clinic or a nurses' training school. Tuberculosis he considered as a field 'of special importance for the Gold Coast' because of the disease's prevalence.[51] He was convinced – obviously contradicting other members of the hospital conference – that the current popularity of the hospital could not be taken for granted in the future. For a small hospital, Brack warned, it would be very difficult to meet the demands facing the general development towards medical specialisation.[52] For Brack, there was no justification for a general district hospital in Agogo, and the future of the Basel Mission's medical work in the Gold Coast did lie elsewhere:

> This fact of isolation is regrettable, especially in view of [the fact] that the country still requires further medical development. Other mission societies have taken up these problems, and in the near future, the building of various mission hospitals in the Gold Coast has to be expected. A mission medical development for the Basel Mission in the Gold Coast can not take place in Agogo. Another, more densely populated area, especially the northern territories would have to be envisaged.[53]

When obviously not all members of the hospital committee shared Brack's estimation concerning Agogo's future, neither did the Mission Committee in Basel. A revision of the hospital's strategic orientation was not considered as urgent. As long as there was a competent physician with training in surgery, as well as a "consciously Christian-missionary character" the work at the hospital could be continued as hitherto, the Committee concluded. The Christian attitude would be an advantage in the competition with even the most modern government hospitals, as patients were aware that '[t]his is a place where I will receive help without having to grease somebody's palm'. And the Committee ruled out the development of a specialisation in tuberculosis as "not possible" because of the proximity to the girls' school.[54]

Brack continued to advocate tuberculosis as a field of activity, claiming that the Gold Coast provided no "advanced" handling and treatment of the disease.[55] And his successor even more strongly advocated the specialisation in tuberculosis. Hans Meister agreed with Brack that the hospital, at least in the long run, would have to look for other functions than those of a general hospital. He was convinced that the future would lie 'above all' in a tuberculosis hospital (even though he also considered a nurses' training school as 'something of high missionary value').[56]

Meister had been superintendent of the Basel Mission's hospital in Moiyen (Meixian), China, for more than ten years, but had to leave after the Basel Mission decided to evacuate its staff from China in 1949.[57] He came to Agogo in 1952 and remained the hospital's superintendent until 1975. In his first annual report from the Gold Coast, Meister's main topic was tuberculosis and he compared the situation to his experience in China. He observed that tuberculosis was more common in China. Apart from the general "constitution" of the population, he attributed this to the 'fact . . . that China stands in a totally different phase of contagion ["Verseuchung"]. China is, like Europe, totally infested with tuberculosis, and therefore a certain relative immunisation can be noted'. As a consequence, he observed three times more different forms of tuberculosis "of the secondary stage" in China than in Agogo. While in China tuberculosis was more common, in the Gold Coast, just because of the "weak infestation", the course of pulmonary tuberculosis was faster and more severe. This was, Meister noted, notwithstanding the 'better nutritional condition on average'.[58] While Meister – in contrast to Golder who had only pointed to the Africans' higher predisposition – at least partially included environmental factors, he still followed the widespread assumption of the "virginity" of most (rural) Africa, at least in relation to China and Europe. This "virgin soil" theory which was predominant in the circle of (colonial) tuberculosis experts until the 1930s,[59] was clearly rejected for the Gold Coast by the tuberculosis survey Golder had referred to, which suggested 'no virgin areas, but high rates of tuberculinization'.[60] In his tentative strategy paper for the hospital a few months later, Meister reported that the 'contagion' with a 'quite vicious' tuberculosis was 'terrifying'.[61]

Meister agreed with Brack that the fight against the disease had hardly commenced, but he also noted that the state was planning to tackle the disease and to build tuberculosis sanatoria and hospitals. Here, Meister saw a good opportunity for the Mission to become active, as a tuberculosis hospital would not only match the agenda of the state but also the missionary aims: first, the mission should care for those who suffered from tuberculosis in the same way they cared for lepers and, second, the tuberculosis patients stayed at the hospital for a longer period, which would make evangelisation more effective. Again, he drew from his experience in China: 'From my experience – I know this from China as well – [the tuberculosis patients] are exceptionally grateful, if not only their body but also their soul is cared for'. Agogo's location, Meister argued, was suitable for a tuberculosis hospital as the '"Davos" of the Gold Coast' because of its altitude and the climate. The long access courses were no problem, as the patients would in any case stay for a longer period. Moreover, Meister argued that the proximity to the girls' school was no problem – if anything, uncontrolled tuberculosis in daily life was more dangerous.[62]

Meister saw two possible forms for Agogo as a tuberculosis centre: As a sanatorium with "conservative therapy" for minor cases with a department for severe, incurable cases or as a hospital for pulmonary and bone tuberculosis with "active therapy". By "active therapy", he meant "modern" drugs as well as surgical methods such as pneumothorax, thoracoplasty and 'maybe even' pulmonary resection. The decision, he believed, was in the first place dependent on the abilities of the doctor. Meister, "undoubtedly", judged the second direction as the right one, even though it would require higher standards concerning quality and quantity of staff. As the tuberculosis treatment would be expensive, Meister apprehended that the hospital's self-supporting policy could not be maintained and help from government might become necessary.[63]

In October 1952, Meister had an encouraging visit by the Gold Coast's first tuberculosis specialist Pointon-Dick, who reckoned the idea of transforming Agogo Hospital into a tuberculosis hospital. His only concern was the remote location as it was desirable that the patients should receive visits and feel comfortable. Meister dismissed the concern, based on 'our experience and our opinion: The Africans don't mind the distance, if they want to make a visit, and anyway too many visits are not desirable'.[64]

Tuberculosis services in the late colonial Gold Coast

In fact, tuberculosis was at that time not only a great concern of the Gold Coast's medical authorities, but also high on the global health agendas. The First World Health Assembly in 1948 declared tuberculosis a "top priority".[65] However, Pointon-Dick's encouragement for Meister's plan was to prove tentative. Soon, the cure and isolation approach with tuberculosis hospitals or sanatoria would give way to other strategies – in the Gold Coast and on a global level.

In 1948, the Gold Coast authorities estimated that there were 6,000 active tuberculosis cases at any time with 12,000 deaths per annum.[66] Tuberculosis was the third highest cause of death reported after the Second World War. Between 1946 and 1950, 73–101 in every 1,000 deaths from all causes were deaths from

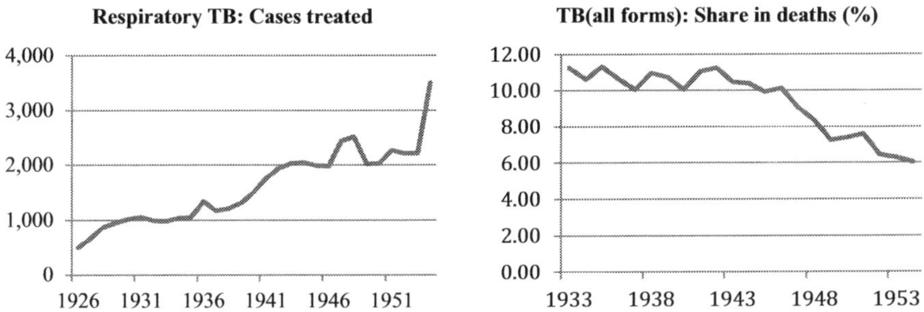

Figure 11.3 Tuberculosis in the Gold Coast

Source: Patterson, *Health in Colonial Ghana*, 129–30, 147, 153.

tuberculosis. Pulmonary tuberculosis was counted responsible for 6.8% of all deaths in hospital.[67] It was only from around 1950, with the introduction of chemotherapy, that the absolute number of deaths from tuberculosis would start to drop considerably (see Figure 11.3).[68]

About 200 of the Gold Coast's 2,000 hospital beds were occupied by tuberculosis patients.[69] In 1952, the "Maude Commission" agreed with the British tuberculosis specialist F.R.G. Heaf's recommendations for West Africa to provide 'simple isolation units in close association with general hospitals'.[70] These recommendations, in turn, corresponded to the policy of the WHO. The organisation, in the early 1950s, stressed the importance of tuberculosis treatment and advised hospitals for tuberculosis patients built in 'the simplest and cheapest style'.[71]

When Pointon-Dick was appointed to the Government of the Gold Coast as a specialist under the Colonial Office to investigate and advise on tuberculosis control in 1952,[72] he established a tuberculosis clinic in Accra, and started to co-ordinate tuberculosis treatment at other hospitals. In 1953, 'steps to organise country-wide tuberculosis services were . . . taken'. In 1954 the Tuberculosis Services were inaugurated, and a Tuberculosis Headquarters for the whole Gold Coast attached to the tuberculosis unit in Accra. A network of service facilities was envisaged. In order to 'bring the clinic and treatment facilities as near as possible to the patient', a hierarchical infrastructure was planned with six regional centres and a system of district tuberculosis centres at hospitals with x-ray and laboratory services as well as local tuberculosis treatment centres (see Figure 11.4).[73]

Modern treatment and tuberculosis hospital

In Agogo, tuberculosis work continued with the former leprosy settlement converted into tuberculosis wards. In 1952, there were 52 in-patients with pulmonary tuberculosis; in 1953, there were 127.[74] Meister reported that the 24 beds, which were "jammed" into the five huts, were permanently occupied. Most patients stayed for two to six months. Meister noted "quite good" success with the "new"

Figure 11.4 The organisation of the Ghana Tuberculosis Services

Source: Koch, 'Tuberculosis in Ghana', 282–9.

drugs Streptomycin and Isoniazid. In 1953, he started with surgical interventions in cases with cavities and conducted two thoracoplasties and one phrenic crush.[75] Five of the 127 in-patients with pulmonary tuberculosis died; 95 were discharged by the end of 1953.[76] However, it is questionable how telling these figures are about success of treatment. After their discharge, the patients were 'cared for at home with drugs and advice' and had to report at the hospital every one or two months.[77] A few years later, a doctor assisting Meister in Agogo noted that the success of 'modern therapy, regularly in combination with surgical measures' were frequently ruined because patients did not come to the monthly controls – often due to financial problems after long-term sickness, as the doctor assumed.[78]

In 1955, Meister advertised the tuberculosis work in Agogo in a circular letter for Swiss audiences. He reported that Agogo provided one of only two hospital departments for tuberculosis in the entire Gold Coast (the other one being the tuberculosis clinic in Accra), and that Agogo was the only place where chemotherapy was combined with surgical operations such as pneumothorax, phrenic crush, thoracoplasty and plombage:[79] The in-patients 'benefit from all modern methods of treatment'.[80] But, Meister wrote, the patients came from the entire country, and as not all could be admitted, 'hundreds' had to be treated as out-patients. They received chemotherapy and were often cured. But, Meister continued, 'these droves of spitting' tuberculosis patients were a threat for the local population, particularly because many of them found accommodation in the town. Meister assured that he had 'controversially' discussed the problem with local authorities and the Ministry of Health. The Minister had thus held out the prospect of 'generous new buildings' for an improvement of the situation – but was busy with other issues. In the meantime, the education of the tuberculosis patients was intensified with presentations about the prevention of infection in the waiting rooms.[81]

Meister undoubtedly gave tuberculosis top-priority in the planning for the hospital. In the annual report for 1955 he explained that the 'problem of the lung tuberculous' was in the focus concerning the organisational matters.[82] Confident with the minister's and the tuberculosis specialist's appreciation of his work (both had visited Agogo in 1955) and, as he assumed, their 'willingness to extend TB work in Agogo with government funds', Meister was confronted with considerations the DMS made on a visit in January 1956: He wished to concentrate the government funds for the reorganisation of tuberculosis work at government institutions – and expected this to reduce the 'unsolicited' rush of tuberculosis patients to Agogo. Meister however did not count on such an effect and assumed determination: Without a positive decision by the Ministry, another sponsor would have to be found who was 'willing to finance a realignment and modernisation' of the tuberculosis work in Agogo.[83]

In order to advance the process (and perhaps startled by the DMS's comments), Meister wrote to the Minister of Health in order to remind him 'of the position regarding TB-Patients at Agogo and the problem of public health involved in it'.[84] He emphasised that the mission had built at its own expense a small building as quarters for out-patients from distant places who had to visit the hospital on several consecutive days for examinations or for weekly Pneumothorax refilling. But now, as Meister explained, some inhabitants and the members of the Local Council had criticised the site of the building due to its proximity to the main road and the schools. Meister indicated to the Minister that Pointon-Dick, at his visit, had agreed 'that it is a concern of Public Health rather than of the Hospital', but

> was not in favour of a TB-Out-Patient Quarters, neither the one which now exists, nor a new one on a new site, which would have to be built and managed by Agogo Local Council. Dr. Pointon-Dick claimed that the best solution to the problem involves retaining in Hospital all Pulmonary TB-Patients whose homes are not near the Hospital and who must be frequently examined or treated.[85]

Apart from the 'public health concern', Meister mentioned two other problems: First, the tuberculosis patients attended the normal out-patients' department and constituted a danger for the other patients; and second, the 'present TB-wards [the lepers' huts] are very primitive indeed' and too far away, 'at the bottom of the hill in the bush'. Accordingly, Meister's request was for a set of new buildings including wards for in-patients (to replace the present 'unsatisfactory TB-settlement'), wards for assessment cases and patients in pneumothorax treatment whose homes were not near (to replace the present 'TB-Out-patients Quarters'), and a separate tuberculosis Out-patients Department with its own waiting room, consulting room, laboratory and dispensary. A suitable site for this building, just next to the hospital, 'in the immediate vicinity of our X-Ray Department and Operating Theatre' had already been identified, and the Agogohene (the local traditional ruler) and members of the Local Council had, '[i]n the course of informal discussions', already 'shown readiness' to provide the site. Only the necessary funds, Meister had to admit, were not available. Therefore he asked the Minister whether he would consider the project worthy of recommendation 'from the viewpoint of the welfare of the unfortunate TB sufferers in this country', and whether the Government would be prepared to contribute the necessary funds 'to enable this Hospital to build and manage a TB-Section in accordance with plans and arrangement agreed upon by your Department'.[86] In other words: Meister wanted to establish, next to the existing 'general hospital', a new tuberculosis hospital or centre, which would only share the operation theatre and the X-ray equipment (as well as the staff) with the former (see Figure 11.5).

Ultimately, Meister's suggestions did not find a sympathetic ear: A year later, he had to report: 'After initial promises, the government has let us down'. The tuberculosis in-patients were still admitted in the 'primitive huts in the bush' and the out-patients did still mingle with other patients and the local population.[87] According to Meister, the government claimed that funds were not available and that free treatment of tuberculosis patients would be a condition for government support. He assumed that collaboration would only be possible if a tuberculosis department were built and administered financially independent from the hospital under the responsibility of the government. He doubted that this would be the 'ideologically' right way to go.[88] In July 1957, the first Ghanaian Minister

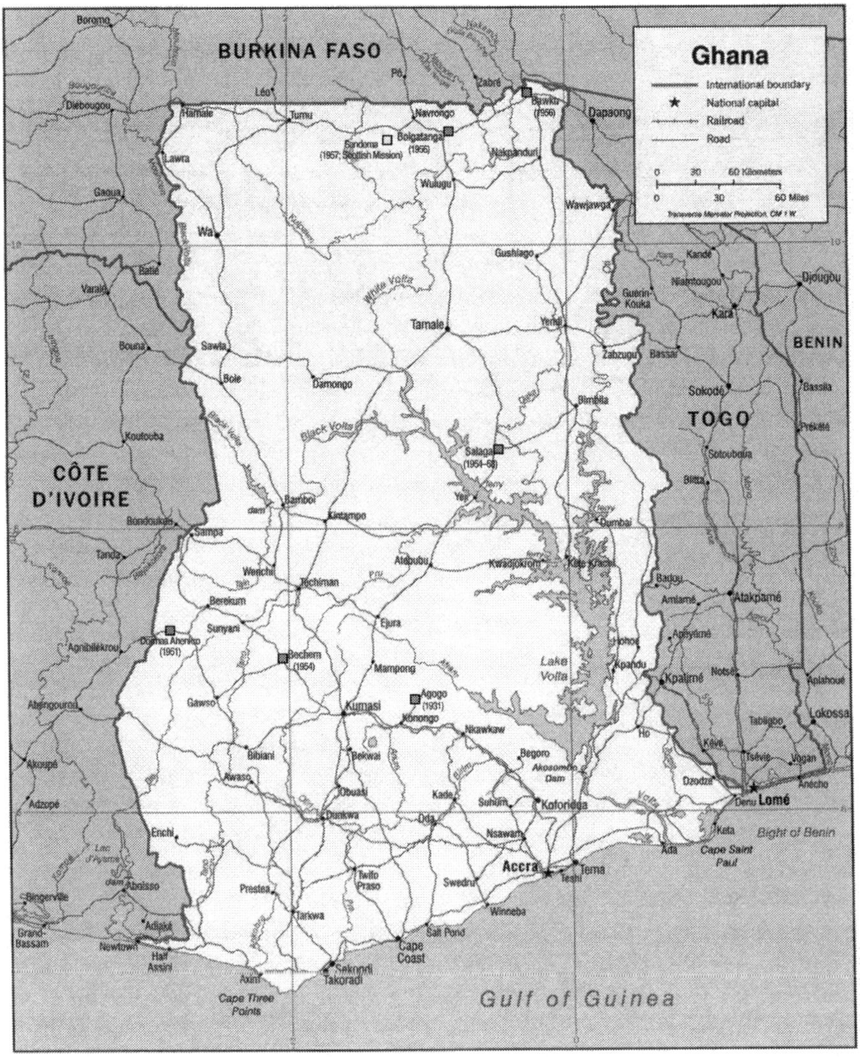

Figure 11.5 Expansion of Medical Services by the Basel Mission

Source: The sites of the Basel Mission's medical facilities in the 1950s (Map: 'U.S. Central Intelligence Agency Maps, 2007'. Accessed 20 February 2013. www.cia.gov/library/publications/cia-maps-publications/Ghana.html; supplemented by the author).

of Health, J.H. Alhassani, visited Agogo. Although commending the tuberculosis work at Agogo, he reaffirmed that the financial situation would not allow the government to support the plans for a tuberculosis centre in Agogo.[89]

By early 1959, Meister had definitively put off the 'project of a new TB Department, for which we had hoped to receive Government grants'. He identified the condition of free treatment of tuberculosis patients as a major obstacle to receive any government support, even in the form of supplying drugs and materials.[90] Free treatment was an issue that had been raised as controversial on several occasions before with.[91] But Meister did not consider the fact that patients in Agogo paid for their treatment 'to be a great evil, because it incites the patients themselves and the medical and nursing staff to more faithful efforts (also in competition with the respective Government services, which are free of charge)' – and for 'needy patients', Meister assured, ways were found to reduce or waive the fees.[92]

Ghana, the WHO and BCG

The question of fees was not the only and hardly the main reason that Meister's plans did not find support among the medical authorities. As A.B.P.W. Koch, Pointon Dick's successor from 1957, noted: 'shortage of hospital beds, x-ray facilities and, above all, of trained medical and nursing staff delayed the much-desired rapid progress' of introducing the tuberculosis services as initially envisaged.[93] This experience – lack of resources and the inability to reach the coverage of the whole population – was shared by other parts of the 'developing' world in the second part of the 1950s. It was to lead Ghana to emphasise the role of 'prevention' – mainly in the form of BCG vaccinations – and, eventually, the WHO to abstain from a vertical, centralised approach (for an overview, see Figure 11.6).[94]

By 1960, the Ghanaian tuberculosis services included three regional and ten district tuberculosis clinics, which were visited monthly by a tuberculosis specialist.[95] A map visualising the future network of the tuberculosis services in Ghana, indicated Agogo as a future 'Local TB Treatment Centre' (and notably not a district centre).[96] In the plans for the Tuberculosis Services, the establishment of ('un-elaborated') isolation units had faded into the background during the 1950s. Koch considered the treatment and isolation of tuberculosis patients as in-patients as not feasible – no more than a close supervision of out-patients, which would have been necessary for the prevention of resistant cultures.[97] Indeed, in 1962, it was reported that 60% of newly registered cases in Ghana 'are lost sight of within six months of starting oral treatment, while 16% are resistant to one of the three commonly used antituberculous drugs and 6% to two of them'.[98] More than on treatment, Koch counted on 'prevention' – on hygiene education and eradication of slum conditions, but more than on anything, he counted on BCG anti-tuberculosis vaccinations:

> BCG vaccination seems to offer a powerful weapon, particularly valuable in less developed countries. Little equipment is needed; costs are not very high; vaccination teams are small and mobile and thus, able to reach almost every village; the vaccination technique is simple and quick, so that large numbers can be vaccinated in comparatively short time.[99]

With this, Koch followed the WHO's tuberculosis strategy. Under the colonial government, no BCG campaign had taken place targeting the general population

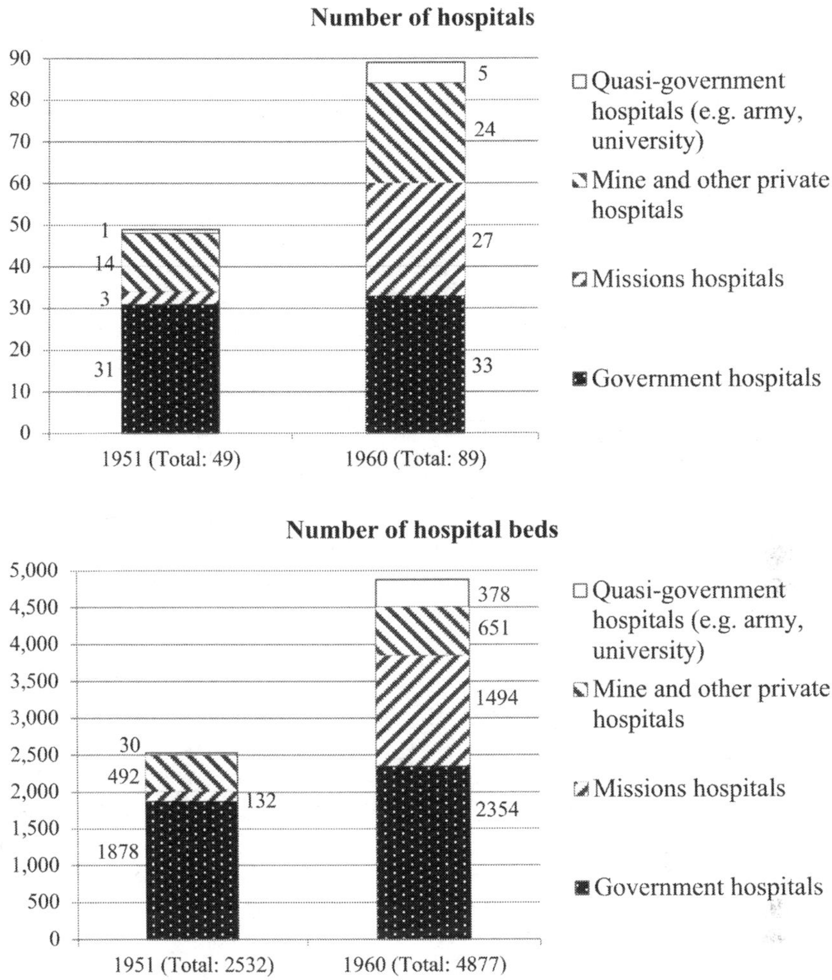

Figure 11.6 Hospitals and hospital beds in the Gold Coast (1951 and 1960)

Source: Addae, *Evolution of Modern Medicine*, 89.

of the Gold Coast.[100] Generally, the British colonial authorities were highly reluctant concerning BCG (in contrast to the French).[101] The WHO, however, had started to shift the emphasis from treatment to prevention, i.e. BCG vaccination, around 1950, and taken over the responsibility of the post-War BCG vaccination programme of UNICEF and the Scandinavian Red Cross societies in 1951.[102] During the 1950s, mass BCG campaigns became the dominant measure in the WHO's tuberculosis work. The reasons why the WHO started to subordinate the initially more established methods of tuberculosis control of 'case-finding, medical care and segregation' to the controversial BCG campaigns advocated by the powerful UNICEF are manifold, as Brimnes demonstrates: the relative weakness of the WHO and its lack of funds; post-war modernism's fascination with technological solutions in general, and vertical approaches for the 'war on disease'; as well as a sense of urgency, as the rise of tuberculosis was seen as a threatening epidemic. BCG vaccination, Brimnes argues, 'was high modernism on the cheap'.[103]

Also in newly independent Ghana, it was 'decided to initiate BCG campaigns on a large scale'. In 1959, the year that Meister buried his tuberculosis project, the first BCG campaign in the Gold Coast was initiated. Some 30,000 pupils in Accra primary and middle schools were tested and vaccinated.[104] In a five-year development plan, the Tuberculosis Headquarters envisaged the establishment of 'tuberculin testing and BCG vaccination centres in association with every hospital and health centre'. Further mobile vaccination teams were to visit remote places. The final aim was 'to vaccinate every new-born child and to tuberculin test and, if necessary, re-vaccinate every child attending school i.e. at about the age of 10 to 12'.[105] However, for a decade, the BCG campaigns were restricted to Accra. Still, in 1964, only six tuberculosis vaccinations were administered in the entire Ashanti Region.[106] In 1969 the regional administrative officer asked the Regional Senior Medical Officer whether arrangements were being made to extend the vaccination of school children against tuberculosis to at least Kumasi and its environs.[107] In the same year, a nation-wide programme aiming at the BCG vaccination of all children under 16 years was started, assisted by the WHO and UNICEF.[108] In 1970, the development plan of the Busia Government stated that the tuberculosis programme, including a BCG campaign 'to cover the entire country' as well as case-finding and home treatment, 'has been retarded by shortages of vaccines and equipment and an insufficient number of doctors'. To that date, 'over 300,000 susceptible people' had been vaccinated and detection and treatment facilities expanded 'from two to nine health posts'.[109] From 1971, BCG was integrated into the measles/smallpox vaccination programme (which had started in 1968),[110] and later in the Expanded Programme on Immunization (EPI), which was launched in Ghana in 1978.[111] The erratic numbers in BCG immunisations in Agogo in the 1970s demonstrate how delicate the supply with vaccines remained.[112]

Even though its effectiveness and value had been heavily contested again and again,[113] BCG was and continues to be one of the most widely used vaccines. Yet when chemotherapy emerged as a relatively effective means for treating tuberculosis, BCG remained the WHO's main instrument in tuberculosis control. In 1964, the Expert Committee on Tuberculosis, despite the slow decline in tuberculosis prevalence in the light of the resources spent on tuberculosis control programmes, still 'firmly believed in BCG, and saw it as the most cost-effective control measure'.[114]

In addition to the rise of BCG, another development in international tuberculosis control efforts had helped to marginalise Meister's plans for Agogo. In the late 1950s, trials in Madras and Bangalore, India, as well as later in Tunis and Nairobi, studied the effect of domiciliary treatment of tuberculosis with different drugs and drug combinations and from different perspectives. The interpretations of the results justified the dismissal of stationary treatment in favour of home treatment. The Expert Committee, in 1964, recommended that 'all financial resources and manpower available for tuberculosis control in the developing countries be confined to organizing efficient ambulatory services and not to constructing new beds'.[115] In combination, domiciliary treatment, the authorisation of BCG vaccination without testing, as well as diagnosis by bacteriological examination of patients with symptoms instead of mass (X-ray) diagnosis laid the ground for the breaking from the vertical approach and the new tuberculosis policy: the integration of tuberculosis services into the general health services.[116] This process of 'integration' was based on 'simplified technology' and entailed the transformation of tuberculosis centres into general health centres and hospitals. Amrith explains the 'strategy of limiting international tuberculosis policy to the question of drugs' and the decreasing inclusion of social factors (even though the Bangalore trials could have underscored the importance of social factors) by the exigencies of cost-effectiveness and the 'shift to the individual level in analysing and understanding tuberculosis' – justified by the Madras and Bangalore studies, 'whatever their mutual contradictions'.[117]

Thorax surgery in Agogo

Meister's vision of 'modern treatment' did not fit into the national Tuberculosis Services' framework and the transnational strategies to control the disease. All the more so, Meister propagated the inclusion of surgical measures in the treatment of tuberculosis. He advocated surgery as a means in tuberculosis treatment not only vis-à-vis the government and the mission audience in Switzerland, but also shared his experience with a professional audience through a presentation at the 'Conference on Tuberculosis and Leprosy of the West African Council for Medical Research' in Nigeria in 1959, and in a corresponding article published in the West African Medical Journal in the same year.[118]

The participants at the conference were physicians from Africa and England. Meister reported on a controversy between 'a group of TB doctors and epidemiologists' on the one hand and a new thorax surgeon at the University of Ibadan. The former argued strongly against the introduction of pulmonary surgery, because they thought it to be too costly. Meister himself, the only contributor who spoke about thorax surgery for tuberculosis, advocated the surgical treatment through thoracoplasty, arguing that satisfying results could be achieved in a small hospital with limited facilities, as in Agogo, and pointed out that the costs were massively lower than estimated by the critics (i.e. £30 instead of £500). Meister was sure to have convinced many of the participants at the conference. According to Meister, there were no thorax surgeries performed in the whole of West and East Africa, except for those in Agogo and Ibadan.[119] This, Meister argues in the article, was due to 'a lack of doctors and of well-equipped hospitals'. But Meister also assumed that some 'doctors and hospitals would have the facilities to engage in this kind of surgical treatment, but they refuse to do so because their nursing staff is said to be inadequate and inexperienced'.[120] Of the 48 tuberculosis patients in Agogo who had thoracoplasties, 44 survived and 'left the hospital after one or two months in improved condition'. Fourteen suffered relapses of the disease, nine were cured and 29 were still under observation (whereas the hospital staff lost sight of some of them).[121] In total, between 1952 and 1958, Meister had treated 57 tuberculosis patients with surgical interventions. He continued with up to 30 surgical interventions against pulmonary tuberculosis per year in the early 1960s.

However, surgical interventions in tuberculosis treatment were, at the time of effective chemotherapy, more and more rarely executed in Western countries[122] and their continuing justification was discussed controversially in medical circles. The proponents saw them as still important supporting measures in severe cases and in cases which did not respond to drugs, and emphasised their role in avoiding the emergence of resistance if cavities fail to close. The opponents emphasised the dangers for complications and the mutilations and other inconveniences the patients had to face.[123] At the same time, in the 1950s and 1960s, surgery was used elsewhere in Africa, including Kenya, Senegal, Liberia and Tanzania.[124] Still, Meister's colleague in Nigeria, Lauckner, was probably representing the contemporarily prevailing attitude towards (more or less) expensive surgical measures concerning tuberculosis. He had given up minor collapse therapy, arguing 'that the results of drug treatment alone (although the danger of resistance exists) are good enough for this to be regarded as a measure of prevention, and that only treatment which is of public health significance can be afforded'.[125]

Beyond tuberculosis

Meister, until his retirement in 1975, continued to perform around ten to 15 surgical interventions against tuberculosis per year (see also Figure 11.7). Until the end of the 1970s, the former leprosy settlement was used for around 100 in-patients per year who were treated with drugs for about three months each. Out-patients were prescribed chemotherapy and 'educated', though follow-ups for home treatment were not organised. However, from the early 1970s, Agogo started to take part in the BCG campaigns for children in the frame of the Under Fives Clinics (which were started in the mid-1960s).

In the context of decolonisation and with the deterioration of the economic situation in the 1960s, the Agogo Hospital lost much of the autonomy it had acquired during the previous five decades. It had to increasingly rely on financial assistance from the state. At the same time, the Presbyterian Church of Ghana (PCG) took over the formal responsibility for the hospital in 1963. Tuberculosis never played the important role Meister envisaged for Agogo Hospital. Instead of an expensive tuberculosis hospital aimed at catering for a relatively small number of patients, a major nurses' training school was established in Agogo from the early 1960s. In contrast to tuberculosis treatment and care, nurses' training had been able to consolidate the interests, values and conceptions attached to health and development of various players including the Basel Mission, the PCG as well as the Ghanaian and Swiss governments. In the early 1970s, the Ecumenical Nurses Training

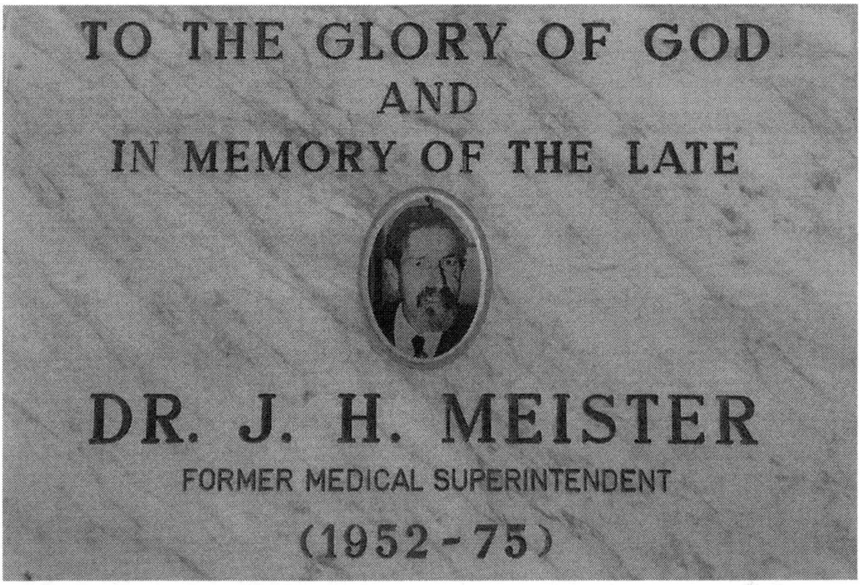

Figure 11.7 Memorial plaque for Hans Meister at Agogo Hospital

Source: Yakubu Ismaila, 2010.

College, a joint venture by the government and the Church Hospital Association of Ghana (CHAG, now Christian Health Association of Ghana) was attached to Agogo Hospital.

Nurses training contributed crucially to the maintenance of the high technical standard the hospital had inherited – despite the hospital's increasing financial dependency on the national state, and in a time when the signs of rural health care showed towards primary health care. In the 1960s and 1970s, the hospital expanded its capacity to more than 200 beds and a nursing staff of 100. The technical standard of the curative services was continuously improved. In addition to the nurses' training centre, the hospital acquired specialisations in ophthalmology, paediatrics and obstetrics. At the same time, in line with the prevalent health care strategies for 'developing countries', Agogo Hospital increasingly engaged in basic and preventive health care and its services were extended beyond the hospital walls. Towards the end of the 1970s, more emphasis was placed on efforts to improve the health situation in the district along the lines of the primary health care paradigm that had come to dominate the international health policies. When primary health care was implemented as a national policy in Ghana, Agogo Hospital became the district hospital and administrative centre for one of the pilot districts. Today it serves as the district hospital for the Ashanti-Akim North District.

Conclusion

The value of the hospital treatment of tuberculosis as such and, especially, the surgical treatment, may be regarded as limited in terms of both public health and the well-being of the individual patient. However, the present chapter does not aim to assess the 'success' of tuberculosis treatment and care in Agogo. Such an assessment would have to take place 'within a historical, social, and perhaps cultural framing of medical success', as Condrau postulates it for judgements on historical treatments.[126] The patients' experience of tuberculosis treatment (surgical and other) and care, as well as the psychological and social consequences for the patients would constitute a highly fascinating subject for further research on Agogo Hospital (e.g. in comparison with other forms of treatment).[127] The same is true for a historically informed epidemiological assessment of the public health effects of the admission of tuberculosis patients in terms of education and especially in terms of isolation (e.g. in the frame of a comparative study).[128]

This chapter has demonstrated how the development of health care delivery is subjected to the constellation of actors at the various levels – local, national, transnational. The medical staff at and around the hospital, the patients, the mission leadership in Basel and on the Gold Coast, the colonial or national authorities all connected different interests, values and ideologies to specific concepts and practices. Tuberculosis treatment inspired by European (middle-class) sanatoria including surgical interventions but also chemotherapy was celebrated as a success in Agogo. The care and treatment of tuberculosis patients corresponded well with the ethical imperatives of the medical mission but also with the major aim of the missionary endeavour: even though the effectiveness of tuberculosis treatment at that time is questionable, tuberculosis patients were seen as a grateful clientele that would, moreover, stay at the hospital for a longer period, which would make evangelisation more effective.

Tuberculosis was also identified as a field in which the hospital could make a welcomed contribution to the colonial health care system, and which would thus serve to justify the maintenance and expansion of (curative) medical services after the Second World War. Indeed, the hospital-based 'active therapy', as promoted at the hospital in the 1950s, can readily be judged as characteristic of the colonial focus on 'modern' curative medicine. But it failed to connect to the currently predominant strategies and policies in the field on an international level and, eventually, to Ghana's national political and medical programme. At the time Agogo Hospital's superintendent Hans Meister made plans for a tuberculosis department, tuberculosis was a great concern for the Gold Coast's medical authorities as well as the WHO and other international organisations. The ways of tackling the disease, however, were subjected to changes in technology and in international policies. Soon, BCG vaccination became the most important tool on the agenda of WHO and UNICEF. In the 1960s, home treatment and the integration of tuberculosis services into the general health services made tuberculosis hospitals obsolete in the predominant health care strategies. These strategies – partly and with delay – entered into the colonial and national health services in the Gold Coast and Ghana. As became apparent, tuberculosis work as conducted in Agogo did not resonate with the development of tuberculosis control policies formulated by national authorities and international organisations. Tuberculosis treatment in Agogo informed by 'modern' curative measures and missionary values did not pass its experimental status; in the context of decolonisation it was overtaken by more 'modern' tools promising to meet the demands of an efficient, just and comprehensive health care system.

Notes

1 Research for this chapter was part of the author's PhD thesis and of the project 'History of Health Systems in Africa' at the Department of History, University of Basel and the Centre for African Studies Basel. It was supported by the Swiss National Science Foundation, the Swiss Network for International Studies, the Freiwillige Akademische Gesellschaft Basel and the Rudolf Geigy Stiftung. The author is grateful to Marcel Dreier and Supriya Guha for their comments and inputs. His sincere thanks go to Patrick Harries for his invaluable inspiration and support.
2 Brassel-Moser, 'Ein Baum – zwei Zweige', 340–52.
3 Cf. Gocking, *The History of Ghana*, 79; Cooper, *Colonialism in Question*, 157; Cooper, *Africa Since 1940*, 36–7; Shipway, *Decolonization and Its Impact*, 116; Clarence-Smith, 'The Organization of "Consent" in British West Africa', 73–4; Packard, 'Post-Colonial Medicine', 100.
4 Cooper, *Africa Since 1940*, 50–3.
5 Maude, *Report*, 63.
6 Addae, *Evolution of Modern Medicine*, 89.
7 Harrison and Worboys, 'Disease of Civilisation', 117–18.
8 Addae, *Evolution of Modern Medicine*, 391.
9 Ibid., 389; Patterson, *Health in Colonial Ghana*, 65.
10 Dumett, 'Disease and Mortality', 215.
11 Ibid., 216–24.
12 'Memorandum: Report on the Sanitation of Ashanti for 1933–1934', 4 January 1934, CSO11/14/215, PRAAD Accra.
13 Addae, *Evolution of Modern Medicine*, 387.
14 It increased until the mid-1940s (from around 100 to around 300). Only from 1950 did it drop dramatically. It has of course to be borne in mind that the increase to some extent has to do with the increasing reporting of deaths and their causes. Patterson, Health in Colonial Ghana, 129–30.
15 Addae, *Evolution of Modern Medicine*, 389.
16 Ibid., 387.
17 Ibid., 390–1; Patterson, *Health in Colonial Ghana*, 66.
18 'DMS to Colonial Secretary', 14 March 1940, CSO11/1/412, PRAAD Accra.
19 Harrison and Worboys, 'Disease of Civilisation', 109.
20 Forster, 'Jahresbericht 1934', 1935, D-4-2-2, Basel Mission; 'Stationskonferenz', 22 March 1937, D-4-5-4, Basel Mission; Süss, 'Jahresbericht 1939', 1940, BM D-4-2,4, Basel Mission.
21 Forster, 'Jahresbericht 1934', 1935, D-4-2-2, Basel Mission.
22 'Stationskonferenz', 20 January 1938, D-4-5-5, Basel Mission.
23 'Inspektor an Station Agogo', 22 February 1938, D-4-5-5, Basel Mission.
24 'Inspektor an Station Agogo', 20 June 1938, D-4-5-5, Basel Mission.
25 'Golder an Kellerhals', 7 May 1948, BM D-4-6,3, Basel Mission.
26 'Golder an Kellerhals', 20 March 1948, BM D-4-6,3, Basel Mission.
27 'Kellerhals an Dora Golder', 31 March 1948, BM D-4-6,3, Basel Mission; 'Golder an Kellerhals', 4 March 1948, BM D-4-6,3, Basel Mission; 'Kellerhals an Golder', 4

October 1948, BM D-4-6,3, Basel Mission; 'Kellerhals an Golder', 29 April 1948, BM D-4-6,3, Basel Mission.
28 'Kellerhals an Golder', 17 August 1948, BM D-4-6,3, Basel Mission.
29 'Kellerhals an Golder', 7 January 1948, BM D-4-6,3, Basel Mission.
30 Todd, 'Tuberculosis Survey'.
31 'Golder an Kellerhals', 7 May 1948, BM D-4-6,3, Basel Mission.
32 'Métraux an Witschi', 11 July 1945, HZ1-N04–01.495, Basel Mission.
33 'Kellerhals an Golder', 19 October 1946, BM D-4-6,3, Basel Mission; 'Golder an Kellerhals', 21 October 1946, BM D-4-6,3, Basel Mission.
34 'Golder an Kellerhals', 15 March 1947, BM D-4-6,3, Basel Mission.
35 'Golder an Kellerhals', 7 May 1948, BM D-4-6,3, Basel Mission.
36 Patterson, *Health in Colonial Ghana*, 66.
37 Maurer, 'Heimgefunden', June 1948, BM D-11,11, Basel Mission.
38 'Golder an Kellerhals', 7 May 1948, BM D-4-6,3, Basel Mission.
39 Ibid.
40 Hardiman, 'Introduction', 33; Worboys, 'Mission and Mandate', 213–14.
41 'Brack an Kellerhals', 4 July 1948, BM D-4-6,3, Basel Mission; 'Kellerhals an Golder', 4 October 1948, BM D-4-6,3, Basel Mission.
42 Brack, 'Bericht 1. August 1948 bis 1. April 1949', 1949, BM D-11,9, Basel Mission.
43 Schellenberg, 'Rundbrief', 5 March 1949, BM D-4-3,5, Basel Mission.
44 Brack, 'Bericht 1. August 1948 bis 1. April 1949', 1949, BM D-11,9, Basel Mission.
45 World Health Organization, *The First Ten Years*, 192–5; Wilcocks, 'Recent Work', 32–3.
46 'Witschi an Brack', 7 November 1949, BM D-4-6,3, Basel Mission.
47 Brack, 'Bericht 1. August 1948 bis 1. April 1949', 1949, BM D-11,9, Basel Mission.
48 'Brack an Witschi', 9 April 1950, BM D-4-6,3, Basel Mission.
49 Fortes, Steel, and Ady, 'Ashanti Survey', 155–6. Agogo was one of the main foci of the Ashanti Survey.
50 Maude, *Report*, 23.
51 'Brack an Witschi', 9 April 1950, BM D-4-6,3, Basel Mission.
52 'Protokoll der Spitalkonferenz', 21 October 1950, BM D-4-6,3, Basel Mission.
53 Brack, 'Jahresbericht 1950', 28 March 1951, PS1-B05-03-10198, Basel Mission.
54 'Witschi an Brack', 24 April 1951, BM D-4-6,3, Basel Mission.
55 Brack, 'Jahresbericht 1951', 1952, BM D-11,9, Basel Mission.
56 Meister, 'Erste Eindrücke', 22 April 1952, PS1-B05-03-10198, Basel Mission.
57 'Witschi an Brack', 23 May 1949, BM D-4-6,3, Basel Mission.
58 Meister, 'Jahresbericht 1952', 17 February 1953, PS1-B05-03-10198, Basel Mission.
59 Cf. Harrison and Worboys, 'Disease of Civilisation'; Wilcocks, 'Recent Work'.
60 Todd, 'Tuberculosis Survey', 200.
61 Meister, 'Erste Eindrücke', 22 April 1952, PS1-B05-03-10198, Basel Mission.
62 Ibid.
63 Ibid.
64 Meister, 'Jahresbericht 1952', 17 February 1953, PS1-B05-03-10198, Basel Mission.
65 Together with malaria, maternal and child health, venereal diseases, nutrition and environmental sanitation. Litsios, *The Third Ten Years*, 39.
66 Todd, 'Tuberculosis Survey', 200.
67 Maude, *Report*, 51.
68 Patterson, *Health in Colonial Ghana*, 129–30.
69 Todd, 'Tuberculosis Survey', 200.
70 Maude, *Report*, 51.
71 British Tuberculosis Association, 'WHO and Tuberculosis Control'.
72 B.C.T., 'Obituary: W. Pointon Dick, M.R.C.S., L.R.C.P'.
73 Koch, 'Tuberculosis in Ghana', 284–6.

74 'Basel Mission Hospital Agogo Annual Report 1952', 1953, PS1-B05-03-10198, Basel Mission; Agogo Hospital, 'Return A (Diseases and Death In-Patients, Diseases Out-Patients) for the Year 1953', 1954, PS1-B05-03-10198, Basel Mission.
75 Meister, 'Jahresbericht 1953', 19 January 1954, PS1-B05–03–10198, Basel Mission; 'Basel Mission Hospital Agogo Statistics of Operations 1953', 1954, PS1-B05-03-10198, Basel Mission.
76 Agogo Hospital, 'Return A (Diseases and Death In-Patients, Diseases Out-Patients) for the Year 1953', 1954, PS1-B05-03-10198, Basel Mission.
77 Meister, 'Jahresbericht 1953', 19 January 1954, PS1-B05-03-10198, Basel Mission.
78 Haaf, 'Bericht Aus Der Medizinischen Arbeit Im Spital Agogo 1956', 1957, PS1-B05-03-10198, Basel Mission.
79 All these surgical interventions aimed at collapsing the parts of the lung with open cavities in order to promote the healing process and limit the spread of the infection.
80 Meister, 'Rundbrief', 26 November 1955, PS1-B05-03-10198, Basel Mission.
81 Ibid.
82 Meister, 'Jahresbericht 1955', 1956, PS1-B05-03-10198, Basel Mission.
83 Ibid.
84 'Meister to Minister of Health', 1 October 1956, PS1-B05-03-10198, Basel Mission.
85 Ibid.
86 Ibid.
87 Meister, 'Jahresbericht 1956', 1957, PS1-B05-03-10198, Basel Mission.
88 Ibid.
89 Haaf, 'Jahresbericht 1957', 1958, PS1-B05-03-10198, Basel Mission.
90 Meister, 'Jahresbericht 1958', 1959, PS1-B05-03-10198, Basel Mission.
91 E.g. Meister, 'Jahresbericht 1956', 1957, PS1-B05-03-10198, Basel Mission; Meister, 'Ärztliche Mission in Ghana', 1959, PS1-B05-03-10198, Basel Mission.
92 Presbyterian Church of Ghana, *Report for 1960*.
93 Koch, 'Tuberculosis in Ghana', 284.
94 Cf. Raviglione and Pio, 'WHO Policies', 776.
95 Koch, 'Tuberculosis in Ghana', 286.
96 Ibid., 285.
97 Koch and Marolda, 'Tuberculosis Prevention in Ghana', 366.
98 Lancet, 'Tuberculosis and Leprosy in Africa', 1116.
99 Koch and Marolda, 'Tuberculosis Prevention in Ghana', 366.
100 Patterson, *Health in Colonial Ghana*, 66.
101 Harrison and Worboys, 'Disease of Civilisation', 118; McMillen and Brimnes, 'Medical Modernization', 182–3; Wilcocks, 'Recent Work', 35–6.
102 World Health Organization, *The First Ten Years*, 191–2.
103 Brimnes, 'BCG Vaccination', 866–7.
104 Koch and Marolda, 'Tuberculosis Prevention in Ghana', 366.
105 Koch, 'Tuberculosis in Ghana', 288–9.
106 Regional SMO J.L.Q. Vanderpuije, 'Ministry of Health Annual Report for Ashanti 1964', 1965, ARG2/8/69, PRAAD Kumasi.
107 'Regional Administrative Officer, Ashanti, to Regional Senior Medical Officer', 23 March 1969, ARG2/14/3, PRAAD Kumasi.
108 'Sen. MO, MFUs: BCG Vaccination against Tuberculosis', 19 September 1969, ARG2/14/3, PRAAD Kumasi.
109 Ghana Government, *Development Plan*, 176–7.
110 Van der Mei, 'The Year 1970 in Children's Ward and Childwelfare Clinic, Agogo Hospital', 1971, PS1-B05–03–10198, Basel Mission. Cf. Akyeampong's critique of the implementation of the WHO's smallpox eradication programme in West Africa 'as testing grounds' and the attachment of measles eradication as an 'encouragement'. Akyeampong, 'Disease in West African History', 204.

111 EPI was initiated in 1973 and adopted in 1974 by the WHO, initially covering tuberculosis, diphtheria, whooping cough, tetanus, measles and poliomyelitis. In 1977, with the support of UNICEF and other funders, it established the goal of immunising all children by 1990. Litsios, *The Third Ten Years*, 191–2. In Ghana there was a pilot project in Central and Northern Regions 1976–8. The programme was launched in 1978, and operational in all districts in 1985. Mensah, 'Factors Contributing to Low Immunization Coverage', 2.

112 Cf. Agogo Hospital annual reports (PS1-B05-03-10198, Basel Mission).

113 Cf. Brimnes, 'BCG Vaccination'; McMillen and Brimnes, 'Medical Modernization'.

114 Brimnes, 'BCG Vaccination', 868, 871.

115 World Health Organization, *The Second Ten Years*, 129.

116 Tubercle, 'Leading Article: Tuberculosis in Africa'; Litsios, *The Third Ten Years*, 196; Raviglione and Pio, 'WHO Policies', 776.

117 Amrith, 'Magic Bullet', 128.

118 McMillen and Brimnes, 'Surgical Treatment'.

119 Meister, 'Bericht über den Besuch der "Conference on Tuberculosis and Leprosy of the West African Council for Medical Research" in Jos vom 17. – 20. Februar 1959', 1959, BM PS1-B05–03–10198, Basel Mission.

120 McMillen and Brimnes, 'Surgical Treatment', 292.

121 Ibid., 293.

122 Zürich, *Zürich Air*, 10–12; Spiegelburg, *New Topic in Tuberculosis Research*, 126; Mackowiak et al., 'Syncopal Episodes', 1801; Chambers, 'Surgical Treatment'.

123 See the controversy among readers of the *British Medical Journal* in 1955: Guyon Trimble, 'Current Therapy'; Livingstone, 'Observations'; BMJ, 'Treatment of Pulmonary Tuberculosis'; Lloyd, 'Treatment of Pulmonary Tuberculosis'; Mitchell, 'Diagnosis and Treatment of Pulmonary Tuberculosis'; Brailsford, 'Treatment of Pulmonary Tuberculosis'; Philip, 'Treatment of Pulmonary Tuberculosis'; Cameron, 'Treatment of Pulmonary Tuberculosis'.

124 Wilcocks, 'Recent Work', 34; Lancet, 'Tuberculosis and Leprosy in Africa'; Dreier, 'Health Care, Welfare and Development'.

125 Cited in Wilcocks, 'Recent Work', 34.

126 Condrau, 'Beyond the Total Institution', 76–7, 88.

127 Cf. ibid., 87–91. On patients who had (successfully) undergone surgical treatment see Schmid and De Haller, 'Late Exudative Complications'; Weissberg and Weissberg, 'Late Complications of Collapse Therapy'.

128 Cf. Wilson's criticism of McKeown. Wilson, 'Commentary'; Condrau and Worboys, *Tuberculosis Then and Now*, 6–7.

Bibliography

Addae, S. Kojo *The Evolution of Modern Medicine in a Developing Country: Ghana 1880–1960* (Edinburgh: Durham Academic Press, 1997).

Akyeampong, Emmanuel Kwaku, 'Disease in West African History', in *Themes in West Africa's History, Western African Studies* (Athens: Ohio University Press, 2006), 186–207.

Amrith, Sunil, 'In Search of a "Magic Bullet" for Tuberculosis: South India and Beyond, 1955–1965', *Social History of Medicine* 17 (2004): 113–30.

B.C.T., 'Obituary: W. Pointon Dick, M.R.C.S., L.R.C.P.', *British Medical Journal* 2 (1956): 305.

BMJ, 'Treatment of Pulmonary Tuberculosis', *The British Medical Journal* 1 (1955): 273–74.

Brailsford, James F., 'Treatment of Pulmonary Tuberculosis', *The British Medical Journal* 1 (1955): 665–6.

Brassel-Moser, Ruedi, 'Ein Baum – zwei Zweige. Zur Geschichte der Basler Mission zwischen 1930 und 1966', *Zeitschrift für Mission* 29 (2003): 334–52.

Brimnes, Niels, 'BCG Vaccination and WHO's Global Strategy for Tuberculosis Control 1948–1983', *Social Science & Medicine* 67 (2008): 863–73.

British Tuberculosis Association, 'WHO and Tuberculosis Control', *Tubercle* 31 (1950): 121.

Cameron, Ewen A., 'Treatment of Pulmonary Tuberculosis', *The British Medical Journal* 1 (1955): 912.

Chambers, John S., 'Surgical Treatment of Pulmonary Tuberculosis – A Decade of Change', *California Medicine* 84 (1956): 388–93.

Clarence-Smith, William Gervase, 'The Organization of "Consent" in British West Africa, 1820s to 1960s', in *Contesting Colonial Hegemony. State and Society in Africa and India*, ed. Dagmar Engels and Shula Marks (London: British Academic Press, 1994), 55–84.

Condrau, Flurin, 'Beyond the Total Institution: Towards a Reinterpretation of the Tuberculosis Sanatorium', in *Tuberculosis Then and Now: Perspectives on the History of an Infectious Disease*, ed. Flurin Condrau and Michael Worboys (Montreal: McGill-Queen's Press – MQUP, 2010), 72–99.

——— and Michael Worboys, eds., *Tuberculosis Then and Now: Perspectives on the History of an Infectious Disease* (Montreal: McGill-Queen's Press, 2010).

Cooper, Frederick, *Africa Since 1940: The Past of the Present* (Cambridge: Cambridge University Press, 2002).

———, *Colonialism in Question: Theory, Knowledge, History* (Berkeley: University of California Press, 2005).

Dreier, Marcel, 'Health Care, Welfare and Development in Rural Africa: The Case of the Catholic Health Services in Ifakara/Tanzania in the 20th Century' (PhD diss., Universität Basel, forthcoming).

Dumett, Raymond, 'Disease and Mortality among Gold Miners of Ghana: Colonial Government and Mining Company Attitudes and Policies, 1900–1938', *Social Science & Medicine* 37 (1993): 213–32.

Fortes, M., R. W. Steel, and P. Ady, 'Ashanti Survey, 1945–46: An Experiment in Social Research', *The Geographical Journal* 110 (1947): 149–77.

Ghana Government, *One-Year Development Plan, July 1970 to June 1971* (Accra-Tema: Ghana Pub. Corp., 1970).

Gocking, Roger, *The History of Ghana, The Greenwood Histories of the Modern Nations* (Westport, CT: Greenwood, 2005).

Hardiman, David, 'Introduction', in *Healing Bodies, Saving Souls Medical Missions in Asia and Africa*, ed. David Hardiman (Amsterdam: Rodopi, 2006), 5–57.

Harrison, Mark and Michael Worboys, 'A Disease of Civilisation: Tuberculosis in Britain, Africa and India, 1900–39', in *Migrants, Minorities & Health: Historical and Contemporary Studies*, ed. Lara Marks and Michael Worboys (London: Routledge, 1997), 93–124.

Koch, A.B.P.W., 'Tuberculosis in Ghana', *Tubercle* 41 (1960): 282–9.

——— and L. Marolda, 'Tuberculosis Prevention in Ghana: A School Children BCG Campaign in Accra', *Tubercle* 42 (1961): 366–70.

Lancet, 'Tuberculosis and Leprosy in Africa', *The Lancet* 279 (1962): 1115–16.

Litsios, Socrates, *The Third Ten Years of the World Health Organization, 1968–1977* (Geneva: World Health Organization, 2008).

Livingstone, James L., 'Observations on the Treatment of Pulmonary Tuberculosis at the Present Time', *The British Medical Journal* 1 (1955): 243–50.

Lloyd, T. W., 'Treatment of Pulmonary Tuberculosis', *The British Medical Journal* 1 (1955): 538.

Mackowiak, Philip A., M. K. Leonard, C. K. Kraft, and R. F. Corpe, 'Man with Syncopal Episodes and Abnormal Chest Radiograph Findings', *Clinical Infectious Diseases* 42 (2006): 1800–2.

Maude, John, *Report of the Commission of Enquiry Into the Health Needs of the Gold Coast* (Accra, Gold Coast: Govt. Print. Dept., 1952).

McMillen, Christian W. and Niels Brimnes, 'Surgical Treatment of Pulmonary Tuberculosis with Limited Facilities', *West African Medical Journal* 7 (1959): 292–8.

———, 'Medical Modernization and Medical Nationalism: Resistance to Mass Tuberculosis Vaccination in Postcolonial India, 1948–1955', *Comparative Studies in Society and History* 52 (2010): 180–209.

Mensah, 'Factors Contributing to Low Immunization Coverage in the Asante Akim North District' (PhD diss., Master of Public Health, University of Ghana, 1998).

Mitchell, Roger S., 'Diagnosis and Treatment of Pulmonary Tuberculosis', *The American Journal of Surgery* 89 (1955): 611–16.

Packard, Randall, 'Post-Colonial Medicine', in *Medicine in the Twentieth Century*, ed. Roger Cooter and John V. Pickstone (Amsterdam: Harwood Academic, 2000), 97–112.

Patterson, K. David, *Health in Colonial Ghana: Disease, Medicine, and Socio-Economic Change, 1900–1955* (Waltham, MA: Crossroad Press, 1981).

Philip, T. V. R., 'Treatment of Pulmonary Tuberculosis', *The British Medical Journal* 1 (1955): 666.

Presbyterian Church of Ghana, *Report for 1960* (Accra: Presbyterian Press, 1961).

Raviglione, M.C. and A. Pio, 'Evolution of WHO Policies for Tuberculosis Control, 1948–2001', *The Lancet* 359 (2002): 775–80.

Schmid, F. G. and R. De Haller, 'Late Exudative Complications of Collapse Therapy for Pulmonary Tuberculosis', *Chest* 89 (1986): 822–7.

Shipway, Martin, *Decolonization and Its Impact. A Comparative Approach to the End of the Colonial Empires* (Malden, MA: Blackwell, 2008).

Spiegelburg, Donald D., *New Topic in Tuberculosis Research* (New York: Nova Publishers, 2006).

Todd, K.W., 'A Gold Coast Tuberculosis Survey 1947', *Tubercle* 29 (1948): 197–200, 207.

Trimble, Harold Guyon, 'Current Therapy in Pulmonary Tuberculosis in the United States', *The British Medical Journal* 1 (1955): 250–3.

Tubercle, 'Leading Article: Tuberculosis in Africa', *Tubercle* 47 (1966): 405–7.

Weissberg, D. and D Weissberg, 'Late Complications of Collapse Therapy for Pulmonary Tuberculosis', *Chest* 120 (2001): 847–51.

Wilcocks, Charles, 'An Analysis of Some Recent Work on Tuberculosis in Africa', *British Journal of Diseases of the Chest* 54 (1960): 31–9.

Wilson, Leonard G., 'Commentary: Medicine, Population, and Tuberculosis', *International Journal of Epidemiology* 34 (2005): 521–4.

Worboys, Michael, 'The Colonial World as Mission and Mandate: Leprosy and Empire, 1900–1940', *Osiris*, 2nd Series 15 (2000): 207–18.

World Health Organization, *The First Ten Years of the World Health Organization* (Geneva: World Health Organization, 1958).

———, *The Second Ten Years of the World Health Organization, 1958–1967* (Geneva: World Health Organization, 1968).

Zürich, Lungenliga, ed., *Zürich Air. Jubiläumsheft der Lungenliga Zürich* (Zürich: Lungenliga Zürich, 2008).

12 When economics went overseas

Epistemic problems in the macroeconomic analysis of late colonial Africa

Daniel Speich Chassé

The peripheries of the European colonial empires were certainly no prominent topic in economic scholarship until the end of World War Two with the possible exception of the theory of international trade. The 'World Economic Surveys' commissioned by the League of Nations in the Interwar Period almost completely focused on entrepreneurial activity among industrialised countries. The British economist James Meade authored one of these reports in 1938. It dealt on 26 pages with the United States, Western Europe and Japan while it dedicated only one single paragraph and one table to the 'primary producing countries' (Australia, Canada, New Zealand, Argentina, Brazil, Chile, Hungary, Romania and Yugoslavia) and one sentence only to the Balkans and to the Dutch East Indies. The Rest of Asia, the USSR and all of Africa went unnoticed.[1] Against this selective spatial pattern of scholarly attention arose, post-1945, a new interest in regions of the world situated on the periphery of Europe's economic interests. This new concern constitutes the topic of this chapter. Focusing on British academic imaginations of Africa it locates macroeconomic research practices in the context of late British colonial policy. New techniques of economic observation, of data gathering, and of quantification arose,[2] and shaped African economies as distinct "others" to the British economy. While economists worked within an assumedly universal framework, they depicted an essentially bifurcated global scenery in which a relatively poor and a relatively wealthy world stood opposed. Non-monetised and largely subsistence-oriented economic collectives characterised one world exemplified by sub-Saharan Africa. Here, British economists diagnosed an under-utilised labour force and significant natural resources and called for the development of the exploitation of these potentials. The other world consisted basically of the industrialised "West" of which, more specifically, the economic structure of Great Britain and the USA formed the templates.

With which observational tools and in what relations to the political demand for expert policy advice have economists started to make sense of African socio-economic interactions around the middle of the 20th century? And to what extent did their discourse structure the conflict between a global North and a global South that unfolded in the wake of decolonisation?[3] The first part of the paper draws attention to the early years in which economic attention was focused on British

Colonial Development in Africa and attempts to locate the individuals behind the knowledge produced at this time. The second part then turns to the construction of development economics after World War Two. In this part, special attention will be given to the techniques of national income accounting in Africa. Macroeconomic statistics strongly shaped the way in which attention was distributed across the globe. Late British colonial policy produced a statistical vision of sub-Saharan Africa that survived decolonisation. It emptied the continent of initiative by pushing local forms of economic life to the margins of its descriptive framework and, in this way, produced an image of Africa as a tabula rasa on which colonial and post-colonial development planners could write as they wished.

Economic universalism

Scholars of colonialism generally agree that late colonialism was characterised by increasing social scientific inquiry and by a growing quest to apply rational techniques of government to the administration of distant possessions.[4] In parallel to the French policy of a "mise en valeur" that was designed in 1923, the British Empire saw the invention of a new policy of "colonial development".[5] In this connection also a new interest in economic knowledge arose. But the latest techniques in macroeconomic analysis, mainly the statistical tools of national income accounting, remained largely ignored. The one exception here is India, where V.K.R.V. Rao computed macroeconomic statistics as early as the 1930s.[6] But an assumingly all-encompassing academic survey on Africa that was initiated by General Smuts and that was conducted under the auspices of the colonial administrator Lord Hailey by a team of researchers including the economist E.A.G. Robinson and the biologist Julian Huxley worked along a completely different epistemic path.[7] In the terms of its use of statistics as a social scientific tool this 1938 volume was still very much closer to the old German tradition of descriptive statistics that had originated in Goettingen in the late 18th century, than to the new quantitative techniques of economic scholarship.[8] Its two dozen chapters first described the peoples of Africa, their languages and their political institutions. It then turned to the colonial practices of domination with special respect to the prevailing systems of taxation and of law. Only in its final chapters did Lord Hailey's African Survey come to the African economy proper by listing potentials of labour force and natural resources. It also included a list of useful infrastructural interventions that had already occurred or were bound to be carried out.

Macroeconomics was absent. In the introduction Lord Hailey positively stated his interest in more closely analysing African conditions:

> A considerable part of the activity of the intellectual world is expended to-day in the study of social institutions, systems of law, and political developments which can now only be examined in retrospect. But Africa presents itself as a living laboratory, in which the reward of study may prove to be not merely the satisfaction of an intellectual impulse, but an effective addition to the welfare of the people.[9]

Historians of anthropology have investigated the metaphor of Africa as "living laboratory" of which colonial scientists and administrators at the time of Lord Hailey made use.[10] The *Africa Survey* was based on anthropology. This discipline promised an approach to the pre-historic human past of which Africa was considered to be a remnant. The interest in such scholarship was two-fold. First, one wanted to enlarge the stock of knowledge of humanity about itself in a disinterested humanistic perspective; and secondly British colonial administrators hoped to refine their tools of economic and social domination by applying the findings from such research.

In this setting anthropologists turned their attention towards Africa as a laboratory where they could find surprising constellations and phenomena. Other social scientific disciplines did not share this epistemic vantage point. As the French anthropologist Georges Balandier observed in 1960, in contrast to anthropology the social sciences of economics, sociology and political science were basically 'unprepared to "go abroad"'.[11] From his perspective, they were fundamentally unable to further develop and adapt their models and epistemic techniques, which they had invented and brought to perfection in the study of industrialised, modern societies, by way of studying African realities. And in fact, when economists started to turn their attention to Africa, the majority did not do so in expectance of as yet undiscovered social institutions or practices. Rather, their intellectual task was to mould African economic interaction into preconceived analytical concepts.

As Western academic endeavours, both anthropology and economics built upon prefigured sets of concepts, terms and assumptions when "going abroad". Both constructed the objects under scrutiny as distinct "others" to the European experience.[12] But anthropology, which was the chief mode of social scientific knowledge production about Africa until 1945, also designed observational techniques which grounded in the substance of local phenomena and tried to arrange them in an interesting way.[13] Economists, in contrast, aimed at understanding African economies almost in unison by measuring the objects under scrutiny against the industrial countries of Europe and America and gauged the respective difference. In so far as the observed constellations did not fit the observational tools, they were described in negative terms. Economic scholarship about Africa cast itself in a language marked by the absence of familiar features, which depicted the African continent – and more specifically sub-Saharan Africa – full of potentials but empty in form. This view was characteristic of colonial social science.

The close links between anthropological inquiry and European colonialism are well explored,[14] and post-colonial theorists have analysed, for example, the colonial dimension of literary studies and historiography.[15] But with economics things are a little different. As indicated above, this discipline was not particularly interested in overseas territories before 1945. Economic arguments always played an important role in legitimising colonial rule, as the extraction of distant riches to the benefit of domestic collectives secured support for any such ventures. But economic scholarship cannot be considered to have been a chief resource in legitimising colonial strategies. There was no need for an economic theory in order to exploit the colonies economically. The one exception here might be Frank William Taussig, an expert in international trade and a towering figure of American economics. In his *Principles of Economics* of 1911 he stated that different human races had different intellectual capacities. Labour could only be productive if workers had the appropriate mental capacity to perform the respective tasks. To Taussig, this was a special African problem:

> Many of the improvements in the arts depend for their application on a good degree of intelligence. The Hottentot cannot use tools even of a comparatively simple kind because his brain power is not sufficiently developed. Negroes are employed in great numbers in the gold mines and diamond mines of South Africa, but for simple pick and shovel work only. For handling and guiding machines skilled and intelligent white mechanics must be employed.[16]

Such a racist observation was helpful in establishing white supremacy in the economic development of Southern Africa.

Post-colonial theorists have identified how a dominant mode of colonial discourse arose as social scientists denigrated the "other" through the construction of a substantial difference between the West and the rest.[17] But this argument was not typical for Taussig's work and was certainly not a common element in Anglo-American economics in the 20th century, which constituted itself during the Inter-war Period as a decidedly universalistic endeavour. As a matter of fact, the rise of development economics after the totalitarian catastrophe of Nazism was closely linked to a crisis in legitimising colonial rule through racial discrimination. As the Africanist Frederick Cooper wrote, after World War Two

> the quest of Britain and France for a legitimate colonialism outside a framework of racial distinction rapidly proved unstable, providing no convincing answer in metropole or colony as to why rule and responsibility of some people over others still made sense.[18]

Turning to economics as an intellectual resource was not helpful for late British colonialism because the fundamental assumption of economics was the equality of all men as producers and consumers of goods and services. The discipline searched for the basic principles and mechanisms in the collective life of all human beings. Moreover, it linked itself to the prospect of overcoming the material inequalities that prevailed under colonial rule and envisioned the possible spread of social security and basic welfare across the planet in the name of economic development.

In the perspective of welfare economics as designed by Arthur Pigou, economics was a social science aiming at generally raising standards of living for all people, regardless of their decent.[19] This universalism made economics the predominant language available on which to build a critical analysis of late colonial Africa. The universalism also made economics a tool to be exploited by the formerly colonised after independence.[20] One African intellectual of the first generation after East African independence, Mahmood Mamdani, recalls:

> Our political consciousness was shaped by a central assumption: We were convinced that the impact of colonialism on our societies was mainly

> economic. We were convinced that political economy was the most appropriate tool to come to analytical grips with the colonial legacy.[21]

Economics came to Africa as a colonial science but it was quickly turned into a tool of anti-colonial struggle. Economics functioned as both a colonial and an anti-colonial science. This ambivalence and the dynamics connected to it are as yet not very well understood.

Mamdani also stressed that the projects of post-colonial nation-building in Africa largely built upon the institutional legacy of colonialism. He especially highlighted the practice of indirect rule which separated settler economies from native economies, thus institutionalising an essential divide between Europe and its "other" on the African continent; even within the newly established African nation states.[22] This brings us back to Malcolm Hailey's *Africa Survey*. Despite its distance from the cutting edge of macroeconomic discussions in the 1930s, the *Survey* had a certain impact upon the unfolding of the new discipline of development economics. It made a description of African economies canonical, which consequently separated all activities and transactions according to their level of modernity (and according to the complexion of the involved actors) into a settler economy on the one hand and a native economy on the other. Such a structure was basic in later conceptions of a "dual economy" in development economic theory especially with respect to the availability of labour.[23] According to Mamdani, this colonial dualism was built into post-colonial states in various ways; and the construction prevailed as new African elites came to occupy the former colonial position without changing the overall socio-economic structure. In fundamentally asymmetrical ways, the pockets of Western economic interaction worked as the template for the development of the native economies, which in their turn were perceived only with respect to their difference to the former.[24]

In the process of Africanisation that followed decolonisation, the (colonial) racial segregation was copied into an emerging class structure in which new African elites took over the positions formerly held by Whites.[25] This gave rise to the sub-Saharan 'Gatekeeper-State'.[26] The colonial imagination of an essentially empty Africa, that needed to be filled with features familiar to Western economic advisors well-served the ideological needs of this new African elite and thus remained in place.[27]

The notion of Africa as being different, in terms of not yet having reached a Western economic form, was prevalent in domestic discourse in many African countries way into the late 20th century in both former British and French spheres of influence.[28] In their empirical social scientific study on the notion of "competence" among different groups of industrialists, entrepreneurs and workers in Abidjan, Bruno Latour and Amina Shabou in 1974 diagnosed a discursive figure of the "manque", a void, a lack of something. The two authors had set out to analyse mental obstacles against the Africanisation of business, i.e. against the quick transfer of leading positions from Europeans to Africans within the Ivorian economy. They found all stake-holders referring to the "manque", an "absence" or "emptiness" associated with things African. Europeans legitimised their presence in the terms of the need to fill this void; while Ivorian elites legitimised their quest for Africanisation by claiming the ability to fill the same void on their own. Latour and Shabou argued that this void was in fact produced by the West precisely through the constant promise of filling it.[29] While offering powerful tools for decolonisation, development economics at the same time prolonged colonial structures of domination into the post-colonial world.[30] In the following section I want to turn to ways in which this ambivalence can be related to the observational techniques of macroeconomics more specifically.

Macroeconomic statistics in Africa

Development economic theory came into existence at a critical moment in the history of economic thought. Its rise was closely linked to what Don Patinkin has called a double revolution of Keynesianism and measurement in economics.[31] New techniques of quantifying macroeconomic interaction and new promises concerning the feasibility of planned social change gained ground towards the end of World War Two in the West. Late British colonial policy took this double trajectory up and gave rise to a new interest of Western economists in the poor parts of the world. Macroeconomics offered several promises. It entailed a universal mode of describing societal interaction irrespective of assumed racial differences. It opened up a perspective of economic development that meant more efficient techniques of exploiting labour and natural resources in the colonial setting and that gave a collective vision to the new "under-developed" states of sub-Saharan Africa.[32] Decolonisation was an age of economic planning, in which statistical observation, macroeconomic modelling and economic policies went hand in hand.

One very prominent point of connection between Keynesian economic policy interventions and statistical quantification was national income accounting. However, all through the 1930s, the 1940s and the 1950s this subfield of economic statistics was the object of a heated debate among economists.[33] The imagination of scholars and politicians was captured by the idea that an all-encompassing statistical system could inform policy decisions through concise figures like a gross or a net domestic product or investment and savings ratios. But an almost Babylonian confusion concerning terminology and basic entities overshadowed such a perspective.[34] Moreover, the tool promised to open up a transparent framework for the comparison of national economies across those divides, that British imperialism had established between the colonisers and the colonised. But different political traditions, racial segregation, differing trajectories of governmental statistics and differences in the economic organisation of societies made such comparative investigations difficult. The colonial construction of Africa as the "other" was not easy to reconcile with the universalism of economic observation. As a result of these difficulties African economies came to be depicted as principally equal to Western economies, but as in substance radically poorer and deviant. To this view, Africa revealed itself as an empty space that needed to be developed. It seems important to separate econometric modelling, economic planning policies and practices of macroeconomic measurement and statistics, in order to fully assess this constellation.

Let us focus on measurement for the purpose of this paper and leave aside the aspects of modelling and planning.[35] Turning to quantification, macroeconomics exclusively focused on what Simon Kuznets called the 'area of the economic principle', i.e. the sphere of monetised market transactions as opposed to 'life at large', thus excluding the private sphere of intra-familial reproduction.[36] Within this supposedly quantifiable realm macroeconomic scholarship posed research questions concerning the economics of wealth, of capital formation and of industrial and agricultural productivity. Looking at sub-Saharan Africa with such lenses, economists diagnosed a lack of monetised transactions, a lack of material wealth, a lack of investment capital and saving capacities, and a lack of market integration. In these respects Africa seemed to be empty. All quantifiable aspects of economic interaction were problematic. And if counted, the resulting figures were scandalously low.

How did such a statistical view become dominant? The first economist to bring the rest of the world to the attention of the West was the statistician Colin Clark with his 1940 compilation of the conditions of economic progress. By bringing his estimated sum totals of national income for all human collectives into a comparative system Clark concluded: 'The world is a wretchedly poor place'.[37] The soundness of his approach was highly contested in the 1940s but it opened up a new avenue of economic scholarship, namely development economics. His figures were of poor quality but they provisionally filled cells in an accounting table that opened up as a new epistemic regime. Just about when Lord Hailey completed the *African Survey*, econometric analysis massively gained ground as a source of governmental knowledge in Great Britain, the United States, the Netherlands and other industrialised countries.[38] Some collaborators in the Survey clearly registered this epistemic change and fully assessed the inadequacy of Hailey's endeavour.

Hailey himself had asked the South African economist Sally Herbert Frankel, the author of a 1929 study on railway investment in Southern Africa, to contribute a chapter on investment to the *African Survey*. In the larger framework of colonial development, which aimed at enhancing the economic productivity of the colonies, the question of capital investment gained importance. The crucial question was, whether returns on investment in government-led colonial activities were high enough to legitimise the imperial project vis-à-vis the British taxpayers.

Frankel started research in 1935 and quickly broadened his topic towards a more general study of the 'effect of modern economic forces on the structure of African society'.[39] While his results could not be included into the *African Survey* because of constraints of the publication deadline, they formed one of the first critical assessments of the power of foreign investment for economic development. Frankel argued that capital alone could not induce growth of productivity or national income. In order to achieve such results, major changes in the socio-economic structure of a given entity had to occur including changes in the mentality of the people. Frankel conceived of economic development as a cultural process – a position which was later taken up by many "heterodox" scholars in development economics.[40] Frankel became professor for Colonial Economics at

Oxford in 1946. As an early member of the Mont Pèlerin Society he subsequently gained the reputation of being a forerunner of the neo-liberal counter-revolution in development economics.[41] But it seems overstressing his point to argue that he was completely against a dirigiste approach.[42] Rather, he invited development planners to not blindly apply simplistic models but to positively take African realities into account. His privileged point of observation became the "Kongwa-Experiment", a grotesque misallocation of investment capital by the British colonial government around 1950 in order to grow peanuts in Tanganyika. It aroused great public interest and hammered one of the nails into the coffin of the British Empire.[43]

Another collaborator of Lord Hailey, the economist E.A.G. Robinson, as early as 1940 clearly sensed that the new quantifying tools of macroeconomics possessed a much more powerful potential than had become evident in the final report of the *African Survey*. In collaboration with the National Institute of Economic and Social Research (NIESR) he invited the young economist Phyllis Deane to compile further figures and to tentatively compose national income accounts for some British colonial possessions during the war. The idea was to convince the Colonial Office of the usefulness of national income accounting for the administration of the colonies. Robinson commissioned Deane to map the Empire economically in order to bring its potential for the extraction of natural resources and hidden labour forces into view.[44]

He designed such a perspective during the years 1940–1, which were an important breach in the history of national income accounting.[45] Based on calculations by Colin Clark, Erwin Rothbarth, Feodora and Richard Stone and James Meade, John Maynard Keynes convinced the British Chancellor of the Exchequer at this specific point in time to make national accounts a government task in domestic policy. In this connection Richard Stone designed a form and a terminology for macroeconomic observation, which was subsequently turned into a global standard by the Organization for European Economic Cooperation (OEEC) and the United Nations.[46] It had some peculiarities. First, Stone conceived of the totality of economic interaction within a given social collective (i.e. the British economy) as an institution that could be compared with a firm. Stone closely cooperated with F. Sewell Bray, a specialist in entrepreneurial accountancy in strengthening this metaphorical transfer.[47] Second, Stone applied the technique of double entry bookkeeping to the statistical registration of British domestic economical interactions. And third he proposed accounting for all flows and stocks in three interlocking balance sheets which were designed according to the General Theory of Keynes. One referred to the sum totals of production, one to all incomes and one to all outlays. Stone baptised his approach "social accounting" as opposed to the entrepreneurial accounting. His triple view resulted in a statistical system which offered a range of possibilities to countercheck single figures and thus to stabilise the plausibility of the final results as hard economic facts.[48] The system encompassed many empty positions which could be imputed by interpolation or otherwise virtually filled. This observational tool was universal in design (i.e. it transported no racial segregations). It designed tables with empty cells that needed to be filled. It constructed an artificial sum total (GDP), which had no real equivalent in economic life but made subsequently an astonishing career in both

Western and African economic policies. And it depicted the British economy as fully compliant while sub-Saharan Africa appeared poor, structurally deviant and empty with respect to important accounting positions.

Stones's tool was applied by Phyllis Deane and thrown over Northern Rhodesia, Nyasaland and Jamaica in her research project that started in 1941. She took up the notion of a laboratory that had informed the *African Survey* and considered her endeavour to be an experiment. In the foreword to the 1948 publication by Deane on 'The Measurement of Colonial National Incomes. An Experiment', E.A.G. Robinson wrote about the new techniques:

> Any test of their application to the measurement of a more primitive national income was . . . difficult, since very few attempts had been made to measure colonial national incomes, and none of them were in a form which readily permitted an already accumulated body of data to be rearranged to see whether it could be used to exploit the advantages of the new techniques.

Robinson highlighted the very newness of the national income approach and contrasted it to older modes of colonial scientific inquiry. He said:

> Indeed, the development of the measurement of colonial national incomes was in itself almost a path-braking task, which was capable of yielding great dividends in knowledge of the economic structure and standards of the colonial territories, the limits of which I myself had learned to appreciate in working with Lord Hailey on his *African Survey*.[49]

But national income accounting did not simplify the colonial access to sub-Saharan Africa and neither did it render an adequate economic picture of the continent. Robinson held high hopes but the experiment failed. The Colonial Office could not be convinced of the usefulness of national income accounting. Instead, Deane herself considered the quality of her results to be rather poor. After the war she embarked on an 18-month field trip to Central Africa, because the published statistics were not helpful in composing true national accounts. Fieldwork seemed necessary in the course of which Deane hoped to gather consistent data in order to compose a neat set of interlocking accounts according to the methodology designed by Richard Stone. But she soon realised that the problems at hand did not only concern the lack of adequate statistical material but were more fundamental. 'The accounting problem is not simply that of the acute scarcity of quantitative data . . . it is also a qualitative problem which brings into question the fundamental validity for primitive communities of the social accounting concepts themselves'. Referring to Arthur Pigou's definition of welfare economics she stated:

> A science which studies 'that part of social welfare which can be brought directly or indirectly into relation with the measuring rod of money' does not appear to have much contribution to make to the problems of a semi-subsistence economy.[50]

At the heart of the problem Deane located differences in the conception and the macroeconomic relevance of the family between Western social structures and African conditions.[51] This difference had consequences for the shape of what was considered "the realm of the economic principle" (Kuznets), or: the economy as the object of economic analysis.

Deane raised the question in what way and to what extent the observational techniques of accounting had to be adjusted to the differing shape of the African object under scrutiny. This problem was debated also with respect to Nigeria. After his studies on the Malaysian rubber economy Peter Thomas Bauer went to Nigeria in 1950 to study the structure of West African trade.[52] At the same time the Colonial Office invited the Cambridge Department of Applied Economics to create an estimate of the Nigerian national income. Again E.A.G. Robinson took over the matter and in cooperation with the Colonial Economic Research Committee asked A.R. Prest and I.G. Stewart to prepare the study. Phyllis Deane, P.T. Bauer, Richard Stone and the anthropologist Meyer Fortes formed a supervising body upon request by Robinson, which was later joined by the black Caribbean economist W. Arthur Lewis.

Prest and Stewart with the resulting 1953 publication opened up an intensive discussion on observational methodology. Their study to some extent reads as a protocol of their successive experience of a fundamental difference between Europe and Africa. Ten main differences were protocolled under the heading 'Dissimilarities of "Western" and Nigerian Economic Conditions'.[53] Not unlike Deane the authors of the Nigerian study wanted to adjust the Western epistemic techniques in view of African specialties in order to reach an adequate observation of both parts of African economies, namely the native subsistence economy and the pockets of Western form. Their highly innovative approach produced some irritation among British colonial officers. The Second Conference of Colonial Government Statisticians in 1953 decided that further on also not directly monetised transactions needed to be accounted for. The issue was political. Drawing the line of economic observation only around the Westernised (i.e. monetised) interactions resulted in incredibly low productivity figures for Africa, because most economic activities in the rural sector were simply ignored. But the statistical construction of per capita incomes divided the product of a partial sector through all heads of population and showed a negatively biassed picture of the levels of wealth. A UN-report on Tanganyika, which was a mandate territory under British rule, took these figures at their face value and scandalised Tanganyikan poverty. Britain reacted in 1955 by commissioning the economist Alan Peacock with a study on the national income of Tanganyika. Three years later Peacock produced results that accounted for all economic activity in the mandate irrespective of their monetary or non-monetary form. Peacock in the introduction stated: 'To suggest that millions of Africans in Tanganyika are on a starvation diet is a travesty of the truth'.[54]

Realising that national income accounting produced an essentially empty picture of sub-Saharan Africa, Prest and Peacock turned to the more positive epistemic techniques of anthropology. Prest had invented a complicated method of quantifying female intra-familial labour by taking brideprices into account. In the same objective Peacock turned to East African ethnographies.[55] He quoted the work of Richard Turnwald and argued that any transaction in some way could be connected to monetised markets.

> So long as the production of any commodity, even cattle blood, has an opportunity cost, then that commodity has a price in terms of other commodities. These commodities in turn will have prices in terms of others, and surely somewhere or other the chain of substitutes will be linked to a commodity which is priced in a market.[56]

To his view there existed no fundamental difference between African subsistence economies and Western capitalism.

Shaping a framework for transnational comparison

In principle, the observational technique of national income accounting allowed for the description of all possible instances. Differences of neither race nor culture limited its epistemic value. But in order to apply it to Africa according to the methodology of Simon Kuznets much work had still to be done. Categories had to be adjusted and the accounting tables needed to be redesigned according to the observed economic realities in the field.

Throwing the observational technique of national income accounting over Africa produced four different stances among economists. The first position occupied by Deane, Prest and Peacock was to revise the categories of observation by recourse to anthropology. They took up Simon Kuznets's conviction that when wanting to compose meaningful national income accounts one had to carefully define the categories according to the cultural and institutional specificities of the social collective under scrutiny. In transposing the Western observational tool to African economies, the pioneering statisticians took brideprices or cattle blood as proxies in order to express the volume of subsistence or semi-subsistence interactions in a monetary form. Some anthropologists highly welcomed this move as it made economic model building one additional tool in the arsenal of ethnographic techniques for the formal description of African economies.[57] Others, however, completely refuted the approach and argued for a substantive difference. To authors like Karl Polanyi or George Dalton the extension and adaption of national income accounting to sub-Saharan African conditions was futile because such endeavours prolonged the Western economic mentality of profit seeking into alien constellations.[58] They called for reconceptualising the basic assumptions of economic analysis.

A second position was voiced by Dudley Seers. He had worked on the Gold Coast (Ghana) and was well familiar with the problems of African national income accounting.[59] His solution was rather modest. He proposed to restrict the quantitative approach and to take into consideration only those interactions which could easily be counted, namely export commodities. All other observational endeavours, to his view, necessarily had to end up in 'the well-known morass which those estimating national incomes of underdeveloped areas either skirt, rush across, or die in'.[60] As a matter of consequence Seers argued that the national income figures for non-Western economies could not be compared with the respective values for industrialised countries. But following Colin Clark, such comparisons became a chief mode in depicting the structures of global inequality between poor Africa and a wealthy Great Britain in the post-colonial world of development. Counting only the undoubtedly countable was an epistemically sound position, but politically, it turned out to be rather unhelpful. With decolonisation also Africans wanted to have their national wealth represented in a globally encompassing system of national income statistics.[61]

Two scholars who have become known as early neo-liberal authors voiced a third stance. Basing upon his work on African trade P.T. Bauer fully rejected the statistical approach of African national income accounting.[62] His argument was that economical scholarship should not rely upon statistical preconceptions but should open up to local surprises. Also, Sally Herbert Frankel argued along the same lines.[63] Macroeconomic statistics and national income accounting to the view of these authors could not render adequate observations of African reality. Frankel and Bauer fully rejected this kind of observation. And they heavily criticised those forms of development economic policy advice which drew clear-cut action plans for the new African governments out of statistical figures about national income and investment and savings ratios.[64] Their critical stance towards planning, which later informed a neo-liberal "counter-revolution" in development economics,[65] was initially voiced as a methodological critique of national income accounting.

However, a fourth view prevailed. By virtue of international organisations like the OEEC/OECD and the UN the accounting system of Richard Stone became a global standard.[66] Further research must show whether there have been any expert economical missions to the global South post-1945 which did not in some way or another refer to the economic indicators provided by national income accounting. Estimating such figures was predominant in international organisations and concurrently became a way of expressing national sovereignty for the new states that emerged out of the British Empire. With a slight touch of cynicism one economist at a conference held in Addis Ababa in 1961 observed:

> Today in many independent countries national accounts are regarded, alongside the national flag and the national anthem, as symbols of independence. This mystical belief can be turned to the planners' advantage, provided national accounts are treated as a means to an end – development – and not as an end in itself.[67]

There is a straight line from colonial economics to post-colonial postures of sovereignty, which Mahmood Mamdani has highlighted. Considering sub-Saharan Africa to be devoid of structures familiar to Western economic eyes with decolonisation became the predominant view of African political problems both for withering colonial officers and new African elites alike.

The figures composed in this view depicted non-Western economies as deviant structures and failed to render their full potential.[68] The observational technique of national income accounting reflected upon African conditions exclusively in negative terms. Poverty, or the absence of Western-style features of wealth became the dominant mode of perception. An economic image of Africa emerged at the moment of imperial decline which immersed global imaginations of Europe and its "other" fully in a specific global allocation of wealth and poverty. Europe and the USA were full of riches, while Africa, in this respect, remained empty and poor. At the same time national income accounting opened up new global avenues of economic policy and set the stage for massive transnational interventions in the name of development aid.

This chapter argues that when economics "went abroad" it took along some peculiar epistemic techniques, which were not necessarily helpful in approaching the problems of African poverty. Development economics grew out of this unhappy conjuncture. It focused on Africa largely in the terms of a void. The notion of a "dual economy" which separated pockets of Western economic interaction from assumedly more genuine African modes prevailed. In a universalistic mode of observation specific asymmetries made epistemic techniques dominant, which denigrated the qualities of the "other" and produced the category of otherness at the same time. National income accounting made Africa poor because it rendered non-Western economic structures negatively and emptied social constellations on the African ground in order to link them up to a new universalistic form of economic knowledge production.[69]

The reason for this specific view was, as I presume, time. Modernisation theory strongly envisioned a future for Africa in which the antagonism between European forms of economic interaction and its "others" would necessarily whither. Or to put it in other words: Empty Africa, in the prevailing perspective of development discourse, would be filled with features familiar to Western observers. Most probably, such a process of globalisation has since 1945 in fact taken place.

Notes

1 Meade, *World Economic Survey*, quoted in Arndt, *Economic Development*, 33.
2 Klein and Morgan, *Age of Economic Measurement*.
3 Chassé, *Erfindung des Bruttosozialprodukts*.
4 Bayly, *Empire and Information*; Bayly, *Indigenous and Colonial Origins*.
5 For the French Empire see Sarraut, *Mise en valeur*; Coquery-Vidrovitch, 'Transfer'. For the British Empire see Constantine, *Making of*; Havinden and Meredith, *Colonialism*. A general overview regarding Africa can be found in Bonneuil, *Development*.
6 Rao, *National Income*.
7 Hailey, *African Survey*.
8 Schlözer, *Theorie*.
9 Hailey, *African Survey*, xxiv.
10 Tilley and Gordon, *Ordering Africa*; Tilley, *Living Laboratory*.
11 Balandier, 'French Tradition', 111; Tilley, 'Introduction. Africa, Imperialism, and Anthropology', in Tilley and Gordon, *Ordering Africa*, 5.
12 The classic study for anthropology is Fabian, *Time*.

13 Berman, 'Ethnography'.
14 As an example, see Harries, *Butterflies.*
15 Said, *Orientalism*; Guha, *History at the Limit.*
16 Taussig, *Principles*, 96.
17 Hall, 'West and the Rest'.
18 Cooper, *Colonialism in Question*, 202.
19 Pigou, *Economics.*
20 This was of course the leitmotif of W. Arthur Lewis's work. Tignor, *Arthur Lewis.* See also Speich, 'Kenyan Style'.
21 Mamdani, 'Beyond Settler', 651.
22 Mamdani, *Citizen and Subject*, 7–8.
23 Lewis, 'Unlimited Supplies'.
24 Mamdani, *Citizen and Subject*, chapter III. Such a dislocation of identity constructions was mirrored into the psyche of the colonial subject by Frantz Fanon. Fanon, *Black Skin.* The French original appeared in 1952.
25 Obama, 'Problems'.
26 Cooper, *Africa since 1940.*
27 Morgan, *On a Mission.* See also Tignor, *Arthur Lewis.*
28 A connection to the specific British policy of Indirect Rule can therefore not be pressed too far.
29 'Les Ivoiriens acceptent tous en profondeur qu'ils manquent, la seule chose qu'ils revendiquent c'est de pouvoir remplir eux-mêmes un peu ce manque -voire l'accélérer-. Jamais on ne soupçonne que l'Occident n'est pas seulement dans le cadre en haut, mais qu'il est le cadre et que c'est lui qui fabrique le manque qu'il feint de remplir'. Latour and Shabou, *Idéologies*, 77.
30 Powerful post-colonial critiques of development economics have been put forward by anthropologists in the 1990s. See Escobar, *Enctountering Development*, chapter 2; Mitchell, 'Fixing the Economy'.
31 Patinkin, 'Keynes'.
32 The continuities from British colonial development to post-colonial development aid are explored in Hodge, *Triumph of the Expert.*
33 Stamp, 'Methods'; Gilbert, 'Measurement'.
34 See Conference on Research in Income and Wealth and the respective publications by the International Association for Research in Income and Wealth which was founded in 1947.
35 This is not to say that a full account of the problems at hand is possible without critical reasoning about the advent of planning post-1945. Boettke, *Collapse.* The conjuncture is obvious in Lewis, *Principles.*
36 Kuznets, 'National Income', 212.
37 Clark, *Conditions*, introduction.
38 On the Dutch experience see Boumans, *How Economists.* On the USA see Yonay, *Struggle and Bernstein, Perilous Progress.*
39 Hailey, *African Survey*, vii. Frankel, *Capital Investment.* See also Frankel, *Railway.*
40 See Hoselitz, *Progress.*
41 Plehwe, 'Origins', 238.
42 Deepak Lal spoke of "dirigiste" approaches in order to name his adversaries. Lal, *Poverty.*
43 Frankel, *Economic Impact*; Hogendorn and Scott, 'Groundnut Scheme'.
44 Here, I draw upon an analysis of the colonial power function of cartographic mapping as shown in Huggan, 'Decolonizing the Map'.
45 Studenski, *Income of Nations.*
46 OEEC, *System of National Accounts*; United Nations Statistical Office, *System of National Accounts.*

47 Stone, 'Use and Development'; Bray, *Review of Singer*. In general see Suzuki, 'Epistemology'.
48 On fact making through accounting see Poovey, *A History*.
49 Deane, *Measurement*, Original emphasis.
50 Deane, *Colonial Social Accounting*, 115.
51 Morgan, *Seeking Parts*. The following section is strongly inspired by Mary Morgan's paper. Inconsistencies and shortcomings are of course all mine.
52 Bauer, *West African Trade*.
53 Prest and Stewart, *National Income*, 4–6.
54 Peacock and Dosser, *National Income of Tanganyika*, 50.
55 Thurnwald, *Economics*.
56 Peacock and Dosser, *National Income of Tanganyika*, 16.
57 Schneider, 'A Model'.
58 Polanyi, 'The Economy'; Dalton, 'Economic Theory'.
59 Seers and Ross, *Report on Financial*.
60 Seers, 'Role of National Income', 166; Seers, *What Are We*.
61 Jerven, *Poor Numbers*.
62 Bauer and Yamey, 'Economic Progress'; Bauer and Yamey, *Economics of Under-developed Countries*.
63 Frankel, *Economic Impact*.
64 W. Arthur Lewis as well as W. Walt Rostow argued that an investment ratio of roughly 15 percent of national income was necessary in order to break the vicious circle of poverty and to make poor economies 'take-off' into sustained growth. Rostow, *The Process*; Lewis' *Theory of Economic Growth*. Bauer criticised both, Lewis and Rostow: Bauer, 'Lewis' Theory'; Bauer and Wilson, 'Stages of Growth'. For Frankel's position towards Lewis see Frankel, 'United Nations Primer'.
65 Toye, *Dilemmas*; Plehwe, 'Origins'.
66 Speich, 'Use of Global Abstractions'.
67 Barkay, 'Statistical Macro-Economic Framework', 85.
68 Rao, 'Some Reflections'. See also Simon Kuznets's critique of Colin Clarks treatment of China in Kuznets, 'National Income', 209.
69 McNeely, *Constructing*.

Bibliography

Arndt, Heinz W., *Economic Development. The History of an Idea* (Chicago: University of Chicago Press, 1987).

Balandier, Georges, 'The French Tradition of African Research', *Human Organization – Journal of the Society for Applied Anthropology* 19, no. 3 (1960): 108–11.

Barkay, Richard M., 'The Statistical Macro-Economic Framework Needed in Development Planning in Africa', in *African Studies in Income and Wealth*, ed. L. H. Samuels (Chicago: Quadrangle Books, 1963), 66–88.

Bauer, Peter T., *West African Trade. A Study of Competition, Oligopoly and Monopoly in a Changing Economy* (Cambridge: Cambridge University Press, 1954).

——— and Basil S. Yamey, 'Economic Progress and Occupational Distribution', *The Economic Journal* 61, no. 244 (1951): 741–55.

———, 'Lewis' Theory of Economic Growth', *The American Economic Review* 46, no. 4 (1956): 632–41.

——— and Basil S. Yamey, *The Economics of Under-developed Countries* (Cambridge: Cambridge University Press, 1957).

——— and Charles Wilson, 'The Stages of Growth', *Economica, New Series* 29, no. 114 (1962): 190–200.

Bayly, Christopher A., *Empire and Information. Intelligence Gathering and Social Communication in India 1780–1870* (Cambridge: Cambridge University Press, 1996).

———, *Indigenous and Colonial Origins of Comparative Economic Development. The Case of Colonial India and Africa* (Washington DC: World Bank, 2008).

Berman, Bruce, 'Ethnography as Politics, Politics as Ethnography. Kenyatta, Malinowski, and the Making of Facing Mount Kenya', *Canadian Journal of African Studies* 30, no. 3 (1996): 313–44.

Bernstein, Michael A., *A Perilous Progress. Economists and Public Purpose in Twentieth-Century America* (Princeton: Princeton University Press, 2001).

Peter J. Boettke, ed., *The Collapse of Development Planning* (New York: New York University Press, 1994).

Bonneuil, Christophe, 'Development as Experiment. Science and State Building in Late Colonial and Postcolonial Africa, 1930–1970', *OSIRIS, 2nd series* 15 (2000): 258–81.

Boumans, Marcel, *How Economists Model the World into Numbers* (London: Routledge, 2005).

Sewell Bray, F., review of *Standardised Accountancy in Germany*, by H.W. Singer, *The Economic Journal* 56, no. 223 (1946): 482–4.

Clark, Colin, *The Conditions of Economic Progress* (London: Palgrave Macmillan, 1940), introduction.

Conference on Research in Income and Wealth, ed., *Studies in Income and Wealth*, vol. 1 (New York: CRIW, 1937).

Constantine, Stephen, *The Making of British Colonial Development Policy 1914–1940* (London: Frank Cass, 1984).

Cooper, Frederick, *Africa Since 1940. The past of the present* (Cambridge: Cambridge University Press, 2002).

———, *Colonialism in Question. Theory, Knowledge, History* (Berkeley: California University Press, 2005).

Coquery-Vidrovitch, Catherine, 'The Transfer of Economic Power in French-Speaking West Africa', in *The Transfer of Power in Africa. Decolonization 1940–1960*, ed. G. Prosser and W. R. Louis (New Haven: Yale University Press, 1982), 105–34.

Dalton, George, 'Economic Theory and Primitive Society', *American Anthropologist, New Series* 63, no. 1 (1961): 1–25.

Deane, Phyllis, *The Measurement of Colonial National Incomes: An Experiment* (Cambridge: Cambridge University Press, 1948).

———, *Colonial Social Accounting* (Cambridge: Cambridge University Press, 1953).

Escobar, Arturo, *Encountering Development: The Making and Unmaking of the Third World* (Princeton: Princeton University Press, 1995).

Fabian, Johannes, *Time and the Other: How Anthropology Makes Its Objects* (New York: Columbia University Press, 1983).

Fanon, Frantz, *Black Skin, White Masks*, trans. Richard Philcox (New York: Grove Press, 2008).

Frankel, Sally Herbert, *The Railway Policy of South Africa* (London: Palgrave Macmillan, 1929).

———, *Capital Investment in Africa: Its Course and Its Effects* (London: Palgrave Macmillan, 1938).

———, 'United Nations Primer for Development', *The Quarterly Journal of Economics* 66, no. 3 (1952): 301–26.

———, *The Economic Impact and Underdeveloped Societies* (Oxford: Blackwell, 1953).

Gilbert, Milton, 'The Measurement of National Wealth: Discussion', *Econometrica* 17 (1949): Supplement Report of the Washington Meeting, 255–72.

Guha, Ranajit, *History at the Limit of World-History* (New York: Columbia University Press, 2002).

Hailey, William Malcolm, *An African Survey: A Study of Problems Arising in Africa South of the Sahara* (London: Oxford University Press, 1938).

Hall, Stuart, 'The West and the Rest', in *Formations of Modernity*, ed. S. Hall and B. Gieben (Milton Keynes: Polity Press, 1992), 275–320.

Harries, Patrick, *Butterflies and Barbarians. Swiss Missionaries in South-East Africa* (Oxford: James Currey, 2007).

Havinden, Michael and David Meredith, *Colonialism and Development. Britain and Its Tropical Colonies, 1850–1960* (London: Routledge, 1993).

Hodge, Joseph. *Triumph of the Expert. Agrarian Doctrines of Development and the Legacies of British Colonialism* (Athens: Ohio University Press, 2007).

Hogendorn, J. S. and K. M. Scott, 'The East African Groundnut Scheme. Lessons of a Large-Scale Agricultural Failure', *African Economic History* 10 (1981): 81–115.

Hoselitz, Bert F., ed., *The Progress of Underdeveloped Areas* (Chicago: Chicago University Press, 1952).

Huggan, Graham, 'Decolonizing the Map: Post-Colonialism, post-Structuralism and the Cartographic Connection', *Ariel. A Review of International English Literature* 20, no. 4 (1989): 115–31.

Jerven, Morten, *Poor Numbers. How We Are Misled by African Development Statistics and What to Do About It* (Ithaca: Cornell University Press, 2013).

Klein, Judy L. and Mary S. Morgan, eds., *The Age of Economic Measurement* (Durham: Duke University Press, 2001).

Kuznets, Simon, 'National Income', in *Encyclopedia of the Social Sciences*, ed. E. R. A. Seligman (New York: Palgrave Macmillan, 1933), 205–24.

———, 'National Income and Industrial Structure', *Econometrica* 17 (1949) Supplement, 205–41.

Lal, Deepak, *The Poverty of 'Development Economics'* (London: Institute for Economic Affairs, 1983).

Latour, Bruno and Amina Shabou, *Les idéologies de la competence en milieu industriel a Abidjan, Sciences humaines, série études industrielles*, vol. 9 (Abidjan: ORSTOM, 1974).

Lewis, W. Arthur, *The Principles of Economic Planning* (London: Allen & Unwin, 1948).

———, 'Economic Development with Unlimited Supplies of Labour', *The Manchester School of Economic and Social Studies* 22, no. 2 (1954): 139–91.

Mamdani, Mahmood, *Citizen and Subject. Contemporary Africa and the Legacy of Late Colonialism* (London: James Currey, 1996).

———, 'Beyond Settler and Native as Political Identities. Overcoming the Political Legacy of Colonialism', *Comparative Studies in Society and History* 43, no. 4 (2001): 651–64.

McNeely, Connie L., *Constructing the Nation-State. International Organization and Prescriptive Action* (Westport: Greenwood Press, 1995).

Meade, James, *World Economic Survey* (Geneva: LoN, 1938)

Mitchell, Timothy, 'Fixing the Economy' *Cultural Studies* 12, no. 1 (1998): 82–101.

Morgan, Mary, *'On a Mission' With Mutable Mobiles*, Working papers on the nature of evidence, LSE Department of Economic History (London: LSE, 2008).

———, *Seeking Parts, Looking for Wholes. History of Observation in Economics Working Paper Series*, vol. 1 (Amsterdam: University of Amsterdam, 2009).

Obama Sr., Barak H., 'Problems Facing Our Socialism', *East Africa Journal* 2, no. 6 (1965): 26–33.

OEEC, ed., *A Standardised System of National Accounts* (Paris: OEEC, 1952).

Patinkin, Don, 'Keynes and Econometrics. On the Interaction Between the Macroeconomic Revolutions of the Interwar Period', *Econometrica* 44, no. 6 (1976): 1091–123.

Peacock, Alan T. and Douglas Dosser, *The National Income of Tanganyika 1952–54* (London: HSMO, 1958).

Pigou, Arthur, *The Economics of Welfare* (London: Palgrave Macmillan, 1920).

Plehwe, Dieter, 'The Origins of the Neoliberal Economic Development Discourse', in *The Road from Mont Pèlerin. The Making of the Neoliberal Thought Collective*, ed. P. Mirowski and D. Plehwe (Cambridge: Cambridge University Press, 2009), 238–79.

Polanyi, Karl, 'The Economy as Instituted Process', in *Trade and Market in the Early Empires. Economies in History and Theory*, ed. K. Polanyi (Glencoe: Free Press, 1957), 243–69.

Poovey, Mary, *A History of the Modern Fact. Problems of Knowledge in the Sciences of Wealth and Society* (Chicago: University of Chicago Press, 1998).

Prest, Alan R. and Ian G. Stewart, *National Income of Nigeria* (London: HSMO, 1953).

Rao, V. K. R. V., *The National Income of British India 1931–32* (London: Palgrave Macmillan, 1940).

———, 'Some Reflections on the Comparability of Real National Incomes of Industrialized and Under-Developed Countries', in *Income and Wealth, Series III*, ed. M. Gilbert (Cambridge: Bowes & Bowes, 1953), 178–210.

Rostow, Walt W., *The Process of Economic Growth* (New York: Norton, 1952).

Said, Edward W., *Orientalism* (New York: Vintage, 1979).

Sarraut, Albert, *La mise en valeur des colonies françaises* (Paris: Payot, 1923).

Schlözer, August Ludwig, *Theorie der Statistik. Nebst Ideen über das Studium der Politik überhaupt* (Göttingen: Vandenhoeck & Ruprecht, 1804).

Schneider, Harold K., 'A Model of African Indigenous Economy and Society', *Comparative Studies in Society and History* 7, no. 1 (1964): 37–55.

Seers, Dudley, 'The Role of National Income Estimates in the Statistical Policy of an Under-developed Area', *The Review of Economic Studies* 20, no. 3 (1952): 159–68.

——— and Claude Richard Ross, *Report on Financial and Physical Problems in the Gold Coast* (Accra: Office of Government Statisticians, 1952).

———, 'What Are We Trying to Measure?', *Journal of Development Studies* 8 (1972): 21–36.

Speich, Daniel Chassé, *Die Erfindung des Bruttosoialprodukts. Globale Ungleichheit in der Wissensgeschichte der Ökonomie* (Göttingen: Vandenhoeck & Ruprecht, 2013).

Speich, Daniel, 'The Kenyan Style of "African Socialism". Developmental Knowledge Claims and the Explanatory Limits of the Cold War', *Diplomatic History* 33, no. 3 (2009): 449–66.

———, 'The Use of Global Abstractions: National Income Accounting in the Period of Imperial Decline', *Journal of Global History* 6, no. 1 (2011): 7–28.

Stamp, Josiah, 'Methods Used in Different Countries for Estimating National Income', *Journal of the Royal Statistical Society* 97, no. 3 (1934): 423–55.

Stone, Richard, 'The Use and Development of National Income Expenditure Estimates', in *Lessons of the British War Economy*, ed. D. N. Chester (Cambridge: Cambridge University Press, 1951), 83–101.

Studenski, Paul, *The Income of Nations: Theory, Measurement, Analysis Past and Present: A Study in Applied Economics and Statistics* (New York: New York University Press, 1958).

Suzuki, Tomo, 'The Epistemology of Macroeconomic Reality: The Keynesian Revolution from an Accounting Point of View', *Accounting, Organization and Society* 28 (2003): 471–517.

Taussig, Frank William, *Principles of Economics* (New York: Palgrave Macmillan, 1911).

Thurnwald, Richard, *Economics in Primitive Communities* (Oxford: Plagrave Macmillan, 1932).

Tignor, Robert L., *W. Arthur Lewis and the Birth of Development Economics* (Princeton: Princeton University Press, 2006).

Tilley, Helen and Robert J. Gordon, eds., *Ordering Africa. Anthropology, European Imperialism, and the Politics of Knowledge* (Manchester: Manchester University Press, 2007).

———, *Africa as a Living Laboratory. Empire, Development, and the Problem of Scientific Knowledge* (Chicago: University of Chicago Press, 2011).

Toye, John, *Dilemmas of Development. Reflections on the Counter-Revolution in Development Theory and Policy* (Oxford: Blackwell, 1987).

United Nations Statistical Office, ed., *A System of National Accounts and Supporting Tables* (New York: UNSO, 1953).

Yonay, Yuval P., *The Struggle over the Soul of Economics* (Princeton: Princeton University Press, 1998).

Index